Pol Hoorelbeke
Process Safety

Also of interest

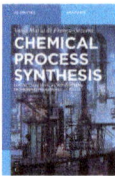

Chemical Process Synthesis.
Connecting Chemical with Systems Engineering Procedures
Bezerra, 2021
ISBN 978-3-11-046825-0, e-ISBN 978-3-11-046826-7

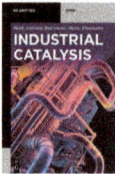

Industrial Catalysis
Benvenuto, Plaumann, 2021
ISBN 978-3-11-054284-4, e-ISBN 978-3-11-054286-8

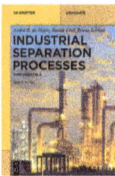

Industrial Separation Processes.
Fundamentals
de Haan, Eral, Schuur, 2020
ISBN 978-3-11-065473-8, e-ISBN 978-3-11-065480-6

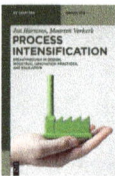

Process Intensification.
Breakthrough in Design, Industrial Innovation Practices,
and Education
Harmsen, Verkerk, 2020
ISBN 978-3-11-065734-0, e-ISBN 978-3-11-065735-7

Pol Hoorelbeke

Process Safety

An Engineering Discipline

DE GRUYTER

Author
Dr. Pol Hoorelbeke
Total S.E.
Health, Safety, Environment Division
Vice President Division HSE Audit & Major Accident Investigations
24 Cours Michelet
92069 Paris
France

ISBN 978-3-11-063205-7
e-ISBN (PDF) 978-3-11-063213-2
e-ISBN (EPUB) 978-3-11-063222-4

Library of Congress Control Number: 2021934878

Bibliographic information published by the Deutsche Nationalbibliothek
The Deutsche Nationalbibliothek lists this publication in the Deutsche Nationalbibliografie;
detailed bibliographic data are available on the Internet at http://dnb.dnb.de.

© 2021 Walter de Gruyter GmbH, Berlin/Boston
Cover image: Digital Vision/Photodisc/Getty Images
Typesetting: Integra Software Services Pvt. Ltd.
Printing and binding: CPI books GmbH, Leck

www.degruyter.com

"I dedicate this book to my lovely wife Martine and my two sons, Günther and Jordi. They are the reasons why I enjoy my life."

Preface

The oil and gas industry processes and transports large quantities of hazardous products in high-tech installations, sometimes in extreme environmental conditions. Despite common agreement in the industry that process safety is of utmost importance, major accidents continue to occur.

There is some discussion about what exactly is covered by the term "process safety." Some perceive it as part of "safety" while others regard it as a "technical discipline."

My main motivation in writing this book is that I am convinced that a better understanding of the subject of process safety can help managers in our industry to reduce the incidence of major accidents.

The first few chapters of the book are dedicated to the history and evolution of Occupational Health and Safety. Since the Industrial Revolution and this for a long time "Industrial Safety" was a synonym for "Occupational Health and Safety at Work." Since WWII a new engineering discipline emerged to deal with the uncertainties associated with new complicated technologies. This engineering discipline is nowadays referred to with terms as "management of process safety" or "management of technological risks." From Chapter 4 onwards, I will deal in more detail with the characteristics of this new engineering discipline. It will be demonstrated that process safety is an engineering discipline that serves the managers who are responsible for industrial safety. But managing process safety is very different from managing what is classically considered as "occupational health and safety." Occupational health and safety is a (legal) relationship between the worker and the employer in which both parties agree to perform the work in a safe way. The managing principles, methods, concepts, etc. to manage occupational health and safety have been established over a period of about 250 years. Process safety, on the other hand, is an engineering discipline that has been developed to deal with the uncertainties of new technologies and processes. The approach is different and should not be confused with managing occupational health and safety.

My vision derives from 34 years of interest in the field of occupational health and safety and process safety. In the first phase of my career, I worked for about 14 years (1987–2002) as a consultant. During this period, my daily task entailed performing all types of studies such as HAZOP studies, reliability studies, Quantitative Risk Assessments, accident investigations, etc. for a range of high-risk industries (onshore as well as offshore) at an international level.

In 2002, I joined an international major energy company (TOTAL) where I went through different positions in HSE functions. I was appointed as Fellow for Process Safety for the Group in 2015. Safety is a value for TOTAL and working in such an exciting company gave me the opportunity to meet people from all over the world and to discuss with thousands of colleagues, safety, in general and process safety, in particular.

https://doi.org/10.1515/9783110632132-202

My academic background (MSc in mechanical engineering, MSc in occupational health and safety and a PhD in Engineering Science) enables me to study the basics of process safety and engineering. I have given more than 100 academic courses and training at different universities and conferences worldwide, which enabled me to have discussions with professors and students from a wide range of backgrounds.

As board member of the EPSC (European Process Safety Center) and FABIG (Fire and Blast Information Group) I have the opportunity to exchange with peers who have been working in the field of process safety for decades.

This background sets the scene for the book. It is a contribution to the international discussion on industrial safety.

Pol Hoorelbeke
December 2020

Acknowledgments

The author gratefully acknowledges the interesting discussions during my career and the help of the people in reviewing the manuscript of the book. In particular, I would like to thank the following.

Professor emeritus Dr. Ir. Marc Van Overmeire and Professor Ir. Paul Olivier introduced me to industrial safety in 1987. Marc, who was also my promotor for my PhD, took the time to read the manuscript.

My many colleagues from Technica and DNV with whom I worked for about 14 years and with whom I was involved in more than 350 assignments in the field of industrial safety.

My many colleagues from TOTAL at various times and places all over the world from whom I learned a lot in theory and in practice. In particular, Xavier Bontemps, Dr. Ir. Dirk Roosendans, Christian Kapp, Paul Vermeiren for their comments on the manuscript and my colleagues Pascal Thiery, Yongjian Yu, Jing Miao and Crystal Chong for the many years of lectures organized by Mme. Miaoxian Wu at the South China University of Technology in Guangzhou, China. Since 2006, more than 1,000 students followed this three days training course.

My friends and colleagues from the European Process Safety Center (Dr. Tijs Koerts, Dr. Hans Schwarz) and from the Fire and Blast Information Group with a special attention for Dr. Bassam Burgan, director of the Steel Construction Institute, who made very useful comments on the manuscript.

My peer colleagues from the headquarters of BP, Chevron, Conoco, ExxonMobil, P66 and Shell with whom I have a chance to share since many years and several times per year experience and good practices in the field of industrial safety. I learned a lot from them.

The almost thousand master students from KU Leuven (Belgium), MINES Paris Tech (France) and from IMT Mines d'Ales (France), who challenge me for more about 20 years during my lectures in addition to the many interesting discussions with Professor Franck Guarnieri and Associate Professor Wim Van Wassenhove of MINES Paris Tech.

I finally want to mention Patrick Pouyanne, Chief Executive Officer of TOTAL. Under his leadership, it became clear to me how important the support of the Chief Executive Officer is in improving Safety. The specialized literature mentions the importance of the CEO sporadically but it was a real enrichment for me to experience this from the front row.

https://doi.org/10.1515/9783110632132-203

Contents

1 Introduction

Occupational health and safety at work and process safety are very closely connected and at the same time very different. Because both subjects are closely connected, people tend to bring them together under the heading of "safety." In many companies, experts in both subjects are part of the HSE department (HSE = Health, Safety and Environment). However, the two subjects are very different:
- The historical motivation to avoid personal or process-related accidents is different.
- The underlying thought process is different.
- The approach is different: preventing person-related accidents requires mainly leadership and knowledge about legislation while preventing technologically related accidents requires mainly scientific competency.

Occupational health and safety at work is a common quest for social justice in the relationship between employers and workers irrespective of the type of industry. The basic premise in occupational health and safety is:
- Workers contribute to their industry by "investing" time, physical capabilities and intellectual capabilities.
- Workers are compensated financially.
- The work undertaken has a potential impact on workers' well-being including their health (positive or negative, real or potential).
- The compensation received by the worker has to balance the worker's investment.

This quest for justice is not new. The code of Hammurabi, a code of law of ancient Mesopotamia dating back to about 1754 BC made reference to "the duty of workers."

Occupational health and safety is part of labor law which regulates the relationship between workers, employing entities, trade unions and the government. The Code of Federal Regulations (CFR) of the United States of America for instance, contains in Title 29, the principal set of rules and regulations issued by federal agencies regarding labor laws. Chapter XVII of CFR Title 29 (version 2014) comprises all the rules and regulations governing occupational health and safety.

The concepts of occupational health and safety (and associated legislation) evolved over time and, in particular, through the first and second industrial revolutions.

The early industries (textile, railroads and mining) that got off to a good start in Europe in the eighteenth century, involved hazardous working conditions. During the first industrial revolution, it was not clear for the society, at least at the beginning, how best to deal with accidents. In many cases, the employer was of the opinion that the worker didn't take enough precautions and that it was somehow the worker's fault that an accident occurred.

But the industrialization also introduced new technologies. Every time a new technology was put to practice, new types of accidents were "discovered." On 10 August

https://doi.org/10.1515/9783110632132-001

1909, about 85 people were killed in a fire in the Paris Metro. The material used in the construction of the wagons was adapted accordingly but "the occurrence of the accident and the lessons learned from the investigation were needed" to improve the design of the wagons. This type of engineering "learn and improve" is different from "occupational health and safety at work" because new technologies bring about unknowns and inherent uncertainties.

Uncertainties due to complicated technologies should not be confused with unknowns that can occur at a certain moment in time, in the field of Occupational Health and Safety (e.g., the uncertainties for exposure to certain products such as asbestos, benzene, etc.). After some time when more knowledge is available, the unknowns are removed and the measures to be taken to avoid any risk are known. Uncertainties associated with complicated technologies can simply not be removed because they are inherent to the fact that technologies can be complex and complicated.

Of course, in both cases (i.e., occupational health and safety at work and industrial safety associated with new technologies), the ultimate goal is to avoid accidents and improve safety. However, "occupational health and safety" and "industrial safety" have different origins and different driving forces.

"Process safety" is an engineering discipline that was developed in the process industries. Its purpose is to avoid losses and to improve reliability and availability of process installations. The use of new technologies and new processes gave rise to accidents that were not known hitherto. Engineers and scientists worked hand in hand to better understand the physics and chemistry behind these accidents. Despite all efforts, uncertainties continue to be an inherent part of the occurrence of these accidents. To deal with these uncertainties, the notion of "risk" was introduced. A broad set of methods and tools were developed to analyze and to assess the risks. Acceptance or rejection of the exposure to hazards is based on risk assessment criteria.

2 The run-up to the Industrial Revolution

2.1 Purpose of this chapter

Occupational health and safety or well-being at work is a subject that arose with the industrialization. The industrialization created the need for a clear defined relationship between the employer and the workers on how to avoid potential negative impacts of the work on the health of the workers. So, why were the rules (legislation) not introduced at the beginning of industrialization? One of the explanations is that this relationship was a new social concept and there was no experience with an industrialized society in which the relationship "employer–workers" was a foundation. The society before the industrialization was built on different beliefs and social structures. The purpose of this chapter is to give some insights on the evolution of the society toward the industrialization.

2.2 The importance of the First and Second Industrial Revolutions

Evolution of European and American societies since around 1700 has provided the basis for the current approach to Occupational Health and Safety worldwide. The First Industrial Revolution saw the transition to new manufacturing processes in Europe and the USA in the period from about 1760 to sometime between 1820 and 1840. This revolution is characterized by such developments as the transition from manual production to machine production, the discovery and introduction of new chemicals, the development of new iron production processes.

The First Industrial Revolution marks a turning point in human history because of its influence on every aspect of daily life. It was driven by a combination of demographics, urbanization and living conditions in towns. The changing needs in Great Britain and a little later in the rest of Europe and the USA, and demographic and societal changes created the seeds for an Industrial Revolution, which in turn changed society in such a way that further industrialization was inevitable. I will focus on Great Britain because of its dominance at that time. During the nineteenth century, the economy of Europe and the rest of the world was dominated by Great Britain. It acted as the workplace of the world. Halfway through the nineteenth century, a quarter of all international trade went through British ports (Evans, 2016). However, the causes and consequences of the First Industrial Revolution became entangled in a vicious circle in which one reinforced the other.

The First Industrial Revolution was followed by a period of rapid industrial development, primarily in Great Britain, Germany and the USA. It was characterized by the construction of railroads, large-scale iron and steel production, widespread

https://doi.org/10.1515/9783110632132-002

use of machinery in manufacturing, greatly increased use of steam power, widespread use of the telegraph, and increased use of energy sources (coal, petroleum and electricity). It also was the period during which modern organizational methods for operating large-scale businesses over vast areas came into existence. The Second Industrial Revolution is generally dated between 1870 and 1914 (the beginning WWI).

The main drivers for the First and Second Industrial Revolutions are:
– Population growth
– People's way of living and in particular:
    – Housing
    – Food
    – Clothing

The evolution of these drivers let to modifications in:
– Societal structure
– Warfare
– Manufacturing of goods and industry in general
– Transportation
– Banking

To make the First and Second Industrial Revolutions possible, science and engineering went through a fundamental transformation.

2.3 Population growth

At the end of the sixteenth century, the world population was estimated around 550 million from which about 110 million lived in Europe. Farmers leased land from the rich while poor people were on the lookout for jobs as wage laborers. The population of Europe grew from about 110 million in 1600 to 742 million in 2000. Figure 2.1 shows the growth (%) of the population in each century. In the seventeenth century for instance, the growth was 13% (number of people in Europe at the end of the seventeenth century compared to the number of people at the beginning of the seventeenth century) while in the eighteenth century, the population in Europe grew by 76% and in the nineteenth century by 122%.

The population growth was not at the same pace in different regions of Europe. The number of people living in (greater) London increased from about 60,000 in 1500 to 600,000 in 1700 and to about 1,000,000 in 1800. The population in London grew, in other words, with a factor of 16.7 between 1500 and 1800. This growth was much faster than in other European cities. In Paris for instance, of the same period the population grew with a factor of 2.8. It was also much faster than in the rest of England, where the overall population growth between 1500 and 1800 was about 3.1 (in France it was 1.6).

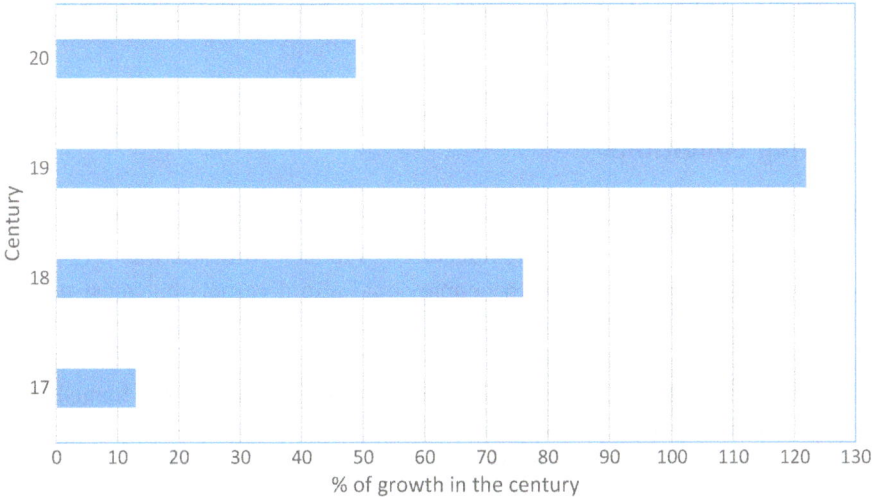

Figure 2.1: European population growth in different centuries.

This rapid urbanization of a number of cities (and in particular, London) combined with the change in living standards probably explains why industrialization started in England.

After 1800, the population of many cities increased rapidly. The population of Glasgow, for example, which was around 77,000 in 1800, grew to 742,000 in 1900. The population of Berlin grew from 172,000 in 1800 to 1,122,000 in 1880. The population of Lisbon, which remained virtually unchanged until well into the nineteenth century, suddenly rose from 232,000 in 1880 to 435,000 just before World War I.

The rapid population growth in major European cities was accompanied by a wave of emigration. It is thought that in the period from 1815 to 1914, a total of around 60 million citizens left Europe (Evans, 2016). More than half (34 million) fled to the USA. All in all, more than a quarter of the natural population growth in Western Europe between 1841 and 1915 was swallowed up by emigration, with a net population decrease of 35 million citizens.

As a result, the population balance in the world began to change and parts of the world became more connected. Halfway through the nineteenth century, the population in the USA was not much larger than that of Great Britain while on the eve of World War I, it had grown to 92 million.

In the nineteenth century, the population of Italy and Spain grew twice as fast as in France while in Great Britain the population grew eight times faster than in France.

The rapid population growth which started in Great Britain and caused industrialization, continued in other parts of Europe. Population growth in Germany was accompanied by turbulent industrialization and lightning-fast urbanization. As a result, the demand for electricity, street lighting, chemically dyed clothing, canned food, etc. increased in turn accelerating industrialization.

2.4 Living conditions

In the sixteenth century, people lived predominantly in small villages and relied on farming. Poor people lived in simple huts of 1 to 3 rooms. Furniture was very basic. The floors were of hard earth. Rich people could afford carpets but most people didn't have them. The flushing toilets were invented by Sir John Harrington in 1596 but most people used chamber pots or chess pits. They produced their own food and made their own clothing, furniture and tools. Most manufacturing was home-based or in small, rural shops, using hand tools and simple machines.

Towns were dirty and smelly because of the absence of sewers and drains. Waste was thrown in the streets. Vermin such as rats were common. Most streets were not paved. When people traveled, they did so on horses and hence paved streets were not needed. Goods were mainly transported by waterways or over land packed on horseback.

In the seventeenth century, many poor people migrated from the countryside to the towns in search of jobs. Outbreaks of plague were common and easily killed large parts (e.g., 10%) of the exposed population. Doctors were rare and expensive and in many cases, they didn't understand the causes of diseases such as smallpox and syphilis.

A "safe working environment," the way we see it today, did not exist. People's primary concern was day-to-day survival. Religious beliefs meant that it was common to consider illness or accidents as "the will of GOD."

John Evelyn (1620–1706), an English writer, called London the "wooden, northern, and inartificial congestion of houses" and warned of the fire hazards due to congestion and an abundance of wood. On Sunday September 2, 1666, shortly after midnight, a fire started at the bakery on Pudding Lane. The Great Fire of London swept through the central parts of London and continued to burn until Thursday, September 6, 1666. The fire consumed 13,200 houses and 87 parish churches. This Great Fire of London was probably the reason why more and more houses were subsequently built of brick or stone. Linen for windows was gradually replaced by glass.

Bricks were not new – man has used brick for building purposes for thousands of years, but demand increased in the nineteenth century. Bricks were made by hand until about 1885. Once the Industrial Revolution took hold, brickmaking machinery was introduced. Consequently, the range of clays that could be made into brick was greatly increased. Handmade brick production capacity was up to 36,000 bricks per week but by 1925, a brickmaking machine made 12,000 bricks a day. As brick structures could be built much quicker and cheaper, they replaced other raw materials like stone or rock. During the building boom of the nineteenth century, when more than 10 billion bricks were produced annually, many American cities like Boston and New York used locally made bricks.

2.5 Food

During the 1500s and 1600s, European traders brought back all kinds of new foods from places they sailed to around the world. The rich began to eat sugar and ginger from India (combining these two new foods together gave us the gingerbread men).

The life of the middle class (e.g., merchants) improved and at the end of the seventeenth century, it is estimated that about half the population of Great Britain could afford to eat meat every day. People also ate more and more with forks. New foods such as bananas and pineapples were introduced. Many coffee houses opened in which merchants met to discuss business.

However, until late in the eighteenth century, most citizens resided in small, rural communities, where their daily existence revolved around farming. Life for the average person was difficult, as incomes were meager and malnourishment was widespread.

The poor made their own food and typically had one meal a day made from mixed grain with water and vegetables. Only the rich (e.g., Tudors in England) could afford sweet food, sugar, imported food and meat.

By the 1800s, Europeans began to eat potatoes, and then tomatoes, both from South America. They cooked these foods in new ways, inventing French fries and potato salad, and adding tomato sauce to pizza and spaghetti and gazpacho. Food scientists developed the chocolate bar in the 1800s. As modern countries formed, governments encouraged their citizens to eat food from different regions. So, instead of only Bretons eating crepes, people all over France began to eat them and instead of only people from Bavaria eating pretzels, they were eaten by people all over Germany (Carr, 2019).

In the 1900s, even more new foods came to Europe from around the world. Immigrants to Europe from Asia and Africa brought with them new foods like couscous, tofu, peanut oil, safflower oil, and curry (Carr, 2019).

The rapid population growth forced the agricultural industry to drastically increase production:
- The potato yield in Germany nearly doubled between 1875 and 1884 and between 1905 and 1914. In 1910, Germany produced a third of all potatoes grown worldwide.
- The production of sugar beet tripled in the same period.
- the cattle stock in Germany increased from 16 million in 1873 to 21 million in 1913, while the number of pigs increased from 12 million in 1892 to 26 million just before WWI.

Inventions brought about the necessary means to realize these efficiency increases: use of phosphates, use of fertilizers (it is generally considered that the modern science of plant nutrition started in the nineteenth century by the work of German chemist Justus von Liebig), milk centrifugal separator (first manufactured by Gustaf de Laval 1845–1913), making it possible to separate cream from milk faster and more easily.

2.6 Clothing

For a long time (up to around 1200), clothing in the Western world changed very little. Twelfth century European fashion was simple and differed only in details from the clothing of the preceding centuries. Apart from the elite, people at that time had low living standards, and clothes were of simple cuts, home or village made. The elite imported silk cloth (silk, and probably cotton, textiles were known in China as early as 3630 BCE). Most people, however, wore only wool or linen, usually undyed, and leather or fur from locally hunted animals (http://www.textileasart.com/weaving.htm).

The thirteenth century saw great progress in the dyeing and working of wool, which was by far the most important material for outerwear. Linen was increasingly used for clothing that was directly in contact with the skin. Unlike wool, linen could be laundered and bleached in the sun.

In the fourteenth century, wool remained the most important material for clothing, due to its numerous favorable qualities, such as the ability to take dye and being a good insulator. Trade in textiles continued to grow throughout the century, and formed an important part of the economy in many parts of Europe.

In the fifteenth and sixteenth centuries, clothes for most people in Europe were simple and practical. Wool remained the most popular fabric for all classes, followed by linen and hemp. Wool fabrics were dyed in rich colors, such as reds, greens and blues (Takeda and Spilker, 2010). The seventeenth and eighteenth centuries are characterized by rapid changes in fashion and the use of new materials. Women, for instance, started to wear frames of wood under their dresses.

Before the eighteenth century, the manufacture of cloth was performed by individual workers in their homes. The rapid population growth (see above) increased the demand for clothes and in the mid-eighteenth century, artisans started inventing ways to become more productive. Silk, wool and fustian fabrics were being eclipsed by cotton, which became the most important textile.

The 1912 book *Manufacturing in Philadelphia, 1683–1912*, listed among the city's textile products carpets, cordage, jute, linen goods, nets and seines, cotton goods including small cotton wares, hosiery and knit goods, shoddy (a recycled wool product), silk fabrics, woolen and worsted products, wool pulling and wool scouring, felt goods, wool hats, and fur felt hats.

John Kay's 1733 flying shuttle enabled cloth to be woven faster and in greater widths. Cotton spinning used Richard Arkwright's Water Frame, James Hargreaves's

Spinning Jenny, and Samuel Crompton's Spinning Mule (a combination of the Spinning Jenny and the Water Frame). This was patented in 1769 and so came out of patent in 1783. The end of the patent was rapidly followed by the erection of many cotton mills. Similar technology was subsequently applied to spinning worsted yarn for various textiles and flax for linen.

2.7 Social structure

Until the sixteenth century, society was characterized by a relationship between landlords and vassals. A Lord is a person with various assets including land and factories. A vassal is a person regarded as having a mutual obligation to a Lord. The vassal used the land or worked in the factory and his labor or service generated wellness for the Lord. Stewards were intermediate persons who ensured that the rules (e.g., paying a rent for the use of the land) established between the Lord and the vassal were followed. The obligations of the vassal could include labor, service and military support by knights in exchange for certain privileges, usually including land held as a tenant. This organization (called feudalism) was a combination of legal and military customs in medieval Europe and flourished between the ninth and sixteenth centuries. Broadly defined, it was a way of structuring society around relationships derived from the holding of land in exchange for service or labor.

An important revolution swept across social structures in the nineteenth century. The aristocracy, the nobility and the rich landowners who had been in power for a long period were gradually replaced by elected members of parliament, who gained legitimacy through elections in which an increasing number of the population participated.

The nineteenth century saw many revolutions against the established order for freedom and political and social democracy. Several waves can be distinguished. A first wave of liberal movements was against the old regime of aristocrats, who wanted to be reinstated after the departure of Napoleon (in 1815). The liberalism centered around the individual freedom, being more important than the state.

A second wave of democratization was followed by social aspiration. In 1801, for instance, the unreformed House of Commons in Great Britain was composed of 658 members with two types of members: County members representing landowners, and Borough members representing the mercantile and trading interests of the kingdom. Hence the power was left in the hands of the aristocracy and the middle classes. The number of voters that elected these people was small. The population in England, Wales and Scotland was about 16.5 million, of which only 439,000 voted.

In France, the situation was worse. In 1831, France had a population of 32 million but there were only 165,000 voters (Sherman, 1971). France followed the USA in the decision on universal adult suffrage. One of the first acts of the Provisional Government in France in March 1848 was the adoption of universal suffrage. The electoral body passed without transition from about 250,000 citizens to 9,500,000 citizens (Rémond, 1974).

In Germany, democratization began in 1871 at the initiative of Bismarck but it was only in 1919 that the Germans would be granted the right to vote. In 1861, the year following the unification of Italy, the legal country had no more than 900,000 voters out of a population of 22 million. Several electoral reforms were adopted (1882, 1912) but universal suffrage was not established until 1919. Between 1848 and 1918, most other countries in Europe adopted provisions that led to universal suffrage: in the Netherlands in 1896, in Belgium in 1893, in Norway in 1905, in Sweden in 1906.

The changes made in the British political system between 1832 and 1884 were important. The electorate increased substantially in size from approximately 366,000 in England and Wales in 1831 to slightly fewer than 8 million in 1885. Parliamentary seats were redistributed to give greater weight to larger towns and cities. The Ballot Act of 1872, which introduced secret ballots, made it far more difficult for voters to be bribed or intimidated.

In many countries, there was still an electoral law, which stipulated that only the rich could be elected and could participate in political life. The Belgian Constitution of 1830 stipulated that the taxpayer must have paid at least 20 francs in taxes in order to have voting rights. In practice, this meant that less than 2% of the population had voting rights. Only in 1893, under pressure from the socialist movement, this electoral system was replaced by universal suffrage for men aged 25 years or more. In 1850, the number of voters in Belgium was 79,998; in 1890, this had only increased to 116,090. After the introduction of the general multiple voting right (in 1948), this became 1,370,687.

The state structures increasingly took over policy. The decline of the aristocracy was an important topic in the press at the end of the nineteenth century. The nobility had obtained its power through the dominance of the lord over his serfs. In the nineteenth century, important functions were transferred to the state, and the nobility became ordinary citizens.

But the difficult living conditions due to industrialization and the feeling of injustice by the working class resulted in a wave of revolutions in 1848, the most widespread in European history. The revolution started in February 1848 in France but very quickly over 50 countries were affected. There was no significant coordination but the major contributing factors were very similar: widespread dissatisfaction with political leadership, quest for participation in government and democracy, demands for freedom of the press, etc.

A sharp fall in grain prices in the seventies of the nineteenth century brought many large landowners into financial difficulties. In Great Britain, the wheat price fell by half between 1871 and 1901 and the area on which wheat was grown was reduced from 14,164 km^2 to 6,070 km^2.

The power of the aristocracy and wealthy landowners was diminishing. Some lost their wealth while others switched to the train of the industrial revolution. Prince Christian Kraft zu Hohenloche-Öhringen, duke of Ujest (1848–1926), for example, employed more than 5,000 people and was the world's largest zinc producer. By 1910, four fifths of his wealth was invested in industry. Prince Guido Henckel von Donners-marck (1830–1916) extended his industrial kingdom into chrome, paper and cellulose and invested in companies in Austria, France, Hungary, Italy and Russia. Neverthe-less, it became increasingly difficult for the aristocracy to dominate the lightning-fast development of industry (Evans, 2016).

2.8 Transportation

Before the industrialization, textiles were made mainly in people's homes (giving rise to the term "cottage industry"), with merchants often providing the raw materi-als and basic equipment, and then picking up the finished product. Workers set their own schedules under this system, which proved difficult for merchants to reg-ulate and resulted in numerous inefficiencies.

Raw materials and finished goods were hauled and distributed via horse-drawn wagons, and by boats along canals and rivers (see Figure 2.2).

Figure 2.2: "Porteurs de farine – scène parisienne 1885" (Louis Carrier-Belleuse 1848–1913).

More people, more crowded towns and changing habits created the need for more food, more textiles, more iron, more coal and more glass. The growth of the popula-tion in particular, in London was so fast that social and political structures that were useful before 1600 were not appropriate for the more complex society.

This complex society needed more materials (iron, glass, and textile), an increased efficiency and productivity in manufacturing and more efficient means for transport and communication. Fulfilling the needs of 60,000 habitants in London (anno 1500) cannot be done in the same way as fulfilling the needs of 1,000,000 habitants in the same London (anno 1800).

New transportation modes were needed. In the Industrial Revolution, inland canals were built in England and later the USA before the development of railways. Matthew Murray proved the viability of the steam engine in 1812. The modern rail system was developed in England, progressing to steam locomotives. George Stephenson, the father of railways, built 16 experimental locomotives for use from the year 1814 to 1826. An underground railway was first built in 1863 in London. In 1880, electric trains and trams were developed.

The transition to new transportation modes required huge amount of workers. Society considered "working" as a duty for everybody. Those who were not old or disabled were "forced" to work.

2.9 Transfer of science and technology

Theories of the Greek philosophers that were held in very high esteem for centuries started to be tested by experiments and were demonstrated to be wrong. Scientists held meetings to discuss their experiments and observations. The Club of the Royal Society for instance was founded in 1662.

Robert Boyle, a great chemist, published in 1642, "The Skeptical Chemist" and Isaac Newton, a great physicist, published in 1687, his "Principia Mathematica." In 1712, Englishman Thomas Newcomen developed the first practical steam engine (which was used primarily to pump water out of mines). By the 1770s, Scottish inventor James Watt had improved on Newcomen's work, and the steam engine went on to power machinery, locomotives and ships.

In the early eighteenth century, Englishman Abraham Darby (1678–1717) discovered a cheaper, easier method to produce cast iron, using a coke-fueled furnace. In the 1850s, British engineer Henry Bessemer (1813–1898) developed the first inexpensive process for mass-producing steel. Both iron and steel became essential materials, used to make everything from appliances, tools and machines, to ships, buildings and infrastructure.

An enormous number of discoveries were made in the field of natural sciences. This was mainly the result of systematically experimenting in the laboratory. A small selection:
- Wilhelm Röntgen discovered the X-rays in 1895.
- Heinrich Rudolf Hertz confirmed the theories of electromagnetic waves.
- Edison invented the electric light bulb in 1879 and the gramophone in 1877.
- Robert Koch discovered the tuberculosis in 1882 and the cholera bacillus in 1883.
- Louis Pasteur found a rabies vaccine in 1881.
- The first film was shown in 1895.

Many results of physics research were quickly applied in technology, either for the development or improvement of new products, or to improve the production process. Soon technical research became a condition for commercial success and therefore an extension of industry.

A worldwide transfer of technology and ideas developed. More and more international conferences and congresses were held. The *Exposition des produits de l'industrie française* was a public event organized in Paris, France, from 1798 to 1849. The purpose was "to offer a panorama of the productions of the various branches of industry with a view to emulation."

These French exhibitions gave rise to the Great Exhibition of the Works of Industry of All Nations or The Great Exhibition in Hyde Park in London, from 1 May to 15 October 1851. It was the first in a series of World's Fairs, exhibitions of culture and industry that became popular in the nineteenth century. World exhibitions in the nineteenth century were largely focused on trade and displayed technological advances and inventions. World expositions were platforms for state-of-the-art science and technology from around the world: London (1851), New York (1853), London (1862), Philadelphia (1876), Paris (1889), Chicago (1893), Brussels (1897), Paris (1900), Buffalo (1901), St. Louis (1904), Ghent (1913) and San Francisco (1915). Posters of the World Exhibitions in Paris and Ghent are shown in Figure 2.3.

Figure 2.3: Posters of Paris and Ghent World Exhibition.

2.10 Banking

The industrialization required new finance mechanisms and new mechanisms for exchange. Banking developed (the Bank of England for instance was founded in 1694) and goldsmiths started to give receipts in the form of notes for deposits of gold. Tradesmen began to exchange these notes as a form of money. Goldsmiths realized that not all of their customers would withdraw their gold at the same time and they issued notes for more gold than they actually had.

Projects became larger, international and required considerable funding. The earliest known state deposit bank, Banco di San Giorgio, was founded in 1407 in Genoa, Italy. However, modern banking practices, including accepting deposits and creating credit while holding reserves at least equal to a fraction of the bank's deposit liabilities emerged in the seventeenth and eighteenth centuries.

Merchants started to store their gold with the goldsmiths of London, who issued receipts ("Banknotes") as payable to the bearer of the document rather than the original depositor. The Bank of England was the first to begin the permanent issue of banknotes, in 1695. By the beginning of the nineteenth century, a bankers' clearing house was established in London to allow multiple banks to clear transactions.

Rothschild family's banking business pioneered international high finance during the industrialization of Europe and was instrumental in supporting railway systems across the world and in complex government financing for projects such as the Suez Canal.

Major nineteenth century businesses founded with Rothschild family capital include:
- Alliance Assurance (1824) (now Royal & Sun Alliance).
- A rail road transport company "Compagnie des Chemin de Fer du Nord" created in Paris in 1845.
- The Rio Tinto mining company founded in 1873, when a multinational consortium of investors purchased a mine complex from the Spanish government in Huelva, and one of the world's largest metals and mining corporations.
- Eramet, a French multinational mining and metallurgy company was founded with the funding of the Rothschild family in 1880.
- De Beers Group, an international corporation that specializes in diamond exploration, mining and diamond retail was founded in 1888 by British businessman Cecil Rhodes, who was financed by the South African diamond magnate Alfred Beit and the London-based N M Rothschild & Sons bank.

Capital (via investments) flowed to foreign countries and overseas territories. In the period 1870 to 1913, 21% of British foreign investments and 16% of German foreign investments went to America (Evans, 2016).

2.11 Resume: population growth, social changes and industrialization – a virtuous circle

This chapter described how society evolved in the period 1700 to 1900. Rapid population growth coupled with increased urbanization brought fundamental changes to the social system.

More food had to be produced and the demand for new materials increased. Global trade accelerated, which in turn increased the demand for better roads, better and faster transport, a new banking system.

These changes formed the basis of the transition to new manufacturing processes in Europe and the USA, now known as the First Industrial Revolution.

This First Industrial Revolution created the demand for more and better machines, new materials, more energy, new products, etc., which resulted in the Second Industrial Revolution. The large land owners invested more and more money in new industries, factories and mines.

The Second Industrial Revolution was a period of rapid industrial development, characterized by the construction of railroads, large-scale iron and steel production, widespread use of machinery in manufacturing, greatly increased use of steam power, widespread use of the telegraph, use of petroleum and the beginning of electrification. It also was the period during which modern organizational methods for operating large-scale businesses over vast areas started to be used.

The Second Industrial Revolution gave rise to large industries that employed many people; these included construction of railways, textile factories, steel factories and coal mines.

The social and economic changes that engulfed Europe in the nineteenth century were exceptional and never seen before. In 1850, the percentage of the population working in agriculture was 75% in Italy, 60% in Germany, 52% in France and 22% in Great Britain. The rapid industrialization brought these figures down at the turn of the century (1900) to 60% in Italy, 35% in Germany, 42% in France and 9% in Great Britain.

The nineteenth century was the century of revolutions. The French Revolution of 1789–1799 was a political uprising that massively reduced the power and privileges of the nobility and the clergy. This political revolution which started in France spread over the entire continent. The traditional land-owner aristocracy was undermined by the forces of economic change and political reforms. A new hybrid social elite emerged based on citizen values such as economy, hard work, freedom, austerity, equality and responsibility. These values dominated society and politics in large parts of Europe and were reflected in sanitation and hygiene, reforms in criminal law and participation of the respectable working class.

The power of the state grew and nobility rights to self-government were abolished. Organized local government replaced noble bailiffs and courts. Nobility groups were pushed aside by elected parliaments, who increased their legitimacy by extending their electoral rights to a larger group of voters.

The knowledge of the potential negative impact of working conditions on people's health was not new at the start of the industrialization:

– In 1566, under the reign of Charles IX (King of France between 1560 and 1574), the status of the roofers of Paris was defined. The ordinance required the implementation of measures to prevent falls from house roofs for the "poor workers."
– In 1700, the Italian doctor Bernardino Ramazzani had published a treatise on the diseases of craftsmen. He studied more than 52 professions and highlighted the relationship between work and the worker's health (Léoni, 2017).
– Chimney Sweeps carcinoma, a sort of skin cancer was already identified in 1775 by Sir Percival.
– Silicosis, black lung and eye problems of mine workers were discussed in the eighteenth century (Oppert, 1866).
– Doctors had already drawn attention from early in the nineteenth century to a number of illnesses in the textile industry such as bronchitis, indigestion, varicose and brown lung disease.

But the combination of the liberal ideology with a rapid change in the social structure of the society resulted in the following situation relative to occupational health and safety at work in the first half of the nineteenth century:

– A very diverse and unregulated working class. The industry was booming and everybody (young or old, man or woman) were allowed to perform the work (there were no legal restrictions).
– A liberal ideology, which was a counter-reaction against the aristocracy. The French Revolution in 1789 forced many aristocrats into exile, relieving them of their lands and power. People were fed up with a government with aristocrats deciding for them what was right and wrong.
– Despite the many occupational accidents and occupational health problems, it was difficult for the government to intervene because of the liberal thinking that prevailed since the end of the eighteenth century.
– The legal framework required to guide the relationship between employer and employee in the new society was not in place at the start of the Industrial Revolution, and because of the previous points it would become a slow learning process with progressive insight.

3 Historical evolution of occupational health and safety in the nineteenth century

3.1 Industrialization in the nineteenth century takes a high toll

3.1.1 Main industries: textile, railroads and mining

The First and Second Industrial Revolution resulted in a society that was very different than what existed till the seventeenth century.

The wide-ranging social impact of both revolutions included the remaking of the working class as new technologies appeared. The changes resulted in the creation of a larger, increasingly professional, middle class, the decline of child labor and the dramatic growth of a consumer-based material culture.

By 1900, the leaders in industrial production were Britain with 24% of the world total, followed by the USA (19%), Germany (13%), Russia (9%) and France (7%). Europe together accounted for 62%. American railroad mileage tripled between 1860 and 1880, and tripled again by 1920, opening new areas to commercial farming, creating a truly national marketplace and inspiring a boom in coal mining and steel production

The mechanization of textile manufacture gave rise to serious labor issues. Work that had previously been done by hand in homes and shops was now centralized and automated in large mills that employed hundreds or thousands, such as those of William Logan Fisher (1781–1862) in Germantown. Working conditions in the mills were often miserable. Employees worked 12- or 14-h days, 6 days and sometimes 7 days a week. Children made up a considerable percentage of the textile workforce and were also subjected to terrible working conditions.

In 1850, Philadelphia had approximately 12,000 textile workers. By 1882, it had over 60,000, employed by nearly 1,000 different firms.

Allegheny County in Pittsburgh, for instance, had in 1908 a population of 1,000,000 of whom 250,000 were wage earners. Seventy thousand were working in the steel mills, 50,000 were working on the railroads and 20,000 in the mines. Hence, these were great employment groups and also great accident groups. In the UK, by 1900 over 620,000 people worked for the railways (about 5% of the total population).

The first railway to be approved by the British parliament was the Stockton and Darlington Railway. The 40-km-long railway, for which Stephenson was the provider of locomotives, was granted permission in 1821 and was constructed in 1825. In 1830, Stephenson engineered the Liverpool and Manchester Railway, in which the locomotive "Rocket" gained worldwide attention, thus starting the Railroad Era. By 1839, the total length of railway lines in Great Britain grew well over 1,500 km, compared to the less than 90 km of 1829.

https://doi.org/10.1515/9783110632132-003

3.1.2 Statistics on accidents for railroad workers

At the beginning of the nineteenth century, there was no legislation that required accident reporting. It is therefore not possible to obtain official statistics on accidents at
work. We can nevertheless form an idea through incomplete figures.

By 1850 about 250,000 workers had laid down 3,000 miles of railways across Britain connecting people like never before. These workers were called (railway) navies.
The railway navies soon came to form a well-paid distinct group performing hard and
dangerous work. They were assembled in huge armies of workers, men and women
tramping from job to job. They worked in harsh conditions and gained reputation for
hard living. Their work conditions were not safe, and housing and sanitation were not
adequate. The death rate during the Woodhead Tunnel construction between 1839
and 1852 was higher than that of the soldiers who fought at the Battle of Waterloo. The
number of navies killed reached about 500 per year in the 1880s and the Woodhead
Tunnel scandal led to a parliamentary enquiry. In 1900 alone, over 16,000 workers
were injured or killed.

Construction work of a single stretch of railway line over a 6-year period (Bronstein, 2008) in the mid-nineteenth century resulted in:
- 32 deaths
- 637 injuries that nowadays would be classified as lost time accidents:
  - 23 compound fractures
  - 74 simple fractures
  - 140 blast burns, severe bruises, cuts and dislocations
  - 400 trapped and broken fingers, injuries to the feet, lacerations of the scalp,
    bruises and broken shins

Once the railways were built and in operation, the railroads continued to subject
workers to dangers that were mainly caused by organizational features such as bad
schedules that created uncertainty about the time that a train was passing by. Brake-
men had to climb on the top of the cars to set the train brakes by hand. Between July
and December 1855, out of a total at-risk workforce of approximately 22,300 workers,
the British Government observed that 63 railway workers were killed and 54 were seriously injured (Clark, 1966).

Statistics on railway accidents show that the incidence of accidents was about
1 worker per 250 in 1847. It decreased to 1 per 1,125 by 1870 (Bronstein, 2008).
About 100 years later (in 1996), the annual risk of fatal injuries is about 1 per
4,200 for locomotive and train crews and about 1 per 5,000 for maintenance personnel (Savage, 1998).

The number of fatalities per 1,000 workers per year for British and American railroad workers in 1889 is given in Table 3.1 (taken from https://eh.net/encyclopedia/
history-of-workplace-safety-in-the-united-states-1880-1970-2/).

Table 3.1: Fatality rates per thousand railroad workers per year in 1889.

	American	British
Railroad workers	2.67	1.14
Trainmen	8.52	4.26

3.1.3 Statistics on accidents in mining

While accident to railway employees were common, coal mine accidents received the greatest amount of attention of all mid-nineteenth-century work-related accidents.

In the counties of Tyne and Wear alone, between 1710 and 1849 at least 120 explosions killed a total of 1,813 miners. A Parliamentary Committee (Parliamentary Papers, 1854) collected information on 90 colliery explosions throughout Great Britain between 1846 and 1852, which killed 1,084 workers. But explosions were not the only cause of fatal accidents in mines. It was not even the major cause.

James Mather (1853) found that 2,143 miners were killed in Great Britain from 21 November 1850 to 31 December 1852 in England, Wales and Scotland (Table 3.2):
- 744 were killed in roof falls (35%)
- 645 were killed in explosions (30%)
- 457 were killed in shaft accidents (21%)
- 297 were killed from other causes (14%)

Table 3.2: Fatalities in mining per district in 1850–1852.

District	Locations of mines	No. of deaths	Percentage of deaths per main cause			
			Explosion	Falls from roof	Shaft accidents	Others
1	Staffordshire, Shropshire, Worcestershire, Lancashire, Cheshire, North Wales	757	25	38	26	11
2	Northumberland, Durham, Cumberland	313	38	26	14	22
3	Scotland	236	33	37	21	9
4	South Wales	362	29	31	19	21
5	Yorkshire, Derbyshire, Nottinghamshire, Leicestershire, Warwickshire	254	41	28	21	10
6	Staffordshire, Monmouthshire, Shropshire	221	23	47	18	12

The data were collected by different inspectors. Mather arranged the data under six separate districts. The detailed data show that the safety problem was widely spread despite some variation in contribution of the main causes.

Of 234 injuries at the Haswell and East Holywell collieries over the period 1849–1851 (Bronstein, 2008), no fewer than 177 (76%) were directly related to falls of coal or stone or to contact with coal. A similar situation was observed at the Ince Hall Coal and Channel Works in Wigan. Over the period 1850–1852, 34 out of 44 fatal accidents were related to roof falls and crushing tubs.

Between September 1862 and July 1863, John Towers recorded:
- 127 workers killed in explosions;
- 20 crushed under wagons or tubs;
- 40 died from falls down the shaft;
- 51 were crushed by falling roofs;
- 5 fell from the cage.

One hundred and twenty-seven workers killed by explosions included, however, 59 workers who were killed by one major accident (the Edmunds Mine explosion of 8 December 1862 which killed 59 and injured another 16 workers).

Table 3.3 gives some fatality rates per thousand workers per year (taken from https://eh.net/encyclopedia/history-of-workplace-safety-in-the-united-states-1880-1970-2/).

Table 3.3: Fatality rates per thousand workers per year.

Period	American Anthracite	Great Britain
1890–1894	3.29	1.61
1900–1904	3.13	1.28

Eastman (1910) gives the number (Table 3.4) of men killed in falls from roofs and coal in France, for each 1,000 workers:

Eastman compares different countries and concludes that the situation in the USA is worse than in European countries (Table 3.5).

3.1.4 Statistics on accidents in textile industry

Accidents were fairly common also in textile factories. They were mainly caused by moving parts of new machinery in combination with poor housekeeping and poor layout. In a 6-month period in 1849, English and Scottish factory commissioners reported 1,114 accidents caused by machinery and 907 other accidents, resulting in 22

Table 3.4: Fatality rates per thousand workers per year due to falls from roofs and coal in France.

Period	Fatalities per 1,000 employees per year
1871–1875	1.26
1876–1880	0.95
1881–1885	0.88
1886–1890	0.63
1891–1895	0.59
1896–1900	0.47

Table 3.5: Five-year average fatality rates per thousand workers per year in coal mines in different countries.

Period	Fatalities per 1,000 employees per year
France (1901–1905)	0.91
Belgium (1902–1906)	1.00
Great Britain (1902–1906)	1.28
USA (1902–1906)	3.39

deaths and 109 amputations. A parliamentary return in 1870 noted the following number of accidents per worker (Bronstein, 2008):
- the cotton industry: 1 accident in 176 employed;
- the woolen industry: 1 accident in 230 employed;
- the worsted industry: 1 accident in 433 employed;
- the silk industry: 1 accident in 1,074 employed.

Certain sectors of the textile industry had a similar accident rate to that experienced on the railroad, although in general textile mill injuries were less serious.

By the 1870s, England required that factories be clean, well ventilated and not overcrowded, and that hoists, exposed gears and other dangerous devices be fenced or railed off. In 1875, the Massachusetts bureau again called for legislation and included a draft law covering machine guarding, fire protection, elevator safety and adequate ventilation. Massachusetts passed the first factory safety and health law in America in 1877 and established an inspection force in 1879. In the neighboring state of Connecticut, there was no factory inspection law until 1887, 10 years after the Massachusetts act. In 1884, the first state west of the Appalachians to pass a factory inspection act was Ohio. With the strong support of organized labor, a Department of Workshops and Factories was created, which included an "Inspector of the Sanitary Condition, Comfort, and Safety of Shops and Factories." The inspector was to visit all factories employing 10 or more persons, to see that machinery guarding, lighting, ventilation, fire exits and so on were adequate. In 1902, Iowa became one of the last northern states to enact factory inspection legislation.

Steam boilers were also a particular source of danger. In the 5-year period 1863–1868, 875 people were killed by exploding boilers (Bronstein, 2008).

Engineer Robert Vinçotte made a complete study of boiler explosions in Europe, which he published in "Chroniques de l'industrie" of 1872. Table 3.6 gives the number of boiler explosions in the period 1864–1869 in Belgium and France (CEOC, 2011). The main cause of the explosions was poor conditions of the boilers (~50% in France and ~73% in Belgium for that period).

Table 3.6: Number of boiler explosions in France and Belgium in the period 1864–1869 (CEOC, 2011).

Year	Belgium			France		
	Explosions	Deaths	Casualties	Explosions	Deaths	Casualties
1864	4	6	0	16	40	18
1865	8	24	27	12	25	25
1866	5	3	1	18	15	22
1867	6	4	6	19	28	36
1868	7	6	1	24	28	35
1869	11	10	6	18	15	20

3.2 Influence of the liberal thought in the nineteenth century

3.2.1 The century of the revolutions

The nineteenth century can certainly be called the century of revolutions. This is very well described by the historian Rémond (1974). Before the 1789 revolution in France, it was accepted by everyone that the aristocracy held the power and that they decided on all important aspects of society.

The outbreak of the French Revolution and the French Revolutionary wars had been received with great alarm by the rulers of Europe's continental powers. The French Republic, under the Directory (a five-member committee that governed France from 2 November 1795, when it replaced the Committee of Public Safety, until 9

November 1799), suffered from heavy levels of corruption and internal strife. It was overthrown by Napoleon Bonaparte in the Coup of 18 Brumaire, and replaced by the French Consulate.

There is a lot of debate among historians as to whether one should speak about French Revolutionary wars followed by Napoleonic wars or refer to the seven major wars between 1792 and 1815 as the Coalition Wars. British historians occasionally refer to the nearly continuous period of warfare from 1792 to 1815 as the Great French War, or as the final phase of the Anglo-French Second Hundred Years' War, spanning the period 1689–1815 (Buffington, 1929; Scott, 1992).

The Napoleonic wars are often categorized into a series of conflicts (called "the coalitions") between France and different Allied powers. Napoleon was defeated by the Sixth Coalition. Coalition troops captured Paris at the end of March 1814 and forced Napoleon to abdicate in early April. He was exiled to the island of Elba, and the Bourbons were restored to power by the Treaty of Paris, signed on 30 May 1814. The treaty recognized the Bourbon monarchy in France, in the person of Louis XVIII, because it was between Louis XVIII (the king of France) and the heads of states of the Coalition great powers. It was also decided to organize the Congress of Vienna, a meeting of ambassadors of European states chaired by Austrian statesman Klemens von Metternich, and held in Vienna from November 1814 to June 1815. The objective of the Congress was to provide a long-term peace plan for Europe by settling critical issues arising from the French Revolutionary Wars and the Napoleonic Wars.

All monarchical Europe flows to Vienna. Fifteen members of royal families sit alongside 200 princes and 216 heads of diplomatic missions. Many lobby groups are also present: the representatives of the Jews of Germany, the Knights of Malta and the abolitionists of the black slave trade, not to mention the inventors of recipes to ensure the peace of the world. There was a lot of fun, many receptions and almost no plenary sessions. Decisions were taken elsewhere.

This restoration of the monarchy is a counterrevolution. But the restoration failed to restore the pre-1789 situation. The monarchy understood that it had to make compromises. A very large part of the population refused to go back to the Ancien Regime.

The administrative organization that was put in place by Napoleon was effective and had to remain in place. The Napoleonic codes recognized the civil liberties and in all the countries which saw social transformations (Netherlands, Belgium, France, etc.), these freedoms were preserved.

The wave of revolutions in the first half of the nineteenth century was a liberal struggle. Liberalism fights against absolutism. Power had to be fragmented, limited, institutionalized and decentralized. The individual had to be at the center and be free.

In the first half of the nineteenth century, this liberal thought spread through all layers of the population. This philosophy had a direct influence on the way in which safety was considered in the workplace. The liberal mainstream thought in particular in the first half of the nineteenth century influenced the relationship between the worker and their employers.

3.2.2 Employers–workers: a new relationship

The old doctrine under common law was that "the King" (in other words, the accepted form of government) "can do no wrong."

The relationship between a worker and an employer was different from the relationship between a lord and a vassal or between a master and a slave. New legal rules to organize the relationship between workers and employers were needed in particular because of the liberal thought.

When the industrialization started, the system in place in Great-Britain and in the USA was "Common Law." Hence, work-related accidents resulted in a request for compensation (by the victims) under common law.

Common law is the body of law derived from judicial decisions of courts and similar tribunals. The defining characteristic of "common law" is that it follows precedent. In cases where the parties disagree on what the law is, a common law court looks to the past precedential decisions of higher courts, and synthesizes the principles of those past cases as applicable to the current facts. Hence, new rules can be developed as new cases are discussed (Ibbetson, 2001). A number of these cases had an important influence on the judgments in occupational accidents.

The following principles were regularly used in the nineteenth century under the common law for judging the employer's liability in workplace accidents:
- **Contributory negligence** allows an employer to be held harmless to the extent that the injured employee failed to use adequate precautions required by ordinary prudence.
- The **Fellow Servant Doctrine** is that an employer can be held harmless to the extent that injury was caused in whole or in part by a peer of the injured worker.
- **Volenti or voluntary assumption of risk** allows an employer to be held harmless to the extent the injured employee voluntarily accepted the risks associated with the work.
- **Free agent doctrine** uses as axiom that the worker is in fact a free person who decides for himself/herself whether he or she wants to take certain risks specific to the work.

3.2.3 Contributory negligence

Contributory negligence was introduced in the Butterfield v. Forrester case in 1809.

Forrester placed a pole on the road next to his house in the course of making repairs to the house. Butterfield was riding at a high speed at approximately 8 pm at twilight and did not see the pole. He struck the pole and suffered personal injuries when he fell off his horse. A witness testified that visibility was 100 yards at the time of the accident and Butterfield might have seen and avoided the pole had he not been riding at such a high speed. There was no evidence that Butterfield had been intoxicated at the time of the accident. At trial, the judge instructed the jury that if an individual riding with reasonable care could have avoided the pole, and if the jury found that Butterfield had not used reasonable care, the verdict should be in Forrester's favor. The jury returned a verdict for Forrester and Butterfield appealed.

The court determined that the plaintiff had failed to use common and ordinary caution, and he was therefore barred from recovery.

Contributory negligence did not have to be gross negligence. It might be a failure to stop a person or a machine, for instance. A parent who allowed a minor child to continue working in a situation that the parent knew to be dangerous was negligent (Bronstein, 2008). If an employer promised to repair something and failed to do so he/she was negligent but if the employer made no such promise, the employee now had to leave the dangerous employment (Bronstein, 2008).

The doctrine of contributory negligence struck a balance between the legal responsibility of the worker and that of the employer (Mitchell and Mitchell, 2010).

3.2.4 Fellow servant rule

In 1837, there was a case "Priestley v. Fowler" in Great-Britain that gave rise to the "*Common Employment Rule*." In 1842, Chief Justice Lemuel Shaw of Massachusetts (USA) used this rule to judge the case "Farwell v. Boston & Worcester Rail Road." In the USA, labor law terminology was the "fellow servant rule."

On 30 May 1835, Charles Priestley, a minor and servant of butcher Thomas Fowler of Market Deeping, was ordered to conduct mutton to market. The meat was placed in a wagon driven by William Beeton, another of Fowler's employees. The van was dangerously overloaded. While traversing the mile south from Peterborough toward Norman Cross, the wagon's front axle cracked, overturning the vehicle. Beeton escaped substantial harm but Priestley was hit by the load resulting in a broken thigh, a dislocated shoulder, and various other injuries.

Priestley was taken to the closest public lodging where he stayed "in a very precarious state," for 19 weeks, during the course of which he was treated by two surgeons. The total cost of Priestley's care and treatment was paid by his father, Brown Priestley. During the Assizes of 1836, Charles Priestley (as a minor through his father) sued his master Fowler for compensation relating to his accident. Priestley pleaded two grounds in support of his claim against his master, a latent defect and the van's overloading. The employer (Fowler) was under a duty "to use due and proper care that said van should be in a proper state of repair" and "not be overloaded, and that the plaintiff should be safely and securely carried thereby."

Acknowledging that the master/servant relationship bound the master directly to "provide for the safety of his servant . . . to the best of his judgment, information, and belief," the Chief Baron emphasized that it could "never" imply an obligation for the master "to take more care of the servant than he may reasonably be expected to do of himself." At the same time, the servant was "not bound to risk his safety in the service of his master" and was free to "decline any service in which he reasonably apprehended injury to himself." This was because servants were in as good, if not better positions, than their masters to appreciate possible hazards.

3.2.5 Volenti non fit injuria or voluntary assumption of risk

Volenti non fit injuria is a common law doctrine which states that no wrong is done to one who consents (volenti non fit injuria).

If someone willingly places themselves in a position, knowing that some degree of harm might result, they are not able to bring a claim against the other party in tort or delict. The similar principle in US law is known as assumption of risk.

In English tort law, volenti is a full defense, that is, it fully exonerates the defendant who succeeds in proving it and this is based on three elements of defense:
–   That the plaintiff perceived the existence of the danger or risk
–   That he or she fully appreciated it
–   That he or she voluntarily agreed to accept the risk

Two examples of voluntary assumption of risks are:
–   An employee jumps from a height of 5 m instead of taking the steps. He/she cannot sue the employer for such an injury because he/she knew what he/she was doing.
–   An employer supplies an employee with a defective piece of machinery, and knowing the machinery is defective, the employee proceeds to use it anyway (albeit carefully).

Voluntary assumption of risk will play a role in the discussion of the nineteenth century doctrine of "Free Agent."

But volenti was not always accepted by the courts. In Smith v Baker [1891], the plaintiff was employed by the defendants on the construction of a railway. While he was working, a crane moved rocks over his head. Both he and his employers knew there was a risk of a stone falling on him and he had complained to them about this. A stone fell and injured the plaintiff and he sued his employers for negligence. The employers pleaded volenti non fit injuria but this was rejected by the court. Although the plaintiff knew of the risk and continued to work, there was no evidence that he had voluntarily undertaken to run the risk of injury. Merely continuing to work did not indicate volens.

3.2.6 Free agent doctrine

In the nineteenth century, it was assumed for a large part of the population that accidents were part of certain activities. When one accepts a job in a mine, one also accepts that an explosion can occur while working in the mine.

Every person who accepts a job knows very well what the risks are and when an accident occurs this person can only blame himself. No one has forced the person to accept the job and associated risks.

This doctrine was called the Free Agent doctrine (Bronstein, 2008). This idea of free labor was supported by economists. A contract between an employer and an employee forms the basis of the economy. Both make clear agreements. If the state intervenes in such a contract, it undermines the basis of a healthy economy. The assertion, commonly made by employers, was that employees were free agents who were to blame for their own accidents and should therefore bear the social cost.

The doctrine of Free Agent came under pressure during the course of the nineteenth century for various reasons. An important reason was the discussion of the meaning of "a free agent."

It was common for whole families to work together underground in the coal mines. Children started work underground when they were around 8 years old, but some were as young as 5. The job of the trapper, often the youngest member of the family, was to open and close the wooden doors (trap doors) that allowed fresh air to flow through the mine. They would usually sit in total darkness for up to 12 h at a time, waiting to let the coal tub through the door. It was not hard work but it was boring and could be very dangerous. If they fell asleep, the safety of the whole workings could be affected.

The older children and women were employed as hurriers, pulling and pushing tubs full of coal along roadways from the coal face to the pit bottom (Figure 3.1).

Figure 3.1: Illustration of a hurrier (https://www.mylearning.org/stories/coal-mining-and-the-victorians/236).

The ideology of free agency helped to prevent male workers from campaigning hard for legislative change. An adult man was a free agent and if he perceived that he needed to work more than 10 h per day for his family, then who was the government to stop him from doing this (Bronstein, 2008).

In 1833, a Royal Commission on the Employment of Children in Factories investigated the use of young workers in industry (Royal Commission, 1833). One of their inquiries was about "could workers who were not free agents under the law be held responsible for workplace accidents?" A father who saw his 11-year-old son die in an accident testified (Royal Commission, 1833,0 p. 869): "it was an accident arising from negligence. He was carelessly cleaning the machine, and was seized by the traps and taken round the drum, and he lived about six hours."

How could one defend that a child knew the risks when the child accepted the job and when an accident occurs the child can only blame himself? The commissioners proposed that, around the age of 14, workers became free agents as a result of changes in the relationship of power within the household.

Also women and in particular pregnant women could not be labeled as a free agent. Neither could slaves be regarded as free agents.

Because woman and child workers were assumed to need more guidance and protection, courts and legislators raised the bar for the employers.

Miss O'Byrne, a female child who was mangled in a machine in the first days of her employment, sued her employer with success in 1858, based on the argument that she was an inexperienced, helpless and ignorant child. In the Hayden v. Smithville Manufacturing Co. (a case in Connecticut, USA, 1861), a 10-year-old boy who had lost his hand in the gearing of an unguarded textile machinery won a claim. Smithville Manufacturing Co. was found negligent by failing to install bonnets around dangerous machinery. The court suggested that the young boy Emory Hayden could only be held to a standard of prudence usual for those of his age. This was of course in contradiction with the notion that a worker accepted all the risks of his employment.

In the nineteenth century, slavery was still legal. Slaves were hired out to work in industries. A hired slave was a rented property who had to be returned to the ultimate owner in the same condition as the moment when the slave was hired. The result was that slaves' owners were able to collect in the courts much more compensation for injured slaves than regular adult male workers were able to collect for their own injuries (Morris, 1996).

Francis Gurney du Pont (1850–1904), grandson of Eleuthère Irénée du Pont de Nemours, kept a notebook in which he recorded all the gunpowder accidents that had occurred at the company since the inception. When a cause was given, it was invariably related to worker error.

There should be no doubt that it is possible for workers to be careless and negligent. But employers in the nineteenth century presumptively assumed that worker carelessness caused workplace accidents and employers never admitted to carelessness or even to taking calculated risks. At the same time, this assumption conflicts with the notion that male workers were capable of keeping the workforce safe.

Slowly but surely opposition to this vision grew. In 1845, the manager of the Govan colliery in Scotland wrote to the mine owner that he could not continue to expose so many lives to apparent danger. The wire ropes that were in use broke without any warning. It still took 3 years to replace the faulty ropes. This is just one of the examples of the awareness that grew indicating that the management and the owners also had a part in making the workplace safe.

White male workers who were considered as free agents and hence could not count on compensation in case of accidents organized themselves in trade unions.

3.3 Influence of economic thought in the nineteenth century

As described above, the direct effects of the industrial revolution included:
- Production of goods and services are set up differently
- The production of goods and services is increasing impressively
- Rapid growth of international and intercontinental trade transport

Economy is the social science that studies the choices people make in the production, consumption and distribution of goods and services. It goes without saying that the industrial revolution had an impact on the evolution of economic thinking. Economic thought can be traced back to more than 2000 years ago but it is generally assumed that "economy" was not a separate academic field until the end of the eighteenth century. The Scottish philosopher Adam Smith (1723–1790) is often noted as the father of modern political economy or "the father of capitalism."

Economic thought in the nineteenth century supports the doctrine that the state should not intervene in the labor market.

Adam Smith published the first version of "An Inquiry into the Nature and Causes of the Wealth of Nations" in 1776 in which he analyzes what determines the prosperity of a nation.

The Wealth of Nations was the product of 17 years of notes and earlier works, as well as an observation of conversation among economists of the time concerning economic and societal conditions during the beginning of the industrial revolution, and it took Smith some 10 years to produce. Smith defended the free market. The study represented a clear paradigm shift in the field of economics.

Alan Greenspan, an American economist who served as Chair of the Federal Reserve of the United States from 1987 to 2006, considers that the Wealth of Nations was "one of the great achievements in human intellectual history" (The Guardian, 2005).

The idea of the free market was, however, not invented by Adam Smith. The idea of a free market and trade is attributed to the French economists François Quesnay

(1694–1774) and Jacques Claude Marie Vincent de Gournay (1712–1759), who launched the term "laissez-faire." These French economists believed, however, that the Wealth of Nations was derived solely from the value of "land agriculture" or "land development" and that agricultural products should be highly priced.

Laissez-faire is an economic system in which transactions between private parties are free from government intervention such as regulation, privileges, tariffs and subsidies. Some basic axioms among others are:

1. The individual is the basic unit in society.
2. The individual has a natural right to freedom.
3. The physical order of nature is a harmonious and self-regulating system.

The doctrine of laissez-faire became an integral part of nineteenth-century European liberalism. Classical liberals were committed to individualism, liberty and equal rights. They believed this required a free economy with minimal government interference.

In Britain, the newspaper *The Economist* was founded in 1843 and became an influential voice for laissez-faire capitalism.

This economic thinking was entirely in line with liberalism and was widely praised and proclaimed well into the nineteenth century. The thinking pattern also supported the doctrine whereby it is assumed that the state should not intervene in the relationship between the employer and the employee. The intervention of the state in the relationship between the employer and the employee would disrupt the economy and should be avoided at all times.

However, the nineteenth century was characterized by successive financial crises. Some of the examples are as follows:

- The Panic of 1825, a stock market crash that started in the Bank of England, has been referred to as the first modern crisis not attributable to an external event. It was part of modern economic cycles that are inherent to capitalism. Seventy banks failed. An infusion of gold reserves from the "Banque de France" saved the Bank of England from complete collapse.
- The Panic of 1837 was a financial crisis in the USA that touched off a major recession that lasted until the mid-1840s. Profits, prices and wages went down while unemployment went up. The recession persisted for approximately 7 years and did not recover until 1843. Banks collapsed, businesses failed, prices declined and thousands of workers lost their jobs. Unemployment may have been as high as 25% in some locations.
- The Panic of 1857 was a financial panic in the USA caused by the declining international economy and overexpansion of the domestic economy. Because of the interconnectedness of the world economy by the 1850s, the financial crisis that began in late 1857 was the first worldwide economic crisis.
- The Panic of 1873 was a financial crisis that triggered an economic depression in Europe and North America that lasted from 1873 until 1877, and even longer in

France and Britain. In Britain, for example, it started two decades of stagnation known as the "Long Depression" that weakened the country's economic leadership. By November 1873, some 55 of the USA nation's railroads had failed, and another 60 went bankrupt by the first anniversary of the crisis. Construction of new rail lines, formerly one of the backbones of the economy, plummeted from 12,070 km of track in 1872 to just about 2,500 km in 1875. In the USA, 18,000 businesses, including 89 railroads, failed between 1873 and 1875. Ten states and hundreds of banks went bankrupt. Unemployment peaked in 1878 at 8.25%, long after the initial financial panic of 1873 had ended.
– The Panic of 1893 was a serious economic depression in the USA that began in 1893 and ended in 1897. Some texts refer to the period as the Great Depression of 1873–1896.

The industrial revolution had also caused a new social class to emerge. This social class (employees in the industry) had a very difficult time. From the middle of the nineteenth century, there was a rise of economists who felt that the existing economic system (a free market without state intervention) had to be questioned. Certain economists began to wonder aloud how to continue to defend a capitalist system, where an increasingly poor social class emerged and where a financial crisis regularly occurred.

To a certain extent, one can compare the situation with what is happening today with regard to climatic change. Some economists see the discussion about climate change as a threat to the economy, while others see it as an opportunity.

Similar discussions took place in the nineteenth century regarding the organization of the labor market. Some economists were convinced that prosperity was the result of a liberal labor market with no rules, where hard work was the engine. The more hours worked, the greater the yield and the more prosperity that would ultimately result. If the state were to intervene in the labor market, this would raise costs and ultimately endanger prosperity.

This very liberal thought was good for the richer class and was therefore maintained. Other economic schools pointed out that the successive economic crises repeatedly reduced prosperity while the working class lived in miserable conditions. Moreover, it was that working class that had to ensure prosperity.

Figure 3.2 shows the GDP (gross domestic product) per capita expressed in 1990 US$ in the period 1820 till 2010 (Bolt, 2014). It is obvious that the GDP growth per capita in the nineteenth century was very slow despite the industrial evolution and the huge number of hours that people worked.

Figure 3.2 shows that the industrial revolution did not bring the average prosperity predicted by the classical economists. The share of the top 10 richest people in the global wealth of a country nevertheless increased in the nineteenth century (Figure 3.3; based on the data from Piketty, 2013).

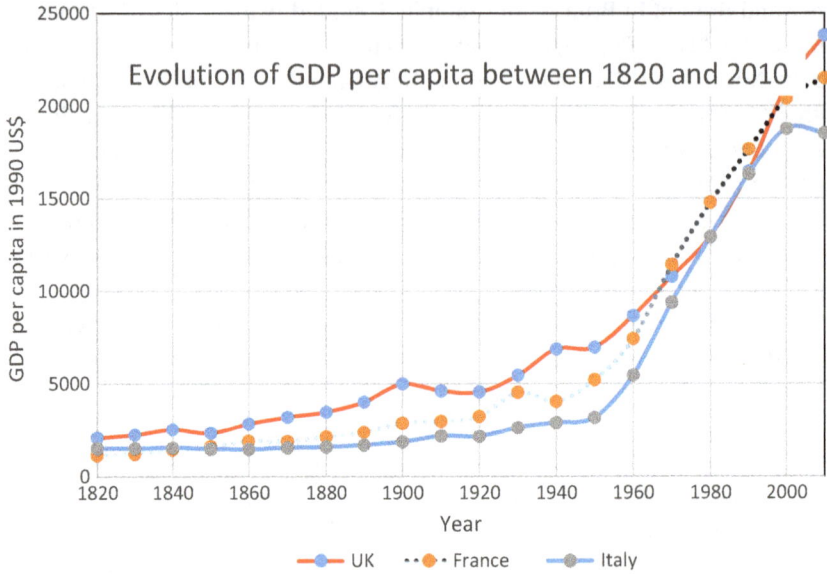

Figure 3.2: Evolution of GDP in some countries between 1820 and 2010.

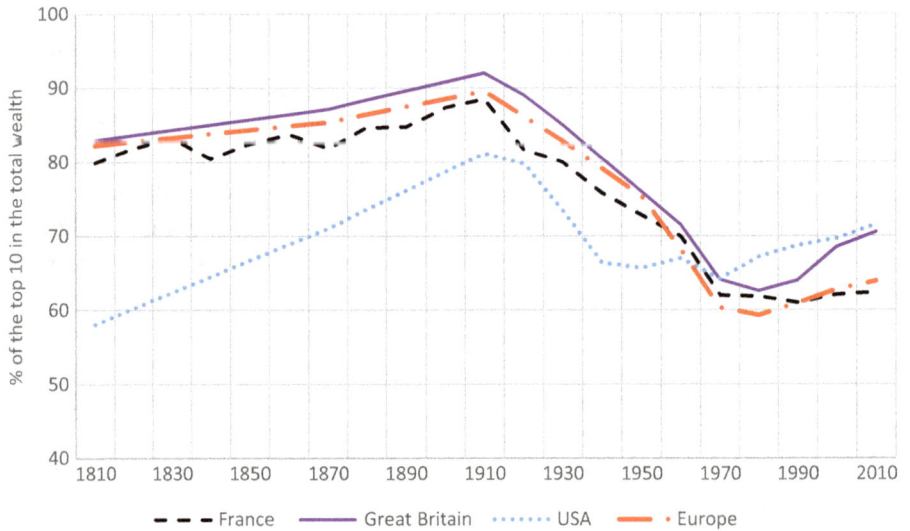

Figure 3.3: Share of the top 10% wealthiest in the total wealth of a country, 1810–2010.

3.4 Utopian socialism

The prevailing ideology in the nineteenth century was that of the classical liberal economists, in which every influence of the state had to be avoided. However, there were also entrepreneurs and economists who believed in a social approach based on the idea that a healthy and happy employee gave a much higher return. Examples are Henri de Saint-Simon, Claude Henri de Rouvroy and Robert Owen.

Claude Henri de Rouvroy (1760–1825) was a French political and economic theorist who claimed that the needs of an industrial class, which he also referred to as the working class, needed to be recognized and fulfilled to have an effective society and an efficient economy.

Robert Owen (1771–1858) was a Welsh textile manufacturer who in the early 1800s became wealthy as an investor of a large textile mill at New Lanark, Scotland. Owen devoted much of his profits to improving the lives of his employees. Owen earned an international reputation by the results of his social and economic ideas that he tested at New Lanark. By taking care of the working conditions of his workers, he won their confidence and improved efficiency at the mill.

Owen's work at New Lanark continued to have significance throughout Britain and in continental Europe. He was a "pioneer in factory reform." Owen raised the demand for an 8-h day in 1810, and instituted the policy at New Lanark. By 1817, he had formulated the goal of the 8-h workday and coined the slogan: "Eight hours labour, Eight hours recreation, Eight hours rest." The ideas of Owen were influential for the social movements of the second half of the nineteenth century.

Karl Marx introduced in his book *For a Ruthless Criticism of Everything* in 1843 the term "utopian socialists" to refer to people like Robert Owen. Friedrich Engels and Karl Marx used the term utopian socialist to make the contrast with their scientific methods for understanding and predicting social, economic and material phenomena by examining their historical trends.

3.5 Renewed political insight at the end of the nineteenth century

3.5.1 Trade unions request for better compensation laws

The basis of the debate in society up to at least mid-nineteenth century was about compensation for the worker who was no longer able to continue his work for health reasons due to the conditions of the work. In a similar way that war deaths are considered as heroic, injured workers in the midst of the industrialization of the nineteenth century might be perceived as participants in, rather than strictly innocent victims of, battle (Bronstein, 2008).

Hence, the important question up to the middle of nineteenth century was more about compensation than about why an accident happened. It was generally accepted that accidents or health problems were inherent to work. But when people became ill or had an accident, there were no clear rules in place dealing with how these people could be helped.

Consequently, brotherhoods and trade unions were established in various places to represent the interests of members. In March 1857, the "Broederlijke Maatschappij der Wevers" was founded in Ghent and a month later the spinners held their "Maatschappij der Noodlijdende Broeders" (Society of the Distressed Brothers) above the font. The first trade unions took a relatively moderate stance and were also open to all workers, irrespective of the political tenor or philosophy of life to which they confessed themselves (Van Onacker, 2009).

The Friendly Society of Operative Stonemasons, Quarrymen and Allied Trades of England and Wales was founded in 1833 as a trade union representing stonemasons and related workers in the UK. It expanded rapidly reaching 4,000 members and 100 branches by 1835. It paid a large sum annually of about £800 for accidents that might have been avoided. The Amalgamated Society of Engineers said in 1876 (Parliamentary Papers, 1876) that it had paid out £26,000 since it was founded in 1851 to members disabled by accident or disease, and cited examples of cases in which the employer should have been liable for compensation.

Robert Knight (1833–1911) was elected secretary-general secretary 1871 of the United Society of Boiler Makers and Iron and Steel Shipbuilders Association. He became very active in the Trade Union Congress, a federation of many trade unions in the UK that was founded in 1860. Knight attacked the doctrine of risk assumption since the risk changed every day as people are hired and fired (Parliamentary Papers, 1876).

By the end of the nineteenth century, trade unions took more and more part in political discussions. At the second Trade Union Congress meeting in 1868, delegates discussed the 8-h working day and election of working people to the parliament.

The combination of the miserable conditions in which the workers found themselves and the financial crisis led to strikes that were brutally tackled. The nineteenth-century governments were formed by Catholics or Liberals. But also in these families there was a growing awareness that the riot among the workers was not due to unwillingness but the result of the difficult circumstances in which this large group of people lived.

Thanks to pioneers such as Frere-Orban, people increasingly opted to avoid riots through prevention. Social legislation had to improve working conditions and dissatisfaction among the workers (Deferme, 2002). Until well into the nineteenth century, social laws usually were a reaction to riots and strikes, but there was a growing awareness that social legislation was needed specifically to protect workers.

3.5.2 The International Workingmen's Association (IWA)

In 1863, French and British workers started to discuss developing a closer working relationship. A French delegation visited London in July 1863, attending a meeting held in St. James's Hall. Here, there was discussion of the need for an international organization, which would among other things prevent the import of foreign workers to break strikes.

In September 1864, some French delegates again visited London with the concrete aim of setting up a special committee for the exchange of information on matters of interest to the workers of all lands. The meeting on 28 September 1864 was attended by a wide array of European radicals, including English Owenites, followers of Pierre-Joseph Proudhon and Louis Auguste Blanqui, Irish and Polish nationalists, Italian republicans and German socialists. The journalist Karl Marx would soon come to play a decisive role in the organization.

The meeting unanimously decided to found an international organization of workers. The center was to be in London, directed by a committee of 21, which was instructed to draft a program and constitution. This executive committee in turn selected a subcommittee to do the actual writing of the organizational program. This subcommittee deferred the task of collective writing in favor of sole authorship by Marx and it was he who ultimately drew up the fundamental documents of the new organization.

The International Workingmen's Association (IWA), often called the First International, was founded by Edward Spencer Beesly, Eugène Varlin, George Odger and Henri Tolain. It organized a yearly congress during which some important resolutions were passed:

- Geneva Congress, 1866. A significant decision at this event was the adoption of the 8-h workday as one of the IWA's fundamental demands.
- Lausanne Congress, 1867. The Congress was attended by 64 delegates from Great Britain, France, Germany, Belgium, Italy and Switzerland. The reports recorded the increased influence of the International on the working classes in various countries. The Lausanne Congress passed a resolution on political freedom, which emphasized that the social emancipation of workers was inseparable from political liberation.
- The Hague Congress, 1873. The Hague Congress decided to relocate the General Council to New York City. The Congress committed the international to building political parties, aimed at capturing state power as an indispensable condition for socialist transformation.

There were various tendencies within social thought, and in 1872 the IWA was split into two groups. From then on, the Marxist and anarchist currents of socialism had distinct organizations at various points including rival internationals.

3.5.3 New compensation laws

Under the old common law, the right of the workman to receive compensation for injury was almost entirely on the demonstration of a "fault" of the employer, of the injured worker himself or of one of his fellow employees.

The new idea for compensation was that compensation was systematic except for accidents caused by serious and willful misconduct of the injured person himself. The advantages of the new compensation doctrine were as follows (Beyer, 1916):

- It is better for the worker because:
  - he/she is sure of receiving compensation in all legitimate cases;
  - the delays and expenses incident to collecting damages through the courts are eliminated, payments being made promptly after the accident at the time when they are most needed;
  - more emphasis is placed on prevention of accidents.
- It is better for the employer because:
  - he/she can count on fixed charges against his business, and make provision for them;
  - he/she is relieved of the trouble and annoyance of lawsuits, and the possibility of excessive verdicts that juries are liable to render.

The first Compensation Act drafted along the new compensation principles was passed by Germany in 1884. Workers, unions and employers quickly saw the advantages of the new compensation laws in particular because of the better relations that were established between the workforce and the employers. Other European countries that enacted Compensation laws were Austria (1887), Norway (1894), Great Britain (1897), France (1898), Spain (1900), Sweden (1901), Russia (1903) and Belgium (1903). The first compensation law enacted by Congress in the USA was in 1908 but the law was limited to certain classes of federal workers. The first state to put in place a compensation law for all workers was New Jersey in 1911. In 1915, another 33 States followed this lead of New Jersey. In 1943, workmen's compensation laws were in force in all states except Mississippi.

3.5.4 A first revolution in thinking about occupational health and safety

The First Industrial Revolution was the transition to new manufacturing processes in Europe and the USA in the period from about 1760 to some time around 1840. This transition included going from hand production methods to machines. Textiles were the dominant industry of the First Industrial Revolution in terms of employment. The number of looms in the UK went from 2400 in 1803 to about 100,000 in 1833 and 250,000 in 1857 (Hills, 1993).

Friedrich Engels (1820–1895) was a German philosopher and journalist, the son of an owner of large textile factories in Salford, England. In 1842, his parents sent the 22-year-old Friedrich to Manchester, England, a manufacturing center where industrialization was on the rise. During his stay in the period 1842–1844, Engels compiled observations and studied contemporary reports. In 1845, he published a book *Die Lage der arbeitenden Klasse in England* in which he focused on both the workers' wages and their living conditions. He argued that the industrial workers had lower incomes than their preindustrial peers and they lived in more unhealthy and unpleasant environments (Engels, 1845).

Engels regularly quotes from reports from parliamentary committees such as the children's Employment Commission's Report. He also makes extensive use of figures. Table 3.7 has been taken from his book. The table must show that there are too many young people and too many women working in the various mines.

Table 3.7: The number of workers employed in the mines in Great Britain, without Ireland, according to a census of 1841 (Engels, 1845).

	Men > 20 years	Men < 20 years	Women > 20 years	Women < 20 years	Together
Coal mines	83,408	32,475	1,185	1,165	118,233
Copper mines	9,866	3,428	913	1,200	15,407
Lead	9,427	1,932	40	20	11,419
Iron	7,773	2,679	424	73	10,949
Tin	4,602	1,349	68	82	6,101
Various minerals are not specified	24,162	6,591	472	491	31,716
Total	**139,238**	**48,454**	**3,102**	**3,031**	**193,825**

According to this census, about 27% of the workers in mines were younger than 20 years and many of them were boys of 12 years old.

The book was translated into English in 1885 by an American, Florence Kelley, authorized by Engels and with a newly written preface by him, it was published in 1887 in New York and in London in 1891.

The miserable living conditions of the working class became a cultural theme in the second half of the nineteenth century. Charles Dickens (1812–1870) describes in his second novel *Oliver Twist; or, the Parish Boy's Progress*, the story of an orphan boy called Oliver Twist. Dickens satirizes the hypocrisies of his time, including child labor, the recruitment of children as criminals and the presence of street children. Dickens' tenth novel *Hard Times – For These Times*, first published in 1854 surveys English society and satirizes the social and economic conditions of the era.

Some of the examples of paintings of poverty of the middle class in the nineteenth century are as follows (see also Howden-Chapman and Johan, 2003):

- Gustave Courbet, *The Stonebreakers*, 1849 and *Charity of a Beggar at Ornans*, 1868
- Adolph Menzel, *Iron Rolling Mill*, or *Modern Cyclops*, 1876
- Vincent van Gogh, *The Potato Eaters*, 1885
- Gustave Caillebotte, *Floor Scrapers*, 1875

From the middle of the nineteenth century, the poverty of the working class was an important theme in all sections of society. It can be compared to a certain extent with the discussions that are taking place today on climate change.

Social pressure, strikes and unions ensured that political reforms took place and that more and more laws were voted and introduced to improve the safety and health of employees.

It can be said that from around 1850, all political parties had understood that a stable, prosperous society could only exist if one could improve the welfare of the working class.

In order to regulate the conditions of industrial employment in Great Britain, the parliament passed the so-called Factory Acts. The first Factory Act was introduced by Sir Robert Peel as the Health and Morals of Apprentices Act 1802. It was largely concerned with the employment of apprentices. This first attempt to improve the lot of factory children is often seen as paving the way for future Factory Acts. It established the principle of intervention by parliament on worker welfare issues against the "laissez-faire" political and economic orthodoxy of that era.

Under the act, regulations and rules came into force on 2 December 1802 and applied to all textile mills and factories employing 3 or more apprentices or 20 employees. The buildings must have sufficient windows and openings for ventilation and should be cleaned at least twice yearly with quicklime and water; this included ceilings and walls. Each apprentice was to be given two sets of clothing, suitable linen, stockings, hats, and shoes, and a new set each year thereafter. Apprentices could not work during the night (between 9 pm and 6 am), and their working hours could not exceed 12 h a day, excluding the time taken for breaks. All apprentices were to be educated in reading, writing and arithmetic for the first 4 years of their apprenticeship. Local magistrates had to appoint two inspectors known as "visitors" to ensure that factories and mills were complying with the act. This first act was followed by the Cotton Mills and Factories Act 1819, the Cotton Mills Regulation Act 1825 and the Labour in Cotton Mills Act 1831.

These acts concentrated on regulating the hours of work and moral welfare of young children employed in cotton mills but their enforcement was left to local magistrates.

In 1833, the government passed a Factory Act to improve conditions for children working in factories. Young children were working very long hours in workplaces, where conditions were often terrible. The basic act was as follows:
- no child workers under 9 years of age;
- employers must have an age certificate for their child workers;
- children of 9–13 years to work not more than 9 h a day;
- children of 13–18 years to work not more than 12 h a day;
- children are not to work at night;
- two-hour schooling each day for children;
- four factory inspectors appointed to enforce the law.

> The first factory inspectors were appointed under the provisions of the Factories Act 1833. Initially their main duty was to prevent injury and overworking in child textile workers. The four inspectors were responsible for approximately 3,000 textile mills and had powers to enter mills and question workers. They were also able to formulate new regulations and laws to ensure the Factories Act could be suitably enforced. Despite serious opposition from contemporary politicians and employers, the factory inspectors were enthusiastic and were able to influence subsequent legislation relating to machinery guarding and accident reporting. (copied from http://www.hse.gov.uk/aboutus/timeline/)

Various factory acts were introduced from the 1840s that regulated the minimum age of the workers and the working time:
- The Mines and Collieries Act 1842 following the terrible revelations of the *Royal Commission on Labour of Young Persons in Mines and Manufactures* (1841) excluded women and girls from underground working, and limited the employment of boys, excluding from underground working those under 10 years.
- A Factory Act (Mills) of 1844, applied only to the textile industry, limited the work of children aged 8–13 to 6.5 h a day, which was considered to be "half-time."
- The Calico Print Works Act of 1845, related to factories printing designs on cotton fabrics, prohibited the employment of children under 8. Those under 13 and women were not to work between 10 pm and 6 am. Children under 13 should attend school for 30 days per half year.
- A Factory Act of 1947 limited women and young persons in textile factories to work not more than 10 h a day or 10.5 if Saturday was a half holiday. This is sometimes known as the Ten Hour Act.
- A Factory Act of 1853 related to mills and required that the work of children aged 9–13 be between the hours of 6 am to 6 pm in summer and 7 am to 7 pm in winter. Similarly, the work of women and young persons was restricted to the hours between 6 am and 6 pm.
- *The Workshops' Regulations Act* for workplaces with less than 50 persons prohibited employment of children under 8 years old. Children aged 8–13 were restricted to half time working. Young persons and women were restricted to a 12 h day with

1.5 hours allowed for meal breaks. Children, young people and women were not to be employed after 2 pm on Saturdays in establishments with more than five employees, and child employees were to attend school 10 h a week.
- *The Factory Act of 1874* forbade employment of children in mills under the age of 9. In 1875, the age limit was raised to 10. Children aged 8 could still be employed in workshops and nontextile factories.

The numerous Factory Acts were repealed and consolidated in the great statute of 1878 a single act of some 107 clauses. The workers fell into four categories:
- "Children" (aged 10–14)
- "Young persons" (aged 14–18)
- "Women" (females aged over 18)
- "Men" (males aged over 18)

The act followed the recommendations of the commission by:
- setting a limit on the hours worked per week by women and young persons in textile factories, nontextile factories and workshops;
- children were not to be employed under the age of 10 and should attend school half-time until 14;
- protected persons (= children, young persons and women) should not be allowed to clean moving machinery;
- the requirement to guard machinery now extended to the protection of men as well as protected persons;
- young persons and children could not work in the manufacture of white lead, or silvering mirrors using mercury;
- children and female young persons could not be employed in glass works;
- girls under 16 could not be employed in the manufacture of bricks (nonornamental) tiles or salt; and
- children could not be employed in the dry grinding of metals or the dipping of lucifer matches.

Subsequent extension, amendment and repeal in such legislation resulted in four additional acts, passed in 1883, 1889, 1891 and 1895, respectively. By the end of the nineteenth century, the social outlook on safety and health at work had changed completely compared to the beginning of the nineteenth century.

Although the term "accident" is still not well defined, it has the popular meaning of "any sudden occurrence, apart from an act of God, illness, or personal violence, by which bodily injury is sustained by any person within the precincts of a factory or workshop, including a workshop employing men only, and the laundries, docks, wharves, quays, warehouses and buildings added for this purpose by the Act of 1895" (Calder, 1899).

Lack of safety in the workplace was seen as unacceptable from a social point of view but also from an economic point of view. The liberal notion that originated in the French revolution and that was booming in the first half of the nineteenth century received a serious headwind from socialist ideology. The various financial crises, the revolutions and the general social dissatisfaction in the nineteenth century made it clear that society could not continue in this way.

The rapidly growing working class who had meanwhile gained more voting rights and participation in society (through political parties and through trade unions) were the driving force behind legislative reforms. The working class had grouped into trade unions who were in contact with each other across borders and evolution in Britain ran parallel with similar actions in other countries.

In the Netherlands, for example, a major uproar broke out in 1886 among workers protesting against miserable working conditions. The government decided to conduct a survey that was not only about child labor but also about the safety, health and well-being of all employees. The report that was published showed a situation that was much worse than expected. The government decided to intervene via the Labor Act 1889 that came into effect on 1 January 1890.

Children under the age of 12 were only allowed to work in the agricultural sector. Women and young people under 16 were allowed to work a maximum of 11 h a day and not on Sundays. Dangerous work was prohibited for these "weaker people." Women were given 4 weeks off after giving birth. An important change was the inspection that was introduced. This improved compliance with the law, but still left something to be desired.

Because the employers themselves did not improve working conditions, the government felt compelled to make it compulsory. The government did not want to impose provisions on working hours for men and therefore the labor law was not amended further, but a new law, the Security Act of 1895, was introduced. The legislation was deliberately kept general, so that legislation could easily be adapted to specific situations through an order in council (van den Bosse, 2012).

In France, the labor inspection was officially created by the 19 May 1874 law, establishing a body of 15 divisionary inspectors, and several departmental inspectors. The 2 November 1892 law in France established a specialized body of civil servants dedicated to inspection of labor conditions. It was first of all charged with the control of the implementation of the 22 March 1841 law prohibiting child labor of less than 8 years old. This law had been enacted following reports by the physician René Villermé. The 1890 law also enacted a maximal length of work for children, women and underaged girls. In 1906, the Ministry of Labor was created. The Labor Code created by the law of 26 November 1912 set for the first time the principles of general safety of premises and protection of workers. It provided for the declaration of any accident at work and any illness as well as measures for improvement in health and safety conditions.

3.6 A more scientific approach to safety at work at the turn of the nineteenth century

3.6.1 Leading studies

At the turn of the century, there were a number of leading publications that can be seen as the start of more modern understanding of the causes of accidents at work. Three of these publications will be discussed in the following chapters:

- **1899**. Inspector John Calder published his book *The Prevention of Factory Accidents* (Calder, 1899). John Calder, Her Majesty's Inspector of Factories of the North of Scotland, published detailed statistics of 725 fatal accidents and 57,423 lost time accidents in the industry in 1898 in the UK.
- **1907**. The Pittsburgh Survey conducted under the direction of Paul Kellogg studied the causes and circumstances of the 526 people killed by work accidents in the period 1 July 1906–30 June 1907 and 509 people injured by work accidents in the months of April, May and June 1907 in the Pittsburgh Steel District of Allegheny County in the USA (Eastman, 1910).
- **1909**. The Fidelity and Casualty Company of New York published a general pamphlet *The Prevention of Industrial Accidents* (Law and William, 1909). It was launched in September 1909 with an initial print run of 30,000 copies and followed in December 1909 with a second print run of 20,000 copies.

These publications showed that people started to look differently at the causes of accidents, which were no longer considered to be solely down to the carelessness of the victim. The Pittsburgh survey puts it as follows:

> If adequate investigation reveals that the most work-accidents happen because workmen are fools, then there is no warrant for direct interference by society in the hope of preventing them. If on the other hand, investigation reveals that a considerable proportion of accidents are due to insufficient concern for the safety of workmen on the part of their employers, then social interference in some form is justified.

If, again, investigation of a large number of cases shows that workmen and their families do not suffer economically from work accidents, and they often make money out of injuries, then we are not warranted in interfering between employers and employees for the sake of further protecting the rights of the latter. But if investigation shows that the majority of work accidents result in serious deprivation to the workers' families and consequent cost to the community, and that the economic loss is inequitably distributed then we shall be warranted in advocating interference to adjust that burden more wisely.

It is interesting to see that the social debate is not so much about the question of whether accidents can and should be avoided. The central question is whether the government should intervene in the employer–employee relationship. The answer to this question must be given on the basis of a consideration of social justice and economic state interests. Today the answer to this question seems trivial. At the beginning of the twentieth century, people still struggled with the idea that the victim should have watched better or simply should not have accepted the work.

3.6.2 John Calder study (1899)

Calder gives an overview of the number of premises that are subject to the Factory Act 1895 (Table 3.8).

Table 3.8: Overview of the number of factories and workers in 1896 in England, Wales, Scotland and Ireland (Calder, 1899).

Classes of work	Registered factories, workshops, departments	Children (11–14)	Young (14–18)	Women	Men	Total
Textile factories	9,951	53,256	236,245	482,030	306,156	1,077,687
Nontextile factories	71,259	7,241	436,502	341,957	1,880,031	2,665,731
Workshops	87,293	3,116	163,982	250,480	237,987	655,565
Total	168,503	63,613	836,729	1,074,467	2,424,174	4,398,983

The number of accidents in these industries in the 3-year period 1896–1898 is given in Table 3.9.

Table 3.9: Overview of the number of factories and workers in 1896 in England, Wales, Scotland and Ireland (Calder, 1899).

Year	Fatal accidents	Nonfatal accidents
1896	596	Not given
1897	655	39,739
1898	725	56,698

The 725 fatal accidents in 1898 equate to a rate of 0.16 fatalities per 1,000 workers per year.

The main causes of the 725 fatal accidents were:
- 233 falls (32%);
- 295 due to contact with machinery (mainly moving parts of machinery – 41%);

- 140 due to material structures and tools, scalds and burns (19%); and
- 63 due to contact with products (hot liquid, molten metal, explosion and steam – 8%).

The 56,698 nonfatal accidents were lost time accidents which equate to 13 lost time accidents per 1,000 employees per year.

Calder classifies the causes of accidents in six categories:

- **Ignorance.** The workpeople are grossly ignorant of the nature of the forces and mechanical arrangements which is in their power either to control or to set free. Nowadays, this category would be classified as "lack of knowledge or lack of understanding of the hazards." Calder mentions a pleasing exception of a large textile factory, where several hundreds of children were employed and educated at the works school. What a child might and might not do in the factory was taught, and each newcomer was, after instruction, only admitted into the factory on probation until the foreman certified that the necessary knowledge for safe and efficient working had been acquired, and was being put into practice. Calder adds that the result was obvious from the remarkable low accident rate among this section of employees.
- **Carelessness.** Sometimes combined with ignorance, sometimes sheer thoughtlessness or folly. Calder concludes that for these cases very little can be done apart from maintaining strict discipline and the adoption of punitive measures.
- **Unsuitable clothing.** This leads to accidents because some machine parts cannot be fenced and the work require operatives to approach the moving parts.
- **Insufficient lighting.** This was an important cause, according to Calder, for serious and fatal falls.
- **Defects of machinery and structures.** Calder mentions more specifically bad belting, pulley and toothed gears, shafting and journals, collars and couplings, weak, badly designed or overloaded parts of machinery, plant and staging giving way, worn-out ropes and damaged and annealed chains breaking, faulty grindstones flying, bursting of jacketed steam pans, closed vessels under pressure but without safety valves or periodical inspection, gangways and stairs of a more or less temporary character.
- **Absence of safeguards.** Calder considers this category as the most important. In his book, he next describes in about 200 pages all types of safeguarding of all types of machines.

Calder identifies the following underlying cause of these shortcomings:

> The cause of this is generally the absorption of the management in the work of the production and the failure to assign the safeguarding of machinery and maintenance of such as a definite duty to some responsible party.

3.6.3 The Pittsburg survey (1907–1908)

The Russell Sage Foundation was established in 1907 for "the improvement of social and living conditions in the United States" by a gift of $10 million from Margaret Olivia Slocum Sage (1828–1918), widow of railroad magnate and financier Russell Sage. In 1907, the foundation funded the Pittsburgh Survey, pioneering sociological study of the working-class conditions in a large US city. Pittsburgh was America's prototypical industrial city with immigrants from Europe trying to escape from poverty and large corporations such as US Steel that dominated local governments. It is widely considered a landmark of the Progressive Era reform movement.

Some 70 investigators began work in 1907 under the direction of survey director Paul Kellog. The research was first published in magazines, including Collier's, in 1908 and 1909, then was expanded into a series of six books (four monographs and two collections of essays) published from 1909 to 1914. Paul Kellogg offered Crystal Eastman her first job, investigating labor conditions for the Pittsburgh Survey sponsored by the Russell Sage Foundation.

Her report, *Work Accidents and the Law* (1910), became a classic and resulted in the first workers' compensation law, which she drafted while serving on a New York state commission. She continued to campaign for occupational safety and health while working as an investigating attorney for the US Commission on Industrial Relations during Woodrow Wilson's presidency. She was at one time called the "most dangerous woman in America," due to her free-love idealism and outspoken nature.

The Pittsburgh Survey analyzed in detail 526 fatal accidents that occurred during one year (from 1 July 1906 till 30 June 1907) and 509 lost time accidents that occurred in a 3-month period (from April 1907 till June 1907).

Table 3.10 gives an overview of the distribution of the 526 fatal accidents among the different occupations in which the victims were employed.

Table 3.10: Overview of the population that was studied in the Pittsburgh Survey (Eastman, 1910).

Occupations in which employed	Size of the population	Number of fatalities	Fatalities per 1,000 employees per year
Railroading	50,000	125	2.5
Mining	20,000	71	3.6
Steel manufacture	70,000	195	2.8
Other occupations	110,000	135	1.2
Total	**250,000**	**526**	**2.1**

The nature and extent of the injury was available in 294 of the 509 nonfatal accident cases:

- Serious permanent injury (e.g., a man lost a leg) in 76 cases (≈25.9%)
- Slight permanent injury (e.g., loss of a finger or a lame leg) in 91 cases (≈30.9%)
- No permanent injury in 127 cases (≈43.2%)

Eighty-four percent of the people who were killed during the survey were not over 40 years while 58% were younger than 30 years. The survey also showed that the general idea that accidents mainly concerned immigrants had to be abandoned because 228 people killed (42.5%) were American born. Table 3.11 gives an overview of the nature of the 195 fatal accidents in the steel industry. Operation of cranes, falls and electrical shocks are still today recurrent causes of accidents.

Table 3.11: Breakdown of the 195 fatal accidents in the steel industry in the Pittsburgh Survey.

Nature of the accident	Number of killed	%
Operation of cranes	42	21.5
Falls	24	12.3
Explosions	22	11.3
Circulation (railroad in yards, trains, operation of rolls)	41	21.0
Loading and piling of steel and iron products	8	4.1
Electrical shocks	7	3.6
Asphyxiation by furnace gas	5	2.6
Miscellaneous	46	23.6
Total	195	100

In Chapter VI of the survey, Eastman investigates the "personal factor in industrial accidents." It was very common in the nineteenth century to assume that most if not all accidents were the result of carelessness of the men. Eastman describes it as follows (Eastman, 1910, page 84):

> So you've come to Pittsburgh to study accidents, have you?" says the superintendent, or the claim agent or the general manager, as the case may be. "Well, I've been in this business fifteen years and I can tell you one thing right now, 95 per cent of our accidents are due to the carelessness of the man who gets hurt. Why, you simply wouldn't believe the things they'll do.

Eastman investigated in detail 410 fatalities (resulted from 377 accidents) for which there were enough data out of the 526 fatalities. She then classified these 410 fatalities by employment and by indications of responsibility. Table 3.12 gives the results (Eastman, 1910). Because there were sometimes joint responsibilities this results in 501 indications:

Table 3.12: A number of 410 work-accident fatalities, classified by employment and by indication of responsibility.

Industries in which the accident occurred	Total indications	Indications of sole or partial responsibility				
		Victim	Fellow workmen	Foremen	Employer	None of these
Steel manufacture	189	45	28	16	48	52
Railroading	91	16	12	9	32	22
Mining	86	31	5	11	14	25
Other industries	135	40	11	13	53	18
All industries	501	132	56	49	147	117
%	100	26.3	11.2	9.8	29.3	23.4

According to Table 3.12, 132 accidents can be attributed directly to the carelessness of the victim, while for 117 fatalities nobody can be blamed. The same study would today, however, be looked at differently. Eastman counts, for instance, in the 117 accidents ("nobody to blame"):
- two great furnace explosions at the Jones and Laughlin Steel Works;
- a mule driver in a mine got down to change a switch, his light went out and in the darkness he was knocked down and killed by the mules; and so on.

Some of the examples of accidents that were attributed to the carelessness of the victim are as follows:
- Garbia Lubitch, a Hungarian, who had been in America 5 months was set alone digging a ditch under the railroad tracks in a mill yard. He was working between the ties, when a train backed down without warning and ran over him. A foreman had told him in English to work at the side.
- Thomas Korenz, a Slav, who had worked as a trestle laborer 14 days and could speak no English, was sent with three of his countrymen to do some work under a car. Later, as a switching engine was about to couple on to the cars on this track, a brakeman was sent to warn any men who might be underneath. The three who could understand hurried out from under the car, forgetting to explain the warning to Korenz. The brakeman, thinking that all was safe, signaled to the engineer to come ahead, and Korenz was killed.
- A boy of 15 years old who had worked 8 h out of a 13-h night turn had a few minutes to rest. He lie down in a wheelbarrow. He fell asleep and was struck and killed by the extending arm of a ladle which the crane man was bringing back to the pit (13 out of the 132 fatalities were boys).

One man, afflicted with epilepsy, fell upon a steam exhaust and inhaled steam until he died. It was considered at that time that the man was responsible for having selected an occupation for which he was unfit. The investigation showed, however, that this man had a wife and four girls under 14 years old. Eastman asked therefore "Are we sure that he is responsible?"

The conclusion of Eastman was that a maximum of 21% of the accidents could be attributed to carelessness.

Eastman describes how the attitude of the men toward safety changed through the establishment of safety committees. The safety inspections and the reports made with recommendations increased the responsibility and commitment of the foremen to keep the plant safe.

The United States Steel Corporation put in place a Central Committee of Safety in 1908. Some of the companies that were brought together in 1901 to form the United States Steel Corporation organized safety departments since 1895. The company Dupont, often considered a pioneer in the field of safety, started Safety Commissions in 1911 (Klein, 2009), that is, about 20 years later than the United States Steel Corporation. This committee was composed of five members representing subsidiary companies operating the largest plants and mills, with an officer of the United States Steel Corporation acting as chairman. It was empowered to appoint inspectors to examine various plants and equipment, and to submit reports of safety conditions, with suggestions for improvement. The committee was further requested to record and disseminate data on regulations, rules, devices and others tending toward safer working conditions in the plants. Meetings of the committee were held about once a month.

In addition to the Central Committee of Safety, the different companies of the United States Steel Corporation had their own safety committee:
- Carnegie Steel Company with 27 plants;
- Illinois Steel Company with 6 plants;
- National Tube Company with 13 plants;
- American Sheet and Tin Plate Company with 14 plants;
- American Bridge Company with 16 plants;
- Tennessee Coal Iron and Railroad Company with 7 plants; and
- American Steel and Wire Company with 32 plants.

Because of the great variety of machinery and the huge challenge to bring this equipment up to the approved standards of safety, and maintaining it in this condition, the responsibility was placed in the plants and special inspectors were appointed and organized in local committees: foremen's committee and workmen's committee.

A foremen's committee usually included the assistant superintendent of the plant, a master mechanic, chief electrician and a department foreman. It was the duty of the foremen's committee to inspect the plant either semimonthly or monthly and to make a written report.

The workmen's committee is entirely distinct and could include, for instance, a machinist, an electrician and a carpenter. At the end of every month, an entirely new committee is appointed. Each committee makes a written report of its inspection, the recommendations have to be numbered and the numbers of any incomplete items are all shown on a monthly statement until they have been carried out. During the month of January 1910, there were approximately 1,500 specific recommendations made by the different inspection committees in the 32 plants of the American Steel and Wire Company. Of these, over 500 were completed entirely before the end of the month, with material ordered and work under way on a great many more.

A certain class of equipment got thorough inspections at frequent intervals by men of special training (e.g., cranes). There were special inspections for boiler plants, engine installations, motor stops and electrical cranes.

In planning a new plant, the drawings were all checked to ensure that the latest safety provisions had been included, and safeguarding of gears, spindles, couplings and so on had to be included in the drawings and all the safety features had to be subjected to the approval of the inspectors. The inspectors had free access at all times to the machinery while it is in the process of construction and erection.

The American Steel and Wire Company also worked on "the human element." They used different ways to work on the psychology of the workers in order to make them realize that it was quite worthy and honorable to be careful and safe. When men received their pay envelope, they found little tips printed on the back of the envelopes (Figure 3.4).

YOU ARE RESPONSIBLE
FOR THE SAFETY OF OTHERS
AS WELL AS YOURSELF.

Figure 3.4: Example of a pay envelope
(Eastman, 1910).

Signs and posters were placed at different locations to warn the worker about the hazards.

Dining rooms were made available where the foremen assembled for lunch with a more or less informal business meeting during which reports of accidents were discussed, general safety recommendations were made and so on.

Following a newspaper account of an accident in an outside company where three men were crushed to death in the air cylinder of an engine, a notice was posted in each of the blowing engine rooms of the American Steel and Wire Company. All companies installed an emergency hospital, where the workers got the first medical assistance by professionals in case of an accident.

Figure 3.5: Typical view of an emergency hospital in one of the mills (Eastman, 1910).

The importance of the Pittsburgh Survey and in particular the report of Eastman cannot be overestimated (Figure 3.5). It was a groundbreaking work. The study demonstrated that:
- The proportion of careless behavior of the worker as the cause of accidents was much lower than what was thought.
- A large number of accidents could be prevented by carrying out systematic inspections involving all layers of the organization.
- Technical measures to prevent accidents were essential.

- Certain equipment required specific inspections that must be carried out by specialists.
- In addition to the technical and organizational measures, special attention had to be paid to the behavior of the workers.

The Eastman study identified at the beginning of the twentieth century, the three important elements that need to be addressed if accidents are to be avoided: *technical measures, organizational measures and human behavior.*

All studies in the course of the twentieth century built on the ideas put forward by Calder and Eastman. They still form the basis of current thinking about occupational safety and are directly related to the vision that people had in the nineteenth century where the carelessness of the slaughterer was central. At the beginning of the twentieth century, accident prevention became a task and responsibility of the managers.

3.6.4 Pamphlet of the Fidelity and Casualty Company, 1909

In the year 1908, the sum of $22,392,072 (equivalent in purchasing power to $624,931,088 in 2019) was paid in premium to insurance companies for liability insurance. Employers' liability insurance was transacted in the USA in 1886 and increased steadily since then. The Fidelity and Casualty Company was of the opinion that much more could be done to prevent accidents. The paper states in its introduction, however, "The prevention of accidents absolutely is of course impossible. But much more can be done to prevent accidents than at first sight appears."

Because at that time statistics regarding the causes of industrial accidents had only been collected in fragmentary form, the Fidelity and Casualty Company referred to statistics collected in Germany by the German Workmen's Insurance and presented in 1904 for the St. Louis Universal Exposition. According to these data, about 57.95% of accidents are due to the negligence of employers and employees and a smaller number, 42.05%, are due to the inevitable risks of employment.

It is interesting to remark that in the beginning of the nineteenth century, the paradigm was that almost all accidents were inevitable risks while 100 years later the percentage of the inevitable risks was about 42%. Today the paradigm is that 100% of the accidents can be avoided, and this paradigm finds its origins in the work of Heinrich (see further). The negligence of employers and employees is further broken down in Table 3.13.

The pamphlet concludes that the prevention lies in the direction, first, of proper design and construction of the plant and appliances, second, of care on the part of employers and employees and, third, of the use of safety devices. Table 3.14 shows the recommendations that relate to the care on the part of employers and employees:

Table 3.13: Causes of evitable accidents according to the German Workmen's Insurance in 1904.

Cause of accident	Percentage of the 57.95% of accidents due to negligence of employers and employees (rounded figures)
Want of skill and carelessness	35
Want of guards	14
Deficient factory arrangements	13
Acting against rules	10
Fault of other (third person)	9
Fault of employers and workmen	8
Not using guards	3
Insufficient instruction	3
Superior force	2
Carelessness	2
Unfit clothes	1

Table 3.14: Suggestions to improve safety according to the Fidelity and Casualty Company study of 1909.

Physical surroundings	Making the physical surroundings of the workmen as comfortable as possible. Plenty of light, good air, safety and comfort pay in a financial sense.
Insufficient lighting	Statistics show that the greatest number of accidents occur during the months of diminishing light.
Overcrowding machinery	Considerations of economy often lead to overcrowding of machinery and hence reduced width of the passageways between the machines.
Slippery floors	Slippery floors constitute an element of danger, especially in conjunction with unguarded machines. A slippery floor may cause a bad fall.
Ignorance	Many accidents are due to the ignorance of the workmen. New hands are put to work on dangerous machines without proper preliminary training of sufficient instructions. All workmen should be carefully instructed, in language they can understand, the proper and safe way to do their work.

Table 3.14 (continued)

Carelessness	The authors refer to statistics in Germany for the year 1897. Out of 45,971 accidents, there were: – 9,363 caused by the "improvidence or inattention of the workmen" – 2,422 cases by "acting in contradiction to instructions" – 861 cases on account of nonuse of existing devices for protection – 533 cases by "frivolous behavior" – 220 cases by "unsuitable clothing" The employer is warranted in making strict rules governing the employees in his/her plant. In fact, it is his/her duty to do so. But the maintenance of discipline, to which punitive measures are added where necessary, many accidents may be prevented. Disobedience of orders should not be tolerated.
Unsuitable clothing	Ragged sleeve ends, loose cravats and coats or overalls not properly buttoned, often catch on moving parts of machinery.
Failure to use safeguards provided	Workmen object to the use of safeguard provided, claiming that they interfere with quantity of output and so cut down their earnings.
Overwork	Fatigue leads to carelessness.
Ventilation	Impure air, gases, vapors, dust and smoke increase chances of accident, in addition to imperiling the health of workmen.
Intoxicants	No man under the influence of liquor, even slightly so, should be permitted to remain in the works.
Supervision and management **Rules**	Managers, superintendents, foremen and others in authority should be persons of experience and given to exercising a high degree of care in all that they do. Utmost care should be taken in selecting them. The education of the workmen in proper and safe practices. Promotion shall be made by the posting of printed rules.
Inspections	Regular and frequent inspections by competent men should be made of all the ways, works, machinery and appliances, so that defects and unsafe conditions may be discovered promptly and remedied.

3.6.5 Paradigm shift on prevention of work-related accidents at the start of the twentieth century

The subject of accidents and their prevention had been recognized as one of great importance in industry since 1830s (Vernon, 1936). However, it is only around the turn of the century that a number of scientific studies (Calder, Eastman, etc.) allowed for a paradigm shift.

These studies laid the foundation for the continuous improvement of safety in the workplace as we know it today. Several theories and metaphors about managing safety would emerge but they would all remain within the main ideas of the studies cited above, that is, the responsibility of the management and the supervision to:

1. improve of physical conditions at the workplace;
2. put in place organizational structures that allow continuous monitoring of the situation and strive for continuous improvement of safety; and
3. better understand the psychology and behavior of the "human element."

The improvement of the physical conditions was mainly oriented toward workplace arrangements (layout and spacing, lighting, ventilation, etc.), housekeeping and safeguards.

The organizational structures were workmen's committees with an active involvements of all levels of the company. These committees discussed unsafe conditions, performed inspections and monitored improvement plans. Rules needed to be enforced and regular inspections were the backbone of the management of safety.

Improvement of the behavior of the workforce entailed improving their knowledge and experience and making them more aware of the hazards via posters and campaigns. Fatigue and lack of mindfulness were identified as potential causes for accidents.

Management and supervisors needed to be selected on the basis of their knowledge and experience. It was considered important that managers, superintendents, foremen and others in authority should be careful and prudent people.

3.7 The evolution of occupational health and safety in the twentieth century

3.7.1 Purpose of this chapter

In the following paragraphs, we will go back to the twentieth century to identify the roots of our current thinking about management of occupational health and safety (OHS). It will be shown that many "modern ideas" about OHS have been proposed and exercised decades ago.

The purpose is to give some indication of the period when people started to publish about a topic, which means that the industry and publishing society were convinced that the subjects (e.g., safety committees and leadership) were important for the prevention of occupational accidents. It also shows that preventing accidents at work is not a simple problem that can be solved by one approach such as "a safety management system" or "a human behavior program." Preventing accidents at work is a subject that has been occupying the industry for more than 150 years and has been extensively and thoroughly considered.

3.7.2 The International Labour Organization

Trade unionism was active during the nineteenth century, and various labor parties and trade unions were formed throughout the industrialized parts of the world. From the mid-nineteenth century onward, the labor movement became increasingly globalized. After World War I (WWI), the allied powers established the International Labour Office (art. 392 of the Treaty of Versailles) as an agency of the League of Nations.

It was French trade unionist Léon Jouhaux who demanded at an international trade union meeting in Leeds that a future peace treaty should contain "Industrial Clauses." The British Trade Union Congress also demanded on that occasion that representatives of the labor movement should participate in an upcoming peace conference.

During WWI, trade unions were crucial for maintaining social peace in order to support war production to the maximum. In exchange, the trade union movement asked for participation and recognition, and in particular the prospect of a better life for worker and soldier once the war was over. No promises without obligation, but a real treaty and a one permanent organization. The revolution in Russia in 1917 with the fall and execution of the tsar followed by the nationalization of all companies alerted the rich that the ghost that Karl Marx had described really existed.

The quest for social justice was not a real surprise. For years, experts and academics had studied labor issues and, apart from that, trade unions and political parties had been campaigning for years to finally come to the realization that some of the core problems could only be dealt with through international agreements.

Whenever demands concerned a shorter working day, a living wage, protection of women and child labor, equal pay for equal work and so on, they clashed on the wall of economic feasibility, because all these demands cost money, and both governments and employers argued that the one-sided introduction would affect competitiveness and thus the quality of life of national economies. And hence an international global approach was needed.

At the time of its establishment, the US government was not a member of International Labor Organization (ILO), as the US Senate rejected the covenant of the League of Nations, and the USA could not join any of its agencies. The USA joined in 1934.

The first annual conference, referred to as the International Labor Conference, began on 29 October 1919 at the Pan American Union Building in Washington, DC, and adopted the first six International Labor Conventions, which dealt with hours of work in industry, unemployment, maternity protection, night work for women, minimum age and night work for young persons in industry. The prominent French socialist Albert Thomas became its first director-general.

3.7.3 Technical standards

Technical standards are continuously developed and improved in all domains. The history and evolution of technical standards is a subject in itself and will not be handled in the frame of this book. The following dates underline the observation that the need for technical standard is not new:

1853 The importance of technical standards was recognized very early. Mathers work (1853) is about the dangers and safety in the mines. He concentrates on a number of critical systems for safety: ventilation, furnace and steam jet and safety lamps.

1880 The American Society of Mechanical Engineers (ASME) was founded in 1880 by Alexander Lyman Holley, Henry Rossiter Worthington, John Edison Sweet and Matthias N. Forney in response to numerous steam boiler pressure vessel failures. ASME, however, extended to all types of areas such as (but not limited) Elevators and Escalators (A17 Series), Overhead and Mobile Cranes and related lifting and rigging equipment (B30 Series), Piping and Pipelines (B31 Series), Bioprocessing Equipment, Valves Flanges and Fittings and Gaskets (B16). ASME has over 110,000 members in more than 150 countries worldwide.

1901 The Engineering Standards Committee was established in London in 1901. It subsequently extended its standardization work and became the British Engineering Standards Association in 1918, adopting the name British Standards Institution (BSI) in 1931 after receiving a Royal Charter in 1929.

1905 Museums were formed to show best practices on how to prevent accidents by mechanical means. The first of these museums was formed in 1893 in Amsterdam. Others were established in Milan in 1894, Munich in 1900, Zurich in 1902, Berlin in 1903, Paris in 1905 and New York in 1907. In 1928, 23 safety museums were known to exist (Vernon, 1936).

1909 Since 1906, the Société Belge des Electriciens has participated in the founding meeting of the International Electrotechnical Commission in London. The BEC was established in 1909 as a result of a growing need for technical specifications for the quality and safety of electrical appliances. In 1923, the BEC created its own national quality brand CEBEC.

1911 The ASME appointed a committee to draft specifications for the safe construction of boilers and other pressure vessels, which in 1914 became the first ASME Boiler Code.

1917 Foundation of the Normenausschuß der deutschen Industrie (NADI, "Standardisation Committee of German Industry"): one of the earliest, and probably the best known, is DIN 476 – the standard that introduced the A-series paper sizes in 1922. In 1975, it was renamed again to Deutsches Institut für Normung, or "DIN" and is recognized by the German government as the official national standards body, representing German interests at international and European levels.

1926   The International Federation of the National Standardizing Associations (ISA) was founded in 1926 with a broader remit to enhance international cooperation for all technical standards and specifications. In October 1946, ISA and UNSCC delegates from 25 countries met in London and agreed to join forces to create the new International Organization for Standardization (ISO). The new organization officially began operations in February 1947.

   "Association française de normalisation" was created in France.

1929   R031 – Prevention of Industrial Accidents Recommendation, 1929 (No. 31) adopted by the Governing Body of the International Labour Office, and having met in its Twelfth Session on 30 May 1929 mentions under point 13:

> 10. It is recommended that the State should establish or promote the establishment of permanent safety exhibitions where the best appliances, arrangements and methods for preventing accidents and promoting safety can be seen (and in the case of machinery, seen in action) and advice and information given to employers, works officials, workers, students in the engineering and technical schools, and others.

It is obvious that the use of technical standards as a vital element in the prevention of occupational accidents is recognized since the early years of the industrial revolution, that is, for more than 150 years. Technical standards and safeguards are considered during the nineteenth century as the main strategy to avoid accidents. Figure 3.6 shows the evolution of standardization organization in the world (taken from Shewhart, 1939).

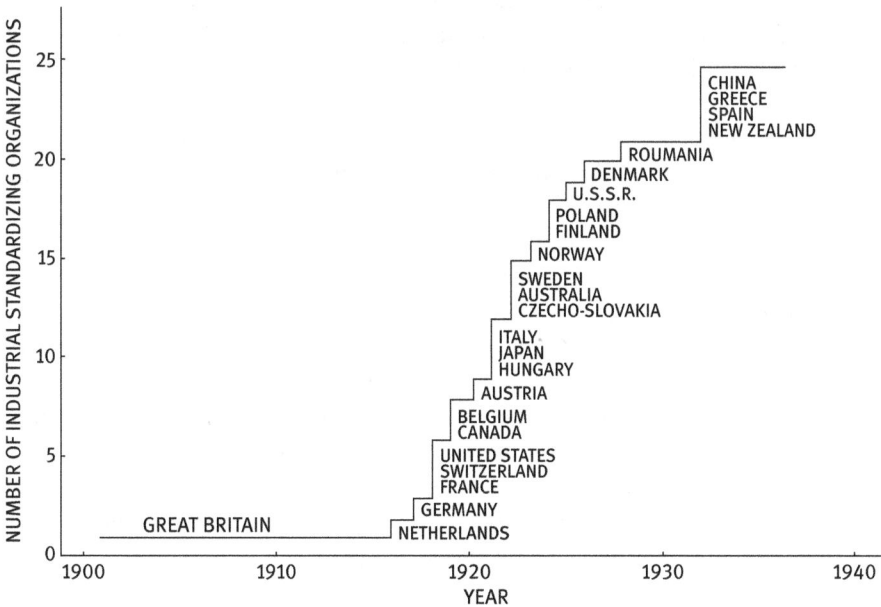

Figure 3.6: Growth of standardization organizations in the world (Shewhart, 1939).

3.7.4 Safety and health standards and rules

1811 By 1811, E. I. du Pont issued official safety rules to help ensure safe manufacture: "As the greatest order is indispensable in the manufacturing as well as for the regularity and the security of works, than the safety of the workmen themselves, the following Rules shall be strictly observed by every one of the men employed in the factory" (Klein, 2009 who refers to "E.I du Pont, Opening Paragraph to 'Rules' For His Manufactory, Hagley Museum and Library, Wilmington, DE, 1811").

1895 Section 8 of the Factory Act of 1895 obliges the "occupier" of a factory to elaborate and to publish special rules for manual labor used in a factory or workshop that is dangerous. Section 9 obliges the worker to follow these rules. Section 10 concerns the obligation for a regular update of the rules and the obligation to give a printed copy of the rules to "any person affected thereby on his or her application."

1909 The use of rules is very common since the beginning of the twentieth century. Law and William (1909) list, for instance, 16 rules for elevator operators, 16 boiler room rules and rules for inspection that machines are safe.

1911 On 25 March 1911, a fire broke out at the Triangle Waist Factory in New York City. Within 18 min, 146 people were dead as a result of the fire. The investigation following the fire revealed great negligence in building safety practices as well as a lack of regulations requiring such practices. In response to the tragedy, workplace safety legislation was introduced and our safety society was born – first in 1911 as the United Association of Casualty Inspectors, then renamed as the American Society of Safety Engineers in 1914 and renamed as the American Society of Safety Professionals (ASSP) in 2018. ASSP has the secretariat for 11 ANSI (American National Standards Institute) committees responsible for more than 100 safety Standards. Today, ASSP has more than 39,000 members in 80 countries around the world.

1934 The Division of Labor Standards, United States Department of Labor, Washington, was created. The objectives of this division are to formulate labor standards and promote the improvement of working conditions.

1943 Blake (1943) dedicates a full chapter of his book to "Safety and Health Standards and Rules."

1948 The Tripartite Technical Conference held in Geneva in 1948 was approved in accordance with a decision of the Governing Body of the International Labour Office, the *Model Code of Safety Regulations for Industrial Establishments for the Guidance of Governments and Industry*. The Model Code was based on a draft prepared by experts over a period of 6 years.

1956 Amendments to certain chapters of the Model Code dealing with subjects of interest to these committees were proposed in 1955 by a panel of experts chosen among the members of the ILO correspondence Committee on Occupational Safety and Health. The Governing Body of the International Labour Office, at its 132nd Session (Geneva, June 1956), authorized the incorporation of these amendments in the *Model Code of Safety Regulations for Industrial Establishments for the Guidance of Governments and Industry* which was first published in 1949 (ILO, 1954). The purpose of the code was to put a guidance at the disposal of governments and industries so that they can fit in framing or in revising their own safety regulations. The Model Code (649 pages) contains 244 detailed regulations (Table 3.15).

Table 3.15: Content of the Model Code of Safety Regulations for Industrial Establishments for the Guidance of Governments and Industry (ILO, 1956).

Introduction		Number of regulations
Chapter I	**General provisions**	6
Chapter III	**Premises of industrial establishments**	
	Section 1 Buildings, structures, workplaces and yards	10
	Section 2 Lighting	3
	Section 3 General ventilation	6
Chapter III	**Fire prevention and protection**	
	Section 1 Building exits	10
	Section 2 Fire fighting facilities	7
	Section 3 Alarm systems and fire drills	2
	Section 4 Storage of explosives and flammable substances	6
	Section 5 Disposal of waste	4
	Section 6 Lightning	4
Chapter IV	**Section 1 General provisions**	5
	Section 2 Prime movers	6
	Section 3 Mechanical power transmission equipment	12
	Section 4 Standard machinery guards	8
	Section 5 Machine guarding, at point of operation	17
	Section 6 Vats and tanks	1
Chapter V	**Electrical equipment**	11
Chapter VI	**Hand tools and portable power-driven tools**	2
Chapter VII	**Boilers and pressure vessels**	
	Section 1 Boilers	5
	Section 2 Unfired pressure vessels	7
	Section 3 Compressors	2
	Section 4 Gas cylinders	1

Table 3.15 (continued)

Introduction			Number of regulations
Chapter VIII	**Furnaces, kilns and ovens**		16
Chapter IX	**Handling and transportation of materials**		
	Section 1	Hoisting equipment other than elevators	11
	Section 2	Conveyors	9
	Section 3	Power trucks and hand trucks	4
	Section 4	Plant railways	5
	Section 5	Piping systems	1
	Section 6	Lifting, carrying, pilling and storage of material	3
Chapter X	**Dangerous and obnoxious substances**		
	Section 1	General provisions	2
	Section 2	Flammable and explosive substances	7
	Section 3	Corrosive, hot and cold substances	4
	Section 4	Infectious, irritating and toxic substances	4
Chapter XI	**Dangerous radiations**		
	Section 1	Infra-red and ultra-violet radiation	2
	Section 2	Ionizing radiation	6
Chapter XII	**Maintenance and repairs**		4
Chapter XIII	**Health protection**		
	Section 1	Sanitation	4
	Section 2	Local exhaust systems	7
Chapter XIV	**Personal protective equipment**		11
Chapter XV	**Selection of workers, medical service, medical aid**		
	Section 1	Selection of workers	2
	Section 2	Medical service	1
	Section 3	Medical service organization and medical aid	5
Chapter XV	**Safety organization**		1
			244

1960 Since the 1960s, industry associations and companies have been creating Health and Safety Executive (HSE) rules. These rules are categorized as general loss control rules, specialized work rules, specialized work permits, permits to operate and the use of signs and color codes. International representatives in cooperation with the National Safety Council National Safety Board selected the following "10 Best General Safety Rules":

1. Follow instructions; do not take chances. If you do not know, ask.
2. Report immediately any condition or practice you think might cause injury to employees or damage to equipment.
3. Put everything you use in its proper place. Disorder causes injury and wastes time, energy and material. Keep your work area clean and orderly.
4. Use the right tools and equipment for the job and use them safely.
5. Whenever you or the equipment you operate is involved in an accident, regardless of how minor, report it immediately. Get first aid promptly.
6. Use, adjust, alter and repair equipment only when authorized.
7. Wear approved personal protective equipment as directed. Keep it in good condition.
8. Do not fool around; avoid distracting others.
9. When lifting, bend your knees, grasp the load firmly, then raise the load keeping your back as straight as possible. Get help for heavy loads.
10. Obey all rules, signs and instructions.

1977 Council Directive 77/576/EEC of 25 July 1977 on the approximation of the laws, regulations and administrative provisions of the member states relating to the provision of safety signs at places of work regulates the safety signs and colors that are mandatory at all places of work in the European member states. The directive recognizes that "the use of uniform safety signs has positive effects both for workers at places of work, inside or outside undertakings, and for other persons having access to such places." The objective of the system of safety signs is to draw attention rapidly and unambiguously to objects and situations capable of causing specific hazards. This directive is revised by the Council Directive 92/58/EEC of 24 June 1992 on the minimum requirements for the provision of safety and/or health signs at work. The annexes of this directive provide detailed information about the minimum requirements for all safety and health signs. Workers and their representatives must be informed of all the measures taken concerning health and safety signs at work and must be given suitable instruction about these signs. This covers the meaning of signs and the general and specific behavior required. The directive defines details about the form (round and triangular), color (red, blue and green) and type of signs to be used.

1980 Companies and industry associations develop booklets or standards with common safe practices for a type of activity or industry. Examples are:
- Code of practice on safety and health in the iron and steel industry adopted at an ILO meeting of experts in 1981. The code was revised in 2005.
- ILO codes of practice on safety and health in coal mines (1986).
- *E&P Onshore Operations Safety Handbook*, API Recommended Practice 54 – Recommended Practice for Occupational Safety for Oil and Gas Well Drilling and Servicing Operations published in 1981. The document was revised in 1992, 1999 and the fourth edition was published in 2019.
- API Recommended Practice 74 – Recommended Practice for Occupational Safety for Onshore Oil and Gas Production Operations. The first edition dates from 2001. The document was revised in 2019.

2000 Over the last two decades, there was an increase in the use of pictograms combined with an effort to keep the number of rules simple and small. In reality, this goes back to where companies started more than 100 years ago which is an attempt to draw the attention of the workers to a small number of rules that cause most of the accidents.

Today, almost all major companies publish "Life Saving Rules," "Golden Rules," and so on. A Google search with the phrase "Life Saving Rules" gives 617,000,000 returns.

Industry trade unions also started to publish "Life Saving Rules." Examples are:
- IOGP (International Association of Oil & Gas Producers) published their "Life Saving Rules" in 2013 and revised them in 2018 (IOGP document Report No. 459).
- API made a summary of their rules from API Recommended Practice 54 and API Recommended Practice 74 and published a booklet in 2018 called "Rules to Live By."
- The World Steel Association published a position paper "SAFETY AND HEALTH IN THE STEEL INDUSTRY" in which they mention the five most common causes of safety incidents and preventative measures:
  1. Moving machinery – isolate, lock or pin all energy sources before any machinery is accessed.
  2. Falling from height – provide regular training, appropriate harnessing equipment and ensure that checks are in place when working at height.
  3. Falling objects – provide regular checks are in place to remove or secure objects in risk areas.
  4. Asphyxiation or gassing – install sensors to test atmospheric condition and provide training on evacuation plans in order to reduce risks of dangerous gases.
  5. Cranes – carry out daily checks before a crane is used to ensure safe and reliable operation.

The importance of (general as well as specific) safety and health standards and rules has been recognized for more than 150 years. Until the early twentieth century, it was common practice to apply a limited number of clear general rules with pictograms in clearly visible places. Although this practice never really disappeared, the number of safety rules and their content became more and more cumbersome. Safety rules became much more detailed.

In the second half of the twentieth century, it was a common practice to have a set of general rules that applied to everyone and a set of very specific rules per activity (welders, builders' scaffolding, etc.). Today, we see an evolution toward a limited set of so-called Life Saving Rules.

One potential problem nowadays is that workers are overwhelmed with rules and they are supposed to strictly follow all the safety rules. New pictograms find their way to the workforce, which may start to confuse the workers (Figure 3.7).

| Warning sign required by the European Directive 92/58/EC to be present at all work places in Europe to warn on the hazard during lifting operations (remark: format, color and sign is mandatory) | Pictogram used as one of the Life Saving Rules of IOGP to alert of the hazard of walking under a suspended load | Pictogram used by an oil company for their golden rule on lifting and hoisting | Pictogram used by API in their brochure "Rules to Live By" (2018) |

Figure 3.7: Different pictograms for one type of hazard.

Within the industry there is a continuous quest for "less rules that are simple, easy to understand and always applicable" and "more comprehensive rules that are detailed enough so that all potential hazards can be avoided." Finding the right balance is a real challenge that is as old as the industry itself. The growing complexity of machines and installations does not make it any easier. It is obvious that overloading people with too many rules and procedures is counterproductive but "ignorance" (which is called today "lack of knowledge or lack of understanding of the hazards") has been recognized as an important cause of accidents since more than 100 years (see, e.g., Calder, 1899). Many companies are evolving from "a sharing culture to a learning culture" but this takes time. A sharing culture is characterized by sending around the information (via posters, safety alerts, accident reports, etc.) while a learning culture pays attention to real learning from positive (best practices) as well as negative events.

3.7.5 Safeguarding of machinery

1844 Chapter XVI of the consolidated Factory Act of 1844 is devoted to the provisions for the prevention of accidents by safeguarding.

1877 In 1877, Massachusetts passed the Nation's first factory inspection law. It required guarding of belts, shafts and gears, protection on elevators and adequate fire exits. Its passage prompted a flurry of state factory acts. By 1890, nine states provided for factory inspectors, 13 required machine safeguarding and 21 made limited provision for health hazards.

1899 Safeguarding is the most important topic in the book of Calder (Calder, 1899):
- Chapter VII: Safeguarding of prime movers
- Chapter VIII: Safeguarding of mill gearing
- Chapter IX: Safeguarding of hoist and lifting tackle
- Chapter X: Safeguarding of dangerous details of machinery

1903 In 1903, the insurance company Aetna created an Engineering and Inspection Division to improve workplace safety.

1904 The Travelers Insurance Company (one of the major America-based insurance companies) organizes a professional corps of safety engineers (Busch, 2018).

1909 The Prevention of Industrial Accidents (Law and William, 1909) gives many examples with nice drawings on how different types of woodworking machines should be safeguarded.

1910 American Steel and Wire Company included the following note in its contracts (Eastman, 1910):

> Safeguarding of gears, spindles, couplings, collars, setscrews, keys, etc. will be covered as fully as possible in the drawings which we furnish, but it is understood that these features shall be subject to the approval of our inspectors, who shall have free access at all times to the machinery while it is in process of construction and erection. Eastman also describes motor stops, automatic engine stops, push bottoms, a switch that can be pulled by means of a rope in case of an emergency and others.

1910 The insurance company Aetna issued in 1910 a book *Safeguarding for the Prevention of Industrial Accidents* (Van Schaak, 1910).

1920 The Travelers Insurance Company publishes the book *Safety in the Machine Shop*.

1940 The American Standards Association published its *Safety Code for Mechanical Power Transmission Apparatus* (very similar to OSHA 1910.219).

1989 The Machinery Directive was first published in 1989. This directive was amended several times and the latest revision – called the new Machinery Directive (Directive 2006/42/EC) – was adopted in June 2006. The safety of the machine can be further increased via the risk analysis which is part of the Directive 2009/104/EC on the use of the work equipment.

Safeguarding of machines has been a concern for more than 150 years. In the beginning of the twentieth century, it was mainly large insurance companies that published books, standards and articles about machine safety and machine guarding but the intent of safeguards for machines at the beginning of the twentieth century was the same as the intent of the Machinery Directive (Directive 2006/42/EC of the European Parliament and of the Council of 17 May 2006). A lot of technical progress was made in improving the safety of the machines but the intention to have safe machines and the commitment to safeguard machines in order to avoid accidents are still the same as 150 years ago.

3.7.6 Safety education, training of people

1853 Chapter IX of the book of Mather (Mather, 1853) is about "Education for the Mines." Mather explains why education is important to ensure safety. Mather mentions:

> Every hour, face to face with some of the most singular and terrific phenomena of nature, a knowledge of their properties and principles will alone enable the miner to deal with them in safety . . . *Sufficient knowledge is not necessary only for many, but for all without exception. An ignorant and careless officer, or one ignorant and reckless man in a pit, may counteract the knowledge and wisdom of the rest, and lay it all lifeless. . . . The men before the Washington explosion warned the officers of the coming danger, which being neglected, 38 lives were lost . . ..*

1899 Calder (1899) demonstrates that ignorance (lack of knowledge) is an important cause of accidents. He mentions the good practice of a large textile factory in which workers were taught about what they might and might not do and that they were only admitted into the factory on probation until the foreman certified that the necessary knowledge for safe and efficient working has been acquired and was being put into practice.

1916 Chapter 46 in the book of Beyer is "Safety education, bulletins, signs, etc." According to Beyer, there is no doubt that many accidents attributed to carelessness can be prevented through safety education (Beyer, 1916, page 359). The methods mentioned by Beyer for the safety education are accident pictures, safety bulletins, safety talks or safety lectures, safety signs and safety slogans. He also listed 28 examples of safety slogans such as:
- "It takes less time to explain why you were late than to make out an accident report not clear"
- "Investigate every accident that happens in your gang, and try to prevent an accident occurring in a like manner"

1929 R031 – Prevention of Industrial Accidents Recommendation, 1929 (No. 31), adopted by the Governing Body of the International Labour Office, and having met in its Twelfth Session on 30 May 1929 mentions under point 13:

> 13. In view of the importance of the work of education referred to in the preceding paragraph, and as a foundation for such education, the Conference recommends that the Members should arrange for the inclusion in the curricula of the elementary schools of lessons designed to inculcate habits of carefulness, and in the curricula of continuation schools lessons in accident prevention and first aid. Instruction in the prevention of industrial accidents should be given in vocational schools of all grades, where the importance of the subject both from the economic and moral standpoints should be impressed upon the pupils.

1940s In all standard books about "industrial safety," this topic is discussed and
1950s considered as important or vital (Blake, 1943, Heinrich and Granniss, 1959, etc.).
1968 Trevor Kletz was appointed the first Technical Safety Advisor of the Heavy Organic Division of ICI (Imperial Chemical Industries) in 1968. ICI was, for much of its history, the largest manufacturer in Britain. Between May 1968 and October 1983, Kletz published every month a *Safety Newsletter* that was sent to all employees. The first topic of his first Newsletter was about "Training." The Newsletter from Kletz demonstrates that training and learning from return of experience was a very topical issue in the industry in the 1960s. The text has been reproduced below:

> Those of us who have children at school know that methods of teaching have changed since we were at school ourselves. The following quotation from an article on education illustrates the change that is taking place. "(There is a) growing concern that children should play a more active part in learning. . . . Class teaching is no longer the appropriate model. . . . By contrast many of the new curricula assume that it is as important for children to find out for themselves as from a teacher, to discuss among themselves as to answer a teacher's questions. Nuffield science and math, for example, put the emphasis on learning by discovery, working in ones and twos; while recent developments in English teaching stress the value of discussion, of the kind of interchange and inquiry that suits small groups of five to fifteen."
>
> Can we learn from this how to improve our safety training?
>
> We cannot let each manager, supervisor and operator blow up his own plant and then discuss the result with him.
>
> But we can get groups of managers, supervisors and operators (mixed or separate) to discuss the causes of accidents that have happened and decide what they would do to prevent a recurrence. They will learn far more in this way than they will learn by listening to someone talking or by reading reports.
>
> I have written up about 50 accident case -histories for discussion in this way. All are illustrated by 35 mm color slides. Many of the case histories deal with fire and explosion hazards but others are of interest to any Works, whatever their processes. Copies of the case-histories and slides are available to those interested and I am willing to take initial discussions myself.

Safety education and training of people has been considered as a main strategy in the prevention of accidents since more than 150 years.

3.7.7 Short service workers

Many companies (e.g., ExxonMobil, Shell), in particular in the USA, have put in place *Short Service Employee (SSE) Management Program* that applies to employees or subcontractors who have less than 6 months experience in the area of work in which they were hired. The purpose of the program is to prevent work-related injuries and illnesses to new hires, temporary workers and subcontractors.

1910 Eastman (1910) pays a lot of attention to what she calls "the greeners." These are young people who just started in a job. They are far less aware of the hazards of the job or from its environment.

1942 Training within Industry Bulletin No. 8 of February 1942 from the United States of Education is about "Introducing the New Employee to the Work Place."

1943 Chapter XXV of the book of Blake (1943) is dedicated to "The New Employee." In 16 pages, Blake defends and explains the importance of a special training for the new employees.

The issue of new employees has been a point of attention for more than 100 years. *All* companies should have a Short Service Employee management program in place.

On 15 July 2009, an explosion occurred in a furnace of a steam superheater in a TOTAL site in Carling. Two people were killed. One of the victims was 20 years old. He was using a piezoelectric ignitor to ignite the gas inside the furnace. He had been only working for a couple of months at the site and he was inexperienced for this task. It is important that companies identify all the tasks that must be strictly reserved for experienced people.

3.7.8 Safety information (posters, bulletins, etc.)

1890 Posters to draw the attention of workers on the importance of safety were extensively used between 16 June 1890 and 7 September 1890 during an exhibition in the Paleis voor Volksvlijt (Palace of Popular Diligence) in Amsterdam to promote health and safety in factories and workshops. The poster says, "Because he was careful, grandfather turned 70 without ever having had one accident" (see Figure 3.8).

Figure 3.8: Examples of posters.

1912 The Safety First movement gets more and more influence when the National Council for Industrial Safety is erected in 1912 by companies and professional associations of engineers (Swuste et al., 2009). Swuste refers to the National Safety Council but it was in 1915 that the Council for Industrial Safety changed its name in the National Safety Council. This Council will act as a coordination center for safety information. In 1925, the National Safety Council will merge with the American Society of Safety Engineers, a society that was erected in 1911 by insurance companies. In the 1930s, the council published six monthly magazines: National Safety News (for industrialists), Public Safety, The Industrial Supervisor, The Safe Worker, The Safe Driver and Safety Education (for teachers).

1932 Posters are widely used in industry. The National Safety First Association, acting in conjunction with the Industrial Welfare Society, provides a weekly service to its members, and in 1932 no less than 200,000 posters were distributed and displayed (Vernon, 1936). It is recommended that one set of posters be used for every 50 employees. Posters should be changed every week and the same new poster should be exhibited at several places in the works (Vernon, 1936). Reitynbarg and Makarow (1933) found out that the average time that people spent on negative posters was 30% higher than the time spent on positive posters.

The use of safety posters (Figure 3.8) to promote safety exists in industry for more than 120 years. There is a general consensus that the use of posters is useful to improve safety. Some sites, however, checked the effectiveness of the use of posters. Surveys in the field observed that people after a couple of days do not pay attention to what is put on the posters.

In one company, the safety poster is renewed monthly. One week after the poster was changed, the employees were asked about the content of the new poster. It appeared that about 30% of the workforce did not even noticed that the poster was changed.

But posters can be very useful when they are part of an active communication campaign. The subject of the poster should be used for discussions during the safety meetings, safety talks and so on. It is also recommended to associate the workers in the definition of the poster campaign.

3.7.9 Job safety analysis

There is a lot of interesting documentation on the Internet about job safety analysis (JSA) or job hazard analysis. See, for instance, the OSHA 3071 publication of the USA Occupational Safety and Health Administration:
-   https://www.osha.gov/Publications/osha3071.html
-   https://www.ccohs.ca/oshanswers/hsprograms/job-haz.html

The importance of JSA is, however, not new.

1900 Job analysis as a management technique was developed around 1900 (Zerga, 1943). It became one of the tools with which managers understood and directed organizations. Taylor made studying the job one of his principles of scientific management (Taylor, 1911).

1943 Gimbel, a safety engineer of General Electric Company, describes in chapter VIII of the book of Blake (1943) the details of what is called in the book *Job Safety Analysis*. A large company brought down its frequency rate (number of disabling injuries per one million hours) from over 50 to under 5 by using *Job Safety Analysis* for all production works. The details on how to conduct a JSA as described by Gimbel are the same today as they were then.

1957 John Grimaldi, former member of the ASME Standing Committee on Safety, presented at the Annual Meeting in Atlanta City methods to perform a JSA.

1959 Heinrich discusses the importance of job analysis and describes the fundamentals of a safety analysis, that is (Heinrich and Granniss, 1959):
   - break down the job or operation into its elementary steps;
   - list them in their proper order; and
   - then examine them critically.

1971 William Fine (1971) proposed a quantitative method to evaluate the seriousness of the risk due to a hazard and to determine the justification for recommended corrective action. This method was taken over by Graham Kinney in 1976 (Kinney and Wiruth, 1976, Graham and Gilbert, 1980) and has since then been used extensively to rank the importance of different tasks or jobs from a risk point of view. The method although developed by William Fine is often referred to as the "Kinney method" (see, e.g., Malchaire, 2002). The method was recommended for a long time as part of critical task analysis programs (e.g., ISRS (International Safety Rating System) edition 6 element 4). When using the method, people should be aware of the difference between criticality of a task and the so-called risk of a task. When a task is extremely performed rarely, its risk goes down (because the risk includes the probability that the task is performed). However, a task can have a very low probability but a high criticality at the moment that the task is performed.

1972 The United States Department of Labor published in 1972 a *Handbook for Analyzing Jobs*.

The importance of job analysis and JSA has been recognized for many decades by all safety professionals. It is fair to say that it is a best practice that is recommended for about 70 years. In many accidents today, however, the investigation reveals that the JSA was not well performed or even not performed at all. In practice, it means that what should be considered by all executors as basic before starting a job is still not recognized as such. JSA is still sometimes considered a theoretical exercise invented by safety people and it is still not always considered an integral part of a job.

3.7.10 Safety departments, safety committees

1885 According to Eastman (1910, page 245), some companies had in 1901 organized safety departments that existed for at least 15 years. The United States Steel Corporation had a central committee of safety and local committees called "foremen's committees." A duty of the foremen's committee is to make an inspection of the plant either semimonthly or monthly and turn in a written report. The recommendations are shown on a monthly statement until they have been carried out.

1892 A safety department was organized at the Joliet Works of the Illinois Steel Company. Because of this early definite start and the fact that its program spread to many other mills, this mill has often been referred to as "the birthplace of the American industrial accident – prevention movement" (Blake, 1943, page 14).

1910 Eastman (1910) describes the organization of the safety committees of the United States Steel Corporation (that included 143 manufacturing plants, employing in all approximately 200,000 people):

- *A central committee* of safety setup in April 1908 was composed of five members representing subsidiary companies. This committee was empowered to appoint inspectors to examine various plants and equipment and submit reports of safety conditions, with suggestions for improvement. Meetings of the committee were held once a month.
- *A local foremen's committee* usually included the assistant superintendent of the plant, the master mechanic, chief electrician and a department foreman or two. It is the duty of the foremen's committee to make an inspection of the plant either semimonthly or monthly and turn in a written report. They also go over the workmen's committee, which reports weekly.
- *The workmen's committee* is entirely distinct and is taken from the rank and file of mill employees. For example, there may be a machinist, an electrician and a wire drawer, a roller and a carpenter. At the end of the month, an entirely new committee was appointed

1929 R031 – Prevention of Industrial Accidents Recommendation, 1929 (No. 31), adopted by the Governing Body of the International Labour Office, and having met in its Twelfth Session on 30 May 1929 mentions under point 8:

8. It is further recommended that the Members should actively and continuously encourage the adoption of measures for the promotion of safety, in particular (a) the establishment in the works of a safety organization which should include arrangements for a works investigation of every accident occurring in the works, and the consideration of the methods to be adopted for preventing a recurrence; the systematic supervision of the works, machinery and plant for the purpose of ensuring safety, and in particular of seeing that all safeguards and

other safety appliances are maintained in proper order and position; the explanation to new, and especially young, workers of the possible dangers of the work of the machinery or plant connected with their work; the organization of first aid and transport for injured workers; and the encouragement of suggestions from the persons employed for rendering work safer; (b) co-operation in the promotion of safety between the management and the workers in individual works, and of employers' and workers' organizations in the industry with each other and with the State and with other appropriate bodies by such methods and arrangements as may appear best adapted to the national conditions and aptitudes. The following methods are suggested as examples for consideration by those concerned: appointment of a safety supervisor for the works, establishment of works safety committees.

1936 Vernon (1936) describes the effect of safety organizations on accident rates. In the period 1927 till 1931, there was a fall of 28% in the frequency rate of the Welsh Plate and Sheet Manufacturer's Association, which employed about 27,000 people. These improvements appear to have been due to the work of the safety committees. At a large biscuit factory, the Safety Committee succeeded in lowering their first-aid cases by 34% between 1927 and 1930. Vernon gave several other examples that clearly demonstrate the positive impact of safety organizations on the accident frequency rate.

1943 Blake (1943) dedicates chapter XXIV to "Safety Organization." The chapter is written by Armstrong, the manager for industrial relations of Westinghouse Electric & Manufacturing Company. Armstrong lists a number of musts before considering the safety organization:
1. Safety "must" have top management approval, sanction and support.
2. Responsibility for safety "must" rest with the supervisory personnel.
3. Safety "must" be given an equally important consideration with other factors of production.
4. Provision "must" be made for prompt action in the elimination of mechanical and personal hazards.

Armstrong distinguishes between three types of safety organizations (depending on the size and type of company):
A. *Line organization*. Those in which safety work is carried on wholly through the line organization
B. *Safety director*. This in which the safety work is directed by a safety director reporting to a major executive
C. *Safety committees*. Those in which safety work is carried on primarily by committees set up for the purpose. In this type of organization, there are different committees: main or governing committee, workmen's committee, technical committee and special purpose committees.

1948 Chapter XVI of the *Model Code of Safety Regulations for Industrial Establishments for the Guidance of Governments and Industry* of the International Labour Office describes the minimum aspects of the safety organization that should be in place in every industrial establishment. It relates to safety rules, discipline, suggestions, workers' safety delegates and their obligations, employer obligations, safety committees, safety officials, accident reports, accident statistics and medical service.

The importance of mixed safety committee (mixed = multidisciplinary) has been recognized for more than 100 years. The positive impact of safety committees on the safety performance is demonstrated and has been quantified (see, e.g., Vernon, 1936). Since 1929, the Governing Body of the International Labour Office has promoted safety organizations and committees between management and workers. In 1948, ILO published a Model Code.

3.7.11 Management, leadership and importance of supervision

The first Factory Act in England was passed in 1802. This Factory Act was revised several times during the nineteenth century but even in the revision of 1895 it is obvious that the importance of leadership of the management is clearly understood. The Factory Acts lists the obligations of the management ("the occupier") but more in a way that the occupier is responsible for the implementation of the legislation. Calder (1899) dedicated a full chapter to the causes of factory accidents but he does not clearly mention the importance of the exemplarity and leadership of the management. This changes very fast in the first decades of the twentieth century.

1916 The role of the superintendent and the foreman in Chemical Works is discussed in Beyer (1916):

> The regulation of the details of the entire question of the protection of workmen is placed under the jurisdiction of each Superintendent, and the responsibility for occurrence of accidents where avoidable by the use of preventive methods enumerated, rests upon him . . ..
>
> You are notified that not only is it your duty as a foreman to see that work under your direction is properly performed, but also that you share with the Superintendent and other members of his staff the responsibility for the safety of the workmen under your immediate charge.
>
> You are directed to acquaint each workman, when placed in your charge, with all of the following rules which apply to the duties of the workman, and to make sure the he understands such of these rules as you read to him, and it is one of your special duties to see that these rules are understood and obeyed.(Beyer, 1916, page 278)

1917　Up to the 1920s, many companies made the safety director responsible for accident reduction (Aldrich, 1997). By the early 1920s, a consensus had emerged that while the safety organization should establish the program, results were the responsibility of operating personnel. The focus shifted from the safety departments to the role of the foreman and supervisor. Safety men, however, realized that line officers would be responsive to safety only if top management was firmly behind the program. Price who had years of experience in implementing safety programs explained in 1917 to the National Safety Council: "The first indispensable feature of a safety program is to convince the general manager – the man who holds the pocketbook. In every case where I have succeeded in getting the general manager with me things have happened" (Price, 1917).

1919　Louis DeBlois, director of safety at Dupont, articulated the key role of top management and of all line managers in front of the National Safety Council. Lammot DuPont and Irénéé DuPont were on the board of directors but they were also directly involved in day-to-day activities of safety committees. DeBlois explained that it was a company axiom that all injuries reflected a failure of operating managers and that top management should state to all line managers that the line management is responsible for all the results and safety must be part of the business of production.

1920–　The idea that safety was integral to good management had become an axiom
1940　to many large corporations. Safety experts encouraged by view published articles in which they tried to demonstrate that safety was synonymous with efficiency and that injuries were more costly than had been thought. The idea of managerial responsibility became widely accepted but its practical implementation was gradual.

1943　The theoretical background of the term "management," which was about resource allocation, production (economics) and pricing issues, was espoused by classical economists such as Adam Smith (1723–1790) and John Stuart Mill (1806–1873). However, it was only around 1920 that the first comprehensive theories of management appeared. Harvard Business School offered the first Master of Business Administration degree in 1921. From then the term "management" became very common. In his book *Industrial Safety*, Blake (1943) devotes under Chapter VII Fundamentals of Accident Prevention a paragraph on "Management must provide leadership":

Under the circumstances, therefore it is absolutely essential that top management as a whole, and the chief operating executive in particular, provide the same kind of leadership in accident prevention as they provide in the field of production. If the manager becomes convinced that accidents can and must be prevented, he will issue the necessary orders that are to be carried down through the organization; he will follow to see that his orders are carried out; he will set a good example for safety; he will do everything which is in power to provide a safe environment in which to work and he will see to it that his assistants do their best to control the behavior of the workers. Without his leadership, the accident-prevention effort in his establishment is bound to become a hit-and-miss affair, and thus will fall far short of maximum results.

1956 Simonds and Grimaldi (1956) published the first edition of their book *Safety Management*. The importance of managing safety in a scientific and similar way to managing the business was well understood at that time. John Grimaldi was a consultant for health, safety and plant protection for General Electric Co. in New York. The book was very successful and had five editions. In its fifth edition, Simonds spent a full chapter on "Organization and Administration Effects":

Safety has been described as everyone's responsibility. As a generalization this is true. It is grossly misleading, however. Because most functions in modern society are fulfilled through an organizational hierarchy, the responsibility for the safety of others increases in significance as the echelons are climbed.

. . .

Where the chief executive is personally interested in safety, the whole organization is aware of it. Progress then is expedited.

1959 In the fourth edition of the book *Industrial Accident Prevention – A Scientific Approach* by Heinrich and Granniss (1959), the basic philosophy of accident prevention explained in chapter 2 of the book is fully in line with the current thinking about accident prevention. The responsibility for occurrence and prevention of accidents (Chapter 2, Section 2.7 of the book) explains the role of different protagonists in the prevention of accidents: the management, the supervisor, the safety engineer and the employee.

1967 Factory Mutual published its second edition of its *Handbook of Industrial Loss Prevention*. The book contains 78 chapters with very detailed technical specifications to avoid property loss due to fire or explosion. In chapter 1, Factory Mutual writes:

Executives of well-managed companies place great importance on the methods used to conserve property, production and above all human lives. They recognize that most disastrous fires today are not the result of weaknesses in the physical safeguards but of failures in the human element. And, almost always, these human errors can be tracked back to a lack of interest by management and to its failure to set the framework for organizing, training and instructing its employees to a comprehensive plan." . . . "Loss prevention and safety are prime responsibilities of operating management. These responsibilities must be accepted by all supervisory employees.

In the nineteenth century, the importance of "managing safety" in a similar way as "managing business or production" was not commonly understood. Scientific management sometimes known as Taylorism was developed by Frederick Taylor during the 1880s and 1890s. It influenced industry as from the beginning of the twentieth century. Many ideas of this theory found their way into the field of occupational safety and became a paradigm since, say, the 1930s.

It is common knowledge today that OHS is to be managed in the same way as production or business in general. The importance of managing OHS and the importance of leadership and supervision are widely recognized since the 1950s.

In many textbooks, the role of the supervisor will be considered of utmost importance. My personal experience is that this is too shortsighted. It puts the responsibility on the supervisor, while the attitude of the worker and the supervisor is influenced by many more people:

1 **The employee**   The person who will have the accident or who will avoid it for the person who is executing the task (could be more than one person). In the nineteenth century, he was considered as the person responsible while responsibility of all other people was considered as marginal. In the twenty-first century, the executor is mainly considered as a potential victim while the underlying cause of the accident is almost always a failure of the management (whoever that might be).

While the responsibility of the victim was exaggerated in the nineteenth century, his/her direct contribution to the accident is underexposed in the twenty-first century.

2 **The CEO of the**   During my career of 20 years in a major energy company, I
company   experienced the influence of several chief executive officers on the organization. All of them without any exception understood the importance of avoiding accidents. All of them considered safety of utmost importance and as a license to operate. The policy of all CEOs I have known was clearly "if we are not capable to do the activity in a safe way we should better not do the business." Although they all had the same principles and policies, the way they conveyed the message in the organization was different. The CEO has to ensure that there is *no room for interpretation about the value of safety*. The organization then moves faster toward zero accident rates than when people think they are allowed to interpret the message about safety.

In many major accidents (Bhopal, Macondo, AZF (AZote Fertilisants), etc.), the highest level of the company is considered as having a responsibility in the accident:

- The Bhopal disaster took place in 1984 in a plant belonging to *Union Carbide* India Ltd in the city of Bhopal, Madhya Pradesh, India. Thousands of people died and thousands more were injured in the disaster. The CEO of the Union Carbide concern, Warren Anderson, was charged with manslaughter by Indian authorities. He flew to India and was promptly placed in custody by Indian authorities, but was allowed to return to the USA. He was declared a fugitive from justice by the Chief Judicial Magistrate, Gulab Sharma, of Bhopal on 1 February 1992, for failing to appear at the court hearings in a culpable homicide case. The chief judicial magistrate of Bhopal issued an arrest warrant for Anderson on 31 July 2009. The USA declined to extradite him. He died at a nursing home in Florida on 29 September 2014 at the age of 92 years.
- On 21 September 2001, there was a major explosion in a fertilizer site AZF that killed 31 people and injured another 2,500. The site started production in 1927 and belonged since 1 November 1987 to the company "Grande Paroisse." In 1990, the French government decided to enter Grande Paroisse under control of Elf Aquitaine. In April 2000, Total Fina acquired Elf Aquitaine to become Total Fina Elf, a major oil company with about 111,000 employees. On 11 December 2011, people invaded the family castle of Thierry Desmaret, CEO of Total Fina Elf, in Montigny. They dismantled doors and windows as a protest against the living conditions of the victims of AZF. On the white stone walls, the word "AZF" was painted with tar.
- On 12 January 2007, *BP* announced that Hayward would replace Lord Browne as chief executive. His replacement was accelerated because of the safety and production issues in Alaska and the explosion at the Texas City refinery that killed 15 people and injured more than 170 others.
- Three years later, on 20 April 2010, an explosion occurred on the Deepwater Horizon oil rig that killed 11 people and led to a major environmental disaster. On 27 July 2010, *BP* announced that Dudley would succeed Tony Hayward as their group chief executive on 1 October 2010. Dudley was also appointed to the board of directors.

3 Business management

Business management will declare that safety is utmost priority (or that safety is a value) but the practical implementation is in the hands of the site management and the people in the field. In the meantime, the business managers ensure that the company performs sound business. For that reason, business managers will tend to put pressure on production, availability, loss control and profit. There is nothing wrong with that because it is indeed one of their important responsibilities. That is why the pressure from the CEO to put safety above all else is vital. There can be no compromise between safety and production.

4 The site manager

The site manager will feel empowered by the top of the company and he/she will convey a similar message to his/her organization. On Sunday 25 March 2012, there was a leak on the Total's Elgin platform 150 miles east of Aberdeen. The manager of the site decided to evacuate immediately 219 people out of 238 workers. The 19 people who were left put the installation in a safe stop and then also left the platform. This decision resulted in a loss of production for many months in addition to costs required to stop the leak and the fines. For the top management of the company, there was no doubt at all: the manager who decided to evacuate took the right decision, irrespective of the (potential) costs. He/she was honored in front of all senior executives of the company because of his/her brave correct decision. These types of clear messages set the trend.

5 The safety engineer

Ensuring a safe environment requires a lot of detailed technical safety-related knowledge. The domain of the safety engineer is very broad. Let us take the simple example of a safety helmet or hard hat. Everybody can easily understand what a safety helmet is and that people should wear the helmet. The line management will demonstrate leadership and ensure that people wear their hard hat. This demonstration of leadership is not new: construction workers of the Golden Gate Bridge in 1933 were obliged to wear hard hats. The safety engineer is the knowledgeable person who needs to study this subject in more detail. Helmets have been used as early as 1898 when the Bullard Company sold protective hats made of leather. But the safety helmet of 1898 is obviously not the safety helmet that complies with the legal obligations of today. Since 1962, almost all safety helmets are made from high-density polyethylene.

Today, in the USA, the hard hats have to comply with the American National Standard for Industrial Head Protection ANSI/ISEA 789.1-2014 (OSHA regulation 1910.135). In Europe, industrial safety helmets have to comply with DIN EN 397. The requirements for the design and manufacture of personal protective equipment to be made available on the market are laid down in the European Regulation 2016/425. The role of the safety engineer is to follow up these evolution of the standards and regulations in the field of occupational safety and to ensure that they are applied in the company.comply with the American National Standard for Industrial Head Protection ANSI/ISEA 789.1-2014 (OSHA regulation 1910.135). In Europe, industrial safety helmets have to comply with DIN EN 397. The requirements for the design and manufacture of personal protective equipment to be made available on the market are laid down in the European Regulation 2016/425. The role of the safety engineer is to follow up these evolution of the standards and regulations in the field of occupational safety and to ensure that they are applied in the company.

6 The line management   The line management (i.e., the managers between the site manager and the supervisor of the employee executor) will feel responsible for the operations. They will feel pressure from all sites (business, cost reduction, availability, environmental issues, safety, etc.). Whether they like it or not, they will have to make daily choices and they will have to deal with potential conflicting messages. When, for instance, "a job has to be done within a certain time frame against a certain budget with a defined number of people," then conditions become the specifications for the work to be done. It is very easy to add on paper that "the work has to be done in a safe manner with respect for the environment, the regulations and standards, . . ." but the specifications will be influential for potential achievements. It is at this point that the positioning of the site manager becomes crucial. The line management will make their choices in function of what they belief is of utmost importance for the site manager.

7 The supervisor

The notion of "supervisor" is often confused with the foreman of the employee or the team that is executing the work. In many cases, there is no supervision at all (e.g., when a worker is working alone) and, in many other cases, the team leader is performing the work together with his/her team. In other words, that person is also not performing "supervision."

There is no doubt at all that the safety of the foreman and of his team is most important for the foreman. Nevertheless, during the execution most of his attention will go to performing the job in accordance with the specifications. In cases where he meets unexpected problems, then his reaction will largely depend on the attitude of the line management which in turn depends on the attitude of the site manager which in turn is largely influenced by the business managers and the CEOs.

When the supervisor is convinced that his/her safety and that of his/her people are of utmost importance, then he/she will stop the work and start talking with the line management. When, on the other hand, the most important perception of the supervisor is that the job has to be done according to the specifications, then he/she and his/her team will try to find a way to work around the hurdles.

3.7.12 Inspection and checklists

A checklist is used to focus human's attention and it helps to ensure consistency and completeness in carrying out a task. Checklists are found in all formats. A typical type of checklist is the so-called punch list, which is used near the end of a large project to verify that the work has been finalized and in compliance with the contract specifications.

Nowadays, checklists are considered as an essential tool for safety inspections to ensure that nothing will be forgotten during the execution of a job that is composed of multiple steps. Checklists are found in different formats at different levels in the company (for maintenance, startup, housekeeping, etc.). Section 6 of the Inherently Safer Process Design of IChemE training (1995) is about checklists.

The use of (What-If) checklists is a specific topic in the Master of Safety Engineering training at the KUL (Hoorelbeke et al., 2019). Many consider 1935 as the year in which checklists were introduced in a systematic way to avoid losses. Although checklists (or inspection lists) are mentioned much earlier, we find in literature as from the 1940s a more systematic promotion for the use of checklists.

1909 Law and William (1909): "Regular and frequent inspections by competent men should be made of all the ways, works, machinery, and appliances, so that defects and unsafe conditions may be discovered promptly and remedied." Law does not make any reference to the systematic use of checklists.

1916 Safety inspections are common practice in industry. Beyer (1916) dedicates chapter 47 of his book to inspection but he gives in many other chapters in his book recommendations on the points to be inspected for certain equipment. For elevators, for instance, he lists cables and connections, emergency clutch and ratchet, guides and cages, automatic stops, gates, signaling apparatus and warning signs, counterweights and attachments, and operating levers.

1935 Many consider 1935 as the year in which checklists started to be used in a systematic way. On 8 August 1934, the United States Army Air Corps (USAAC) tendered a proposal for a multiengine bomber to replace the Martin B-10. Boeing was in competition with their Model 299 B-17. On 20 August 1935, the prototype flew from Seattle to Wright Field in 9 h and 3 min with an average cruising speed of 252 miles per hour (406 km/h), much faster than the competition. Development continued on the Boeing Model 299, and on 30 October 1935, Army Air Corps test pilot Major Ployer Peter Hill and Boeing employee Les Tower took the Model 299 on a second evaluation flight. The crew forgot to disengage the "gust locks," which locked control surfaces in place while the aircraft was parked on the ground, and after takeoff, the aircraft entered a steep climb, stalled, nosed over and crashed, killing Hill and Tower (other observers survived with injuries). The investigation revealed that the captain had left the elevator lock on. During a think-tank session, it was determined that pilots needed a checklist. The B-17 was the first aircraft to get a checklist (Figure 3.9).

1943 Inspection of equipment, installations, plants and others is an important topic throughout the book of Blake (1943). The use of checklists is mentioned, and an example of a checklist is given (for plant housekeeping) but it is not recommended as common practice (Figure 3.10).

Safety inspections are common practice in industry since the beginning of the twentieth century. The systematic use of checklists to perform safety inspection, however, seems to be in industry as from the 1940s.

3.7.13 Human behavior

3.7.13.1 The syndrome of accident proneness
Human behavior has been considered as a main cause of accidents since the beginning of industrialization. "Human behavior" as a cause of accidents has been a subject of debate and controversy for about 200 years and it is probably the most studied

APPROVED B-17F and G CHECKLIST

REVISED 3-1-44

PILOT'S DUTIES IN RED
COPILOT'S DUTIES IN BLACK

BEFORE STARTING
1. Pilot's Preflight—COMPLETE
2. Form 1A—CHECKED
3. Controls and Seats—CHECKED
4. Fuel Transfer Valves & Switch—OFF
5. Intercoolers—Cold
6. Gyros—UNCAGED
7. Fuel Shut-off Switches—OPEN
8. Gear Switch—NEUTRAL
9. Cowl Flaps—Open Right—
   OPEN LEFT—Locked
10. Turbos—OFF
11. Idle cut-off—CHECKED
12. Throttles—CLOSED
13. High RPM—CHECKED
14. Autopilot—OFF
15. De-icers and Anti-icers, Wing and
    Prop—OFF
16. Cabin Heat—OFF
17. Generators—OFF

STARTING ENGINES
1. Fire Guard and Call Clear—LEFT Right
2. Master Switch—ON
3. Battery switches and inverters—ON &
   CHECKED
4. Parking Brakes—Hydraulic Check—On-
   CHECKED
5. Booster Pumps—Pressure—ON &
   CHECKED
6. Carburetor Filters—Open
7. Fuel Quantity—Gallons per tank
8. Start Engines: both magnetos on
   after one revolution
9. Flight Indicator & Vacuum Pressures
   CHECKED
10. Radio—On
11. Check Instruments—CHECKED
12. Crew Report
13. Radio Call & Altimeter—SET

ENGINE RUN-UP
1. Brakes—Locked
2. Trim Tabs—SET
3. Exercise Turbos and Props
4. Check Generators—CHECKED & OFF
5. Run up Engines

BEFORE TAKEOFF
1. Tailwheel—Locked
2. Gyro—Set
3. Generators—ON

AFTER TAKEOFF
1. Wheel—PILOT'S SIGNAL
2. Power Reduction
3. Cowl Flaps
4. Wheel Check—OK right—OK LEFT

BEFORE LANDING
1. Radio Call, Altimeter—SET
2. Crew Positions—OK
3. Autopilot—OFF
4. Booster Pumps—On
5. Mixture Controls—AUTO-RICH
6. Intercooler—Set
7. Carburetor Filters—Open
8. Wing De-icers—Off
9. Landing Gear
   a. Visual—Down Right—DOWN LEFT
      Tailwheel Down, Antenna in, Ball
      Turret Checked
   b. Light—OK
   c. Switch Off—Neutral
10. Hydraulic Pressure—OK Valve closed
11. RPM 2100—Set
12. Turbos—Set
13. Flaps ⅓—½ Down

FINAL APPROACH
14. Flaps—PILOT'S SIGNAL
15. RPM 2200—PILOT'S SIGNAL

Figure 3.9: Checklist for Boeing B-17 airplane.

"cause of accidents." But human behavior is at the same time a consequence of something else (environmental factors, technical or organizational deficiencies, etc.) and
when human behavior is a contributor to an accident it should be investigated more
in depth instead of classifying it as a "cause."

PLANT HOUSEKEEPING INSPECTION CHECK LIST

COMMITTEE................................ DATE ....
GROUP .... DEPARTMENT ....

1. *Buildings:*
   a. Are walls clean for this department? ....
   b. Are windows clean for this department? ....
   c. Are walls free of unnecessary hangings? ....
   d. Is proper light provided? ....
   e. Are platforms in good condition? ....
   f. Are stairs clean and well lighted; have they standard rails and standard treads? ....

2. *Floors:*
   a. Is floor surface good for this department? ....
   b. Is it swept clean, free of loose materials, and is it clean in the corners, back of radiators, along the walls, and around the columns? ....
   c. Is it free of oil, grease, etc.? ....
   d. Are operating floors, or work positions, free of loose stone, scrap, metal or other materials? ....
   e. Is the building free of unnecessary articles? ....
   f. Are receptacles provided for refuse? ....

3. *Aisles:*
   a. Are aisles free of obstructions? ....
   b. Is there safe and free passage to fire extinguishers, fire blankets, and stretcher cases? ....
   c. Is there safe and free passage to work positions? ....

4. *Machinery and Equipment:*
   a. Is it clean and free of unnecessary material or hangings? ...
   b. Is it free of unnecessary dripping of oil or grease? ....
   c. Is position around it clean and free of rags, paper, etc.? ..
   d. Are lockers and cupboards clean and free of unnecessary material, both on top of them and inside of them? ....
   e. Are benches and seats clean and in good condition? ....
   f. Are drinking fountains clean? ....
   g. Are toilet rooms clean and well ventilated? ....
   h. Are proper guards provided and in good condition? ....

5. *Stock and Material:*
   a. Is it properly piled and arranged? ....
   b. Is it loaded safely and orderly in ships, cars, trucks, etc.? ....

6. *Tools:*
   a. Are they properly arranged in place? ....
   b. Are they free of oil and grease? ....
   c. Are they in good working condition? ....
   d. Are tool rooms orderly and clean? ....

7. *Grounds:* (Fifteen feet from outside wall or to first railroad track)
   a. Is yard outside building free of refuse such as fruit peelings, scrap, wood, iron, etc.? ....
   b. Were winter hazards checked? ....

Figure 3.10: Checklist for plant housekeeping (Blake, 1943).

1910   Eastman (1910) deals with what she calls "the human element":

> From statistics which have been prepared in this country (= USA) and in Germany, it
> would appear that about one-third of the total number of industrial accidents are attrib-
> utable in whole or in part directly to carelessness of negligence on the part of the work-
> ers themselves. In other words, a considerable percentage of the accidents which occur can
> be charged to the human element and cannot be prevented by mechanical safeguards. If
> they are to be materially reduced they require other treatment. The problem here is largely
> a psychological one, and we are working on it in a number of different ways."

1914   In the WWI period, psychologists in Britain and Germany independently and
simultaneously originated the idea of accident proneness (Unfallneigung).
The idea is that some people are more prone to accidents than others.

1919   Greenwood and Woods (1919) performed a statistical analysis of accidents
in a munitions factory. They demonstrated that the occurrence of accidents
could not be explained by a normal probability distribution and hence they
concluded that some people have a greater predisposition than others to
suffer accidents. This theory was called accident proneness.

1926   Accident proneness in the railway service was investigated by Schmitt who
kept 486 persons under observation for a year (Schmitt, 1926).

1936   Chapter II of the book of Vernon is about "Individual Differences in
Susceptibility" to accidents. He describes in about 20 pages the studies
that have been conducted since 1919 on accident proneness. His main
conclusion is that there is a need for further investigation into the changes
in accident frequency induced by transferring the accident-prone persons to
other occupations.

2008   The discussion about accident proneness is ongoing. Burham (2008) gives
an overview of the different studies since the original article of Greenwood
and Woods and explains why psychiatrists did not adopt the idea.

Since about 100 years, studies have been conducted to demonstrate the existence of
accident proneness. This means that some persons are accident prone and that even
a full realization of the importance of the avoidance of accidents does not prevent
such persons from having accidents. That is why people should go through a strict
selection process and when accident-prone workers are identified they should be
transferred to other occupations. Vernon describes in 1936 (Vernon, 1936) how psychological tests can be used to detect accident proneness.

3.7.13.2 Scientific management approach to human performance improvement

The early discussions among psychologists on accident proneness in combination
with a scientific approach to improve productivity (Taylorism) formed the basis of
the Human Factors Society of America Society which was founded in 1956.

1913   John B Watson published an article "Psychology as the Behaviorist Views It." Watson argued that mental activity could not be observed and hence he rejected the studying of consciousness. Watson was largely influenced by the work of Ivan Pavlov (1849–1936) on classical conditioning, which focuses on using preceding conditions to alter behavioral conditions. Pavlov who won the Nobel Prize in Physiology or Medicine in 1904 is considered as the founder of modern behavior therapy. Some very simple concepts such as the ABC model (antecedents, behaviors and consequences) find their roots in the behaviorist theories of Watson.

1910+   According to Drury (1915) during a meeting in 1910 in the Interstate Commerce Commission, the word "scientific management" was introduced by Louis D. Brandeis to explain a system, in which high wages could be paid while having low labor cost. He alleged that scientific management would overcome railroad inefficiencies. Scientific management, sometimes known as Taylorism, was one of the earliest attempts to apply science to the engineering of processes of management. Scientific management aims at improving human performance.

1940 till Human performance was extensively studied during World War II by the
1945   US army and the British Forces in particular with regard to the effects of cockpit design features on the errors made by pilots.

        Maslow's hierarchy of needs (published in 1943) orders our needs (from most basic to least basic) as biological, *safety*, belonging, esteem and self-actualization. In other words, biological and safety needs are critical to survival, and the rest are more or less gravy. In other words, biological and safety needs are critical to survival, and the rest are more or less gravy.

1948   The Ergonomics Research Society, which was formed in 1948 as a successor to the operations research groups, was used to great effect by the British Forces during World War II.

1955   Planning activity for the Human Factors Society of America Society began in 1955 in Southern California by a joint committee representing the Aeromedical Engineering Association of Los Angeles and the Human Engineering Society of San Diego. The members of the committee were involved in human factors work in industry, universities and government laboratories. The formal concept for and features of a human factors organization were developed by a founding committee in 1956 and approved at a joint meeting of the two local organizations. The founding of the Society with its Constitutional Convention and First National Meeting in Tulsa, Oklahoma, was on 25 September 1957. That meeting was held in conjunction with the Fifth Annual Human Engineering Conference of the Office of Naval Research, which had served as an additional influence leading to a national society (Stuster, 2006). Human factors psychologists and engineers are concerned with anything that affects the performance of system operators – whether hardware, software or live ware. Every year there is a national congress.

Human Factors and Ergonomics Society now has 67 active chapters throughout the USA, Canada and Europe, 46 of them were student chapters. The society's technical groups now number 24.

1960s onward   Since the 1960s the literature on human performance is overwhelming and the domain becomes very broad. The most often cited domains of specialization are physical, cognitive and organizational ergonomics. Cognitive ergonomics focuses on mental processes such as perception, memory, information processing, reasoning and motor response as they affect interactions among humans and other elements of a system. Organizational ergonomics is concerned with the optimization of sociotechnical systems, including their organizational structures, policies and processes (Salvendy, 2012).

1972   At the request of the United States Atomic Energy Commission, the WASH 1400 study is started. The study will take 3 years and a total of 60 people during 80 man-years of effort will make an estimate of the risks of the nuclear industry. Sandia National Laboratories (SNL) personnel were involved in the reliability analysis. Swain and Guttman from SNL are the human reliability analysts. They develop human error probabilities.

1976   The WASH 1400 study was published (in 1975), but during this study the need arose for a more systematic approach for the analysis of human reliability. Research is conducted as from September 1976 and a first draft *Handbook on Human Reliability Analysis* is available. The handbook is extensively used by practitioners in human reliability analysis.

1981   First publication of the textbook *Principles of Neural Science* was edited by Eric R. Kandel and others. Kandel won the 2000 Nobel Prize in Physiology or Medicine for his research on the physiological basis of memory storage in neurons. The understanding of the biological basis of learning, behavior and perception is considered as the "ultimate challenge" of the biological sciences.

1983   Comments and suggestions for improvement are incorporated in the revised *Handbook of Human Reliability Analysis with Emphasis on Nuclear Power Plant Applications* which becomes NUREG/CR-1278. The handbook presents "methods, models and estimated human error probabilities to enable qualified analysts to make quantitative or qualitative assessments of occurrences of human errors that may affect the availability or operational reliability of engineered safety features and components in nuclear plants" (Swain and Guttmann, 1983).

1983   Jens Rasmussen from Riso National Laboratory in Denmark publishes an article in which he distinguishes three types of behavior (skill based, rule based and knowledge based) that interact with each other. The SRK model finds its way to industry and it is still used today as a model to explain and understand some behaviors (Rasmussen, 1983).

1986   Consultancy companies in the field of quantitative risk assessments establish a "Human Factors Unit" to include human factors in their major risk studies.

1989   Licht and Polzella prepared a report in response to a request by the Human Factors Committee of the National Research Council to aid its deliberations regarding a standardized definition of human factors (Licht and Polzella 1989). They compiled 90 definitions. They found that there is a history of debate over the comparability of the terms human factors, human factors engineering and ergonomics. Especially in technical documentation, new words (or combinations of words) are constantly being introduced and it is not clear if the "new" expression means the same as the term they have seen before, almost the same, or something completely different. In reviewing the different terminology used, there appeared to be three broad categories of definitions: human factors, human factors engineering and ergonomics.

1990   The Human Factors Unit in the Stockport office of Technica "is led by Jeremy Williams who has over 20 years of experience in the application of human factors technology in the industrial field" (from an article in *Technica Times*, Winter 1990). This demonstrates that the human factors approach is well established. Consultants deliver services to onshore and offshore companies on the design and assessment of workplace for safe and efficient performance of tasks.

2000   Since 2000, there is an increase in articles about the use of neuroscience in the field of safety and more in particular for understanding and managing the neural processes in the brain that precedes and leads to that behavior.

2004   Whittingham publishes a comprehensive book that gives a good overview of the status of the different approaches of "human errors." It includes a comprehensive discussion of errors in practice, latent errors and violations, human reliability analysis, human error modeling and human error in event sequences (Whittingham, 2004).

2010   Since 2010, we see an increase in the use of the term "neurosafety," which is the use of neuroscience in the field of safety. The assumption is that creating sustainable safety cultures needs better understanding and managing the neural processes in the brain rather than to focus on behavior.

2012   The domain of human factors ergonomics is becoming so broad that the fourth edition of the *Handbook of Human Factors and Ergonomics of Salvendy* contains 61 chapters (1,700 pages) and that it was written by 131 experts.

The "human factor" is by far the factor most studied in the field of OHS since at least 150 years. This is logic because it all started with humans, and humans are at the heart of the industry.

A lot of progress has been made in the field of ergonomics and in the field of the relation man–machine.

A lot of research has been done to understand human behavior, to improve human performance and to reduce human errors. These subjects are, however, much more complicated. Many ideas and theories have been proposed and tested. Some of these theories pop-up as a general accepted paradigm for a certain period of time after which the theory is replaced by a new paradigm. Some concepts (e.g., skill–rule–knowledge-based behavior and ABC theory) found their way to the industry because they are simple to understand and to some extent the concepts help in explaining the workforce how behaviors can be influenced. The field of "human behavior" is without any doubt the domain where new ideas are proposed (e.g., neurosafety). These new ideas are interesting but their translation in pragmatic useful methods and tools is still a challenge.

Methods for quantification of human error have been developed but these methods have a limited use in the field of OHS. They are more intended for the use in quantitative major risk assessments.

3.7.13.3 Industrial, work and organizational psychology

The industrial, work and organizational (IWO) psychology is a discipline in psychology based on the science of human behavior relating to work and applies psychological theories and principles to organizations and individuals in their places of work as well as the individual's work life more generally.

The roots of IWO psychology trace back nearly to the beginning of psychology as a science, when Wilhelm Wundt founded one of the first psychological laboratories in 1879 in Leipzig, Germany. Also the publications of Cattell are referred to as the starting point for what is called today IWO psychology. Between 1883 and 1886, James Cattell (1860–1944), an American professor in psychology, published nine articles discussing human reaction time rates and individual differences. The historical development of IWO psychology was paralleled in different regions of the world (the USA, the UK, Germany, etc.).

In July 1892, the American Psychological Association (APA) was founded at Clark University.

The APA is today the largest scientific and professional organization of psychologists in the USA, with over 118,000 members, including scientists, educators, clinicians, consultants and students.

The "industrial" side of IWO psychology originated in research on individual differences, assessment and the prediction of work performance. Industrial psychology crystallized during WWI. After the war, the growing industrial base in the USA was a source of momentum for what was then called industrial psychology. Research in IWO psychology covers a very broad group of subcategories. Some of them are explained below.

Job analysis   Job analysis is a family of procedures to identify the content of a job in terms of activities involved and attributes or job requirements needed to perform the activities. Theories about the psychology of task analysis and task observation found their way to the domain of OHS. Job analysis as a management technique was developed around 1900 (Zerga, 1943). It became one of the tools with which managers understood and directed organizations. Taylor made studying the job one of his principles of scientific management (Taylor, 1911). But this early interest in job analysis disappeared as the human relations movement focused on other issues. It was not until the 1960s that psychologists and other behavioral scientists rediscovered jobs as a focus of study in organizations.

Practical applications of these theories in industry in the domain of safety are JSA, task risk analysis, job observations and behavior observation programs. In many cases, the theory has been translated into very practical methods. A common cause in accidents is the lack of a correct JSA. The underlying cause is usually that the implementers have insufficient knowledge of the reason for these techniques and therefore consider them more as a burden than a necessity. The appointment of a coordinator specialist who monitors the correct use of the various methods is not an unnecessary luxury. This person must ensure that the use of the various techniques (JSA, task observation, etc.) is done correctly and that all appropriate levels in the organization are aware of the reason why the technique is used.

Personnel recruitment and selection

Personnel selection is the methodical process used to hire (or, less commonly, promote) individuals. Although the term can apply to all aspects of the process (recruitment, selection, hiring, acculturation, etc.) the most common meaning focuses on the selection of workers. All safety management systems include personnel recruitment and selection as vital in the management of safety. Safety professionals, therefore, started to include personnel selection and hiring in their safety management systems. Following the first studies on accident proneness in the beginning of the twentieth century (e.g., Greenwoods and Woods, 1919; Farmer, 1927), the idea that some people were by nature more susceptible to accidents spread throughout the industrialized world. One very important strategy of defense was to test people before hiring them. Some psychologists by the 1920s wrote about the potential for management in separating out from some factory tasks those workers inclined to suffer accidents. "Hiring and Placement" has been one of the elements of ISRS since its earliest versions in the early 1980s (Bigelow and Robso, 2005). The purpose is to ensure that the person has the medical and physical capabilities to perform the job safely.

According to the Society for Industrial and Organizational Psychology, all types of psychological tests often seek information on a candidate's leadership and teamwork skills, interpersonal skills, extraversion and creativity. Questions about education, training, work experience and interests are used to predict success on the job.

Cognitive ability tests, also called aptitude tests, typically use questions or problems to measure a candidate's ability to learn quickly, and use logic, reasoning, reading comprehension and other mental abilities that are important for success in many different jobs.

Personality tests typically measure traits related to behavior at work, interpersonal interactions and satisfaction with different aspects of work.

Performance management and performance appraisals

Performance appraisals are a part of career development and consist of regular reviews of employee performance within organizations.

Performance management is not a new concept. It has been there since more than 100 years but the way psychologists think about it has changed massively:

A distinct and formal management procedure used in the evaluation of work performance, appraisal really dates from the time of the Second World War. http://www.whatishumanresource.com/history-origin-ofperformance-appraisal

By the mid-1950s, companies were using personality-based systems for measuring "Human Performance." The traditional emphasis on (financial) reward outcomes was progressively rejected. In the 1950s in the USA, the potential usefulness of appraisal as a tool for motivation and development was gradually recognized.

By the 1960s, the influence of the management by objectives movement meant that performance appraisal developed a greater emphasis on goal-setting and the assessment of performance-related abilities.

The use of psychometrics as part of the appraisal process emerged as a trend in the 1970s and gained momentum over the next two decades. The relative objectivity of psychometrics made them more acceptable in the new litigious environment. Research and development of all types of assessments such as written tests, aptitude tests, physical tests, psychomotor tests, personality tests, integrity and reliability tests, work samples and simulations were proposed.

Through the 1980s and 1990s, the concept of performance management provided a more holistic approach to generating motivation, improving performance and managing human resources.

In recent years, performance management has evolved even further, with many companies pulling down the traditional hierarchy in favor of more equal working environments. This has led to an increase in performance management systems that seek multiple feedback sources when assessing an employee's performance – this is known as 360-degree feedback. The majority of research on performance appraisal is drawn from a Western context and it is therefore not always possible to assume that the findings from a piece of performance appraisal research can be generalized across all national and organizational cultures.

The subject is so broad that in 1962 the International Society for Performance Improvement (ISPI) was erected. ISPI is dedicated to improving individual, organizational and societal performance.

In the workplace, ISPI represents more than 10,000 international and chapter members throughout the USA, Canada and 40 other countries. ISPI's mission is to develop and recognize the proficiency of our members and advocate the use of Human Performance Technology (Pershing, 2006, page 1311).

Many safety professionals, nowadays, are very interested in this part of the IWO psychology. The number of occupational accidents has been reduced significantly over the last decades, and a further reduction by technical means or organizational means (e.g., modern safety management systems) seems a real challenge. A number of safety professionals, therefore, focus again on "the improvement of the performance of the workers." The importance of the "human element" in the debate is not new at all.

Over the last 50 years, the focus in the field of safety evolved from "human performance" (in the 1960s) to "human reliability" (e.g., Bell et al., 2009) via "human error" (Kletz, 1985) and "human factors" (Dekker, 2005) back toward "human performance" (Conklin, 2019).

Occupational health and well-being

The link between industrial working conditions and mental health was studied since early twentieth century. Important founders of occupational psychology in Britain are Charles Myers and the Industrial Fatigue Research Board:

Myers established in 1921 the National Institute of Industrial Psychology. It was to be run as "a scientific nonprofit association," dependent for its operation primarily on fees earned for diagnostic investigations and advisory work carried out for industrial and commercial firms to improve working conditions and human performance.

Meanwhile, the Industrial Fatigue Research Board (later renamed the Industrial Health Research Board) was established almost contemporaneously to continue in peacetime the investigation of industrial fatigue and factors affecting the personal health and efficiency of workers that had begun in 1915 with the Health of Munition Workers Committee. Attempts to meet the armed forces' ever-increasing demand for munitions by extending the hours worked by women in the munitions factories, such that 90 h a week was common and 100 h not unknown, had led to decreased productivity and increased sickness absence and accidents with the result that the committee was charged with the systematic investigation of these problems.

Industrial psychology has since then been expanding at a rapid rate all over the world. Since the start back in the 1920s, there have been many discussions about the purpose of industrial psychology as a science. Some saw it as a scientific domain that was needed to investigate how the employee could be deployed even more efficiently while others consider it as the scientific basis for promoting well-being at work. Drever in his book entitled *The Psychology of Industry*, published in 1921, compared the two spheres of human performance activity as follows:

The aim of scientific management "is confessedly to increase profit and output" and therefore considers psychological phenomena from the viewpoint of management' but industrial psychology is strictly impartial. It concerns itself with the facts, and its investigations and results are equally at the service of employer and employee. On the whole its tendency has perhaps been to support the worker and his claims, since the worker is the effective agent in nearly every process it investigates, and an understanding of the facts is impossible without understanding the point of view of the worker, as well as the psychological processes involved in the work itself.

Remuneration and compensation

IWO psychologists have studied for decades the importance of different types of compensation. Rynes et al. (2004) give an overview of different studies since 1957. Herzberg et al. (1957) reviewed 16 studies and showed that pay ranked sixth. Lawler (1971) reviewed 49 studies and showed that pay ranked third across studies. Tower (2003) surveyed more than 35,000 US employees and found that importance of pay varies by objective.

The IWO psychology research in this field is relevant to Occupational Safety and Health specialists because it provides scientific research on how organizations can improve occupational safety and health by means of economic incentives. The European Agency for Safety and Health at Work, (2010) published a comprehensive review of the subject.

The scope of the report is on economic incentives in OSH, defined as external economic benefits offered to employers to motivate them to invest in safer and healthier workplaces. The incentives in OSH described in this report are thus external and economic. External means that these incentives are established by organizations outside the enterprise, usually public administration bodies or insurers; these incentives may act at national, regional or sector level. With regard to the economic aspect of incentives, there are two major categories:

1. Financial incentives (positive or negative), such as insurance-related incentives (e.g., variable premiums), funding schemes and tax-based incentives (tax reduction or specific taxes)
2. Nonfinancial incentives, including recognition schemes such as awards, aiming at positive recognition but not having substantial direct financial implications.

There has been a lot of debate as to whether financial incentives are useful at company level to improve occupational safety and health performance. Some companies take into account the number of accidents or the reduction in the number of accidents from one year to another year in calculating the bonus of the managers. OSHA in the USA issued a warning about these types of incentive programs (https://www.osha.gov/as/opa/whistleblowermemo.html):

Finally, some employers establish programs that unintentionally or intentionally give employees an incentive to not report injuries. For example, an employer might enter all employees who have not been injured in the previous year in a drawing to win a prize, or a team of employees might be awarded a bonus if no one from the team is injured over some period of time.

In addition, if the incentive is great enough that its loss dissuades reasonable workers from reporting injuries, the program would result in the employer's failure to record injuries that it is required to record under Part 1904. In this case, the employer is violating that rule, and a referral for a recordkeeping investigation should be made.

If employees do not feel free to report injuries or illnesses, the employer's entire workforce is put at risk. Employers do not learn of and correct dangerous conditions that have resulted in injuries, and injured employees may not receive the proper medical attention, or the workers' compensation benefits to which they are entitled. Ensuring that employees can report injuries or illnesses without fear of retaliation is therefore crucial to protecting worker safety and health.

OSHA explained in 2018 that it has no objection at all against incentive programs. They can even be important to promote workplace safety and health but the employer has to demonstrate that he/she is serious about creating a culture of safety and not just the appearance of reducing rates (https://www.osha.gov/laws-regs/standardinterpretations/2018-10-11):

One type of incentive program rewards workers for reporting near-misses or hazards, and encourages involvement in a safety and health management system. Positive action taken under this type of program is always permissible under § 1904.35(b)(1)(iv). Another type of incentive program is rate based and focuses on reducing the number of reported injuries and illnesses. This type of program typically rewards employees with a prize or bonus at the end of an injury-free month or evaluates managers based on their work unit's lack of injuries. Rate-based incentive programs are also permissible under § 1904.35(b)(1)(iv) as long as they are not implemented in a manner that discourages reporting. Thus, if an employer takes a negative action against an employee under a rate-based incentive program, such as withholding a prize or bonus because of a reported injury, OSHA would not cite the employer under § 1904.35(b)(1)(iv) as long as the employer has implemented adequate precautions to ensure that employees feel free to report an injury or illness.

Productive behavior

Productive behavior is best defined as any behavior by an employee who has a positive impact on the goals and objectives of an organization. All organizations will define productivity, profitability and loss control as their main objectives. A central question for the management and the safety professionals is how the workers can be kept motivated in applying all safety rules. IWO psychologists have been studying individual psychology and group dynamics for many decades in order to understand how productive behavior can be achieved and optimized and how counterproductive behavior (is it antonymous?) can be avoided. Organizational constraints have been identified and studied as aspects that interfere with or prevent good job performance. Some of them are very relevant for productive safety performance: job-related information, tools and equipment, budgetary support, task preparation, time availability and the work environment.

According to De Fruyt and Jesus (2003), it was demonstrated in several studies across different countries (USA and Europe) that the most important traits to predict work performance are conscientiousness and neuroticism (emotional stability). Traits not only predict different facets of job performance, but they also affect a range of additional work outcomes including job satisfaction, job commitment, voluntary turnover, absenteeism and deviant behaviors.

According to Ones and Viswesyaran (1997), there are five reasons why conscientiousness predicts job performance. According to these authors, "conscientious individuals":

– spend more time on the task they are assigned (rather than daydreaming and engaging in other unproductive activities), which results in greater productivity;
– spend more time on the task, allowing them to acquire more job knowledge;
– go beyond role requirements in the workplace;
– set goals autonomously and persist in following them;

- avoid counterproductive behavior; and
- selection of workers or managers on "conscientiousness" is not done in industry.

Other domains that are extensively studied by IWO psychologists are training and training evaluation, motivation in the workplace, job satisfaction and commitment, occupational stress, occupational safety, organizational culture, group behavior team effectiveness (team composition, task design, organizational resources, team rewards and team goals) and leadership (leader-focused approaches, contingency-focused approaches, follower-focused approaches and organizational development).

The influence of IWO psychology on the well-being of the workers has been very important because the IWO psychologists delivered for more than 100 years now the background and insights that OHS performance results from the way that workers are managed.

3.7.13.4 Some "old–new" simple approaches

When an accident happens, the management will often be confronted with questions such as:
- Why do people not follow the safety rules?
- Why do people omit to perform a task analysis?
- Why do people take undue risks?
- Why did the victim bypass the safety measure?

Safety professionals will try to find easy answers. Ideas and concepts from the IWO psychology will be transformed into pragmatic simple tools and theories.

A number of organizations, mainly in the USA, are currently working on human performance programs based on the research of Dr. Todd Conklin. He is described as:

> One of the most influential, most innovative, and most controversial thinkers in occupational safety and health these days is Dr. Todd Conklin, who's famous for his human and organizational performance (HOP) approach to safety matters.

Let us have a closer look at some of these "innovative" HOP principles:

HOP principle	Basic fundamentals in accident prevention since more than 80 years
Human error is normal	This statement is one of the fundamentals of accident prevention for more than 100 years. "Carelessness is an unavoidable characteristic of our human makeup. In recognition of this human imperfection, many accidents are unpreventable and the inevitable by-product of industry" (Beyer, 1916).

(continued)

HOP principle	Basic fundamentals in accident prevention since more than 80 years
	The theme of this book is that it is difficult for engineers to change human nature and therefore, instead of trying to persuade people not to make mistakes, we should accept people as we find them and try to remove opportunities for error by changing the work situation, that is, the plant or equipment design or the method of working. (Kletz, 1985)
Blame fixes nothing	Blake (1943, Chapter X Accident Investigation):
	The purpose of accident investigation is to discover the causative factors, the hazardous conditions and practices that brought the accident about, so that proper action may be taken to prevent a recurrence. The need is for full information as to causes – all the correctible causes that led to the accident, not just the major cause. This point brings out the importance of eliminating the factor of fixing blame. If part of the purpose is to fix blame, or if workers think it is, vital information will often be withheld or facts will be distorted.
	Kletz (1992, page 73):
	Even here before blaming people for so-called violations, we should make sure that the rules were clear, that the need of them had been explained and that no one had turned a blind eye on the previous failures to follow them.
	Chapter 13.3 of the book of Kletz is about "Blame in accident investigations" but he also writes "Of course, if a man makes mistakes repeated errors, more than a normal person would make, or shows that he is incapable of understanding what he is required to do, or is unwilling to do it, then he may have to be moved."
Learning is vital	Education, training and learning have been fundamental to accident prevention for more than 100 years. See, for instance, chapter 46 in the book of Beyer (1916):
	Safety inspectors are frequently asked the following question: How can you prevent this type of accidents? . . . There is an answer and it is found in two words – "safety education. (page 359)
Context drives behavior	The idea is that behavior is an outcome of organizational context. This was, for instance, well discussed by Heinrich and Granniss (1959). In the middle of the 1920s, a series of theorems were developed and illustrated by a "domino sequence." According to Heinrich, accidents are caused by the faults of persons *which are inherited or acquired by the environment.*

(continued)

HOP principle	Basic fundamentals in accident prevention since more than 80 years
How you respond to failure matters	Blake (1943, chapter X Accident Investigation):
	At times temptation to punish particularly thoughtless or inconsiderate action is difficult to resist. Experience, however, seems clearly to be on the side of limiting disciplinary action of any kind to instances of action so objectionable that the fellow worker themselves favor punishment.

The so-called innovative ideas from the HOP approach are very basic ideas that have been known by psychologists and safety professionals for about 100 years. In a large number of European countries (e.g., France and Belgium) this is common sense. In some other regions of the world (Asia and Middle East), a "blame" is a cultural element. When something goes wrong somebody must receive the blame. The HOP approach will depend on the region seen as common sense or revolutionary. *This observation does not detract from the wonderful work that many consultants such as Dr. Conklin deliver. Their support to industry really helps to improve safety.*

A similar observation can be made about the "SAFETY DIFFERENTLY" movement that started with a book by Professor Sidney Dekker (2014). The movement claims to be different from the "old approach to safety" (Table 3.16).

Packing existing ideas in new packaging can help to initiate a new dynamic. It is at the same time important that management and people in the field realize that these approaches are not new and that there is no "single solution" to the problem of prevention of occupational accidents. The cultural aspects also need to be taken into account. The perception and feeling of people in Asia, Africa or Europe in case of a workplace accident is not the same. This does not mean that the people accept accidents, it simply means that they will look at it from a different cultural background.

3.7.13.5 Human behavior is not a root cause

"Human error or inappropriate actions" are regularly mentioned among the apparent causes and root causes of accidents. In industry however human error is not considered as a root cause. *"human error is not a root cause: it is a symptom of underlying causes that must be identified among the elements that form a work situation."*   (TOTAL, 2019)

The knowledge that human failure is a symptom and that root causes must be sought has been known by industry for a long time.

"People say that accidents are due to human error, which is like saying falls are due to gravity" (quote of Trevor Kletz in the 1980).

Table 3.16: Comparison between the old and new approaches according to SAFETY DIFFERENTLY movement.

"Old approach" according to SAFETY DIFFERENTLY	SAFETY DIFFERENTLY	Comment
Workers are considered the cause of poor safety performance. Workers make mistakes, violate rules and ultimately make safety number look bad. This means workers represent a problem that an organization needs to solve.	People are not the problem to control, they are the solution. Learn how your workers create success on a daily basis and harness their skills and competencies to build a safer workplace.	Both statements are in essence the same. Since more than 100 years, there is a consensus that people are the key to success. Safety activities that have been put in place since the beginning of the twentieth century aim to improve their skills and competences to build a safer workplace.

The SAFETY DIFFERENTLY movement starts with a wrong statement about "how people are seen in industry." The mainstream approach in industry for more than 100 years is that "people are key to success." A small number of safety experts do study the "poor performance of humans" but this is a minority. |

(continued)

Table 3.16 (continued)

"Old approach" according to SAFETY DIFFERENTLY	SAFETY DIFFERENTLY	Comment
Because of this organizations intervene to try and influence workers' behavior. Managers develop strict guidelines and tell workers what to do, because they cannot be trusted to operate safely alone.	Rather than intervening in worker's behaviors, intervene in the condition of their work. This involves collaborating with frontline staff and providing them with right tool and environment to get the job done safely. The key here is intervening in workplace conditions rather than in working behavior.	Since the 1920s, all safety professionals agree that the machines and environment have to be adapted to humans in order to achieve "a maximum of efficiency coupled with a minimum of health hazards, absence of fatigue and a guarantee of the sound health and all round personal development of the working people" (Myasishchev's and Bekhterev at the First Conference on Scientific Organization of Labour in 1921).
		It is a legal duty of the line management to provide the workers with correct tools and to put them in safe conditions before starting the work. But it is also a duty of the management to provide clear instructions and guidelines and to require the workers to follow strictly these instructions. Managers have the duty to perform inspections and checks.
		Managers and workers meet in the field on many occasions and have to find a good balance between clear instructions and inspections and giving the worker the empowerment to take decisions in executing the work safely.

Organizations measure their safety success through the absence of negative events.

Measure safety as a presence of positive capacities. If you want to stop things from going wrong, enhance the capacities that make things go right.

Occupational health and safety-related activities are result-oriented activities that strive for zero harm. In order to achieve that we need to avoid accidents and in order to avoid accidents it is important to tackle the causes of accidents before they happen. Measurement of the number of accidents and quantitative monitoring of all substandard conditions and acts have been crucial to improving occupational health and safety. Since the 1920s, it has become obvious that, in addition, we also need to measure leading indicators and to promote safety via good examples. This is the basis of the safety management systems that have been put in place since the 1970s. Continuous improvement according to modern safety management (e.g., ISRS) is about improving the effectiveness of the safety management activities (safety meetings, training, rewarding, housekeeping, etc.). But the bottom line remains to achieve zero harm.

This balanced approach between measurement of lagging and leading indicators is in place in industry for many decades.

Nevertheless, it remains a tricky issue. Trevor Kletz, the first Technical Safety Advisor of the Petrochemical Division of ICI Ltd, wrote in his *Safety Newsletter* number 10 of May 1969 (Figure 3.11).

10/13 "HUMAN FAILING"

There were 30 minor accidents in April in one of the Division's Works. Over half were recorded by the supervisor as caused by "human failing". It is an easy way out for the supervisor or manager as then he has to do nothing except tell the man to be more careful.

These accidents were discussed individually with the supervisors and in all but two cases they agreed that there was something they could do to prevent the accident happening again.

If you don't believe that over 90% of all accidents can be prevented by better management let me know and we can discuss a few examples.

20 May 1969

Figure 3.11: Extract from ICI Safety Newsletter May 1969.

In his *Safety Newsletter* number 12 of August 1969, he repeated:

> I have often commented on our readiness to write down "human failing" as the cause of fires and accidents (see Newsletter 10, Item 13). Over half our accidents are given this cause, though there is nearly always something that managers or supervisors could do to make a repetition less likely. If we say that an accident can be prevented by better design, better training, better methods of working, better systems of inspection and so on then we are led on to do something to prevent it happening again, But if we say that an accident is due to "human failing", then there is nothing we can do to prevent it occurring again, except telling someone to take more care.

The term "human error" embraces various kinds of human inappropriate actions. A classification helps to identify the best methods to prevent the error (Kletz, 1991):
– mistakes (the person does not know what to do because of poor training or instruction);
– violations (the person does not want to do it because of wrong motivation);
– mismatches (the person is not able to do it); and
– slips or lapses of attention (inevitable from time to time).

Violations are not always negative! If instructions are wrong (due to ignorance or slips), violations can prevent accidents.

An interesting field guide to human error investigations was published by Dekker (2002).

A comprehensive overview of *Human Performance Improvement* is given in two handbooks issued by the US Department of Energy that can be downloaded for free on http://www.hss.energy.gov/nuclearsafety/ns/techstds/:

- DOE-HDBK-1028-2009, *Human Performance Improvement Handbook, Volume 1: Concept and principles*, 2009 (175 pages):
  This *Human Performance Improvement Handbook* consists of five chapters entitled: "An Introduction to Human Performance," "Reducing Error," "Managing Controls," "Culture and Leadership", and "Organizations at Work." It is a set of concepts and principles associated with a performance model that illustrates the organizational context of human performance. The model contends that human performance is a system that comprises a network of elements that work together to produce repeatable outcomes. The system encompasses organizational factors, job-site conditions, individual behavior and results. The system approach puts new perspective on human error: it is not a cause of failure, alone, but rather the effect or symptom of deeper trouble in the system. Human error is not random; it is systematically connected to features of people's tools, the tasks they perform and the operating environment in which they work.
- DOE-HDBK-1028-2009, *Human Performance Improvement Handbook, Volume 2: Human Performance Tools for Individuals, Work Teams, and Management*, 2009 (137 pages). That volume describes methods and techniques for catching and reducing errors and locating and eliminating latent organizational weaknesses.

3.8 Evidence-based safety

For most of history, professions have based their practices on expertise derived from experience passed down in the form of tradition. Some of these practices continue to exist, although there is no evidence that these practices are justified. Since the 1970s, there is an increasing literature about evidence-based medicine which can be described as "the conscientious, explicit and judicious use of current best evidence in making decisions about the care of individual patients." Evidence-based medicine found its way to other domains and since about 30 years there is literature about evidence-based practices in all types of domains: evidence-based dentistry, evidence-based design, evidence-based education, evidence-based legislation, evidence-based management, evidence-based nursing, evidence-based toxicology and so on. In other words, the phrase "evidence-based" is a buzzword in many fields.

The movement toward evidence-based practices attempts to encourage and, in some instances, to force professionals and other decision makers to pay more attention to evidence to inform their decision making and to avoid making decisions based on tradition and intuition.

Wang et al. (2017) give a historical overview how the concept found its entry in the field of safety. Evidence-based safety is considered as a subcategory of evidence-based management. Wang makes a comparison between:
- experienced-based safety management;
- theory-based safety management;

- behavior-based safety management;
- standard-based safety management;
- risk-based safety management;
- accident-based safety management; and
- countermeasure-based safety management.

There is no doubt that all these approaches have their merits because they are all based on some "evidence." However, one should realize that all these approaches also have drawbacks and that the word might be recent but the underlying approaches are not new. Hence, the discussion is mainly a philosophical discussion within the profession of safety specialists with little added value for the safety of the people in the field.

A major critic on "evidence-based safety" is as follows:

- Scientific evidence can be in error. Weigman (2005) in her report *The Consequences of Errors* writes: "[. . .] when models or dogmas are established on the basis of erroneous science, they are all the more difficult to eliminate." A well-known example is the error that was made in 1890 when the German scientist, Gustav von Bunge, found the iron content of spinach to be 35 mg per 100 g. It was overlooked that he had analyzed powdered dried spinach. The fictional American cartoon character Popeye the Sailor was getting strength by eating spinach. The error and the myth survived until 1937.
- When scientific evidence is established, one has to understand why the results were obtained. Safety is a very complex matter and the different elements are interacting dynamically. Hence, when a safety method is tested it could give good results but because of reasons that are not understood and hence the wrong conclusions could be drawn.
- It is not because there is no scientific evidence that something does not work. Sometimes good results are obtained based on experience and intuition without a good explanation. Theories like "high reliability organizations" (HRO) were explained after the observation that some organizations were very efficient. The organizations were not put in place based on evidence but based on experience. The notion HRO arose when people started to study these organizations and they started to question why these organizations had good results.

A more pragmatic approach than evidence-based safety could be a "result-based safety" approach. This approach would identify organizations that have a good safety performance (even without any scientific evidence), trying to understand why they have good results and checking whether the good practices can be extrapolated.

3.9 Tribute to Heinrich and Bird

3.9.1 Influential authors on occupational safety

Herbert William Heinrich (1886–1962) and Frank E. Bird Jr. (1921–2007) are two influential authors of the twentieth century. Their main contribution to the safety practice and theory is that each of them brought together long-standing ideas and presented them in a more holistic framework.

Both gentlemen were important because for decades they were the references for safety thinking. Certain concepts such as Heinrich's domino model or Bird's pyramid have become household names in the world of safety science.

The books and articles of both Heinrich and Bird do not, however, contain much new material compared to what other contemporaries wrote. Books by, for example, Beyer (1916), DeBlois (1926), Vernon (1936) and Blake (1943) indicate that the concepts and ideas of Heinrich and Bird were mainstream as from the 1920s.

Busch (2018) who wrote a very interesting master thesis about Heinrich tries to explain why Heinrich is often credited for ideas that already existed and were available in the literature.

The main legacy from Heinrich and Bird are as follows:

- William Heinrich made a clear difference between the notion of "injury" and the notion of "incident." The occurrence of injury is rare and thus more a question of (bad) luck. In some cases, it could be a fatality while in other very similar cases it could be a first aid. An incident, however, is caused and hence can be prevented. It is therefore important to concentrate on the incidents and their causes and not on the number of injuries. The causes are further classified as substandard acts and substandard conditions.
- According to Heinrich "ALL" accidents (98% according to Heinrich) can be prevented. This was a rupture with the discussions before 1920s when it was common to state that the preventable accidents were somewhere in the order of 30–50%. Heinrich overemphasized the contribution of substandard acts (compared to substandard conditions) but it all depends on the definition of the contribution of humans. Substandard conditions can in some way often be considered as caused by substandard acts. The term substandard acts will later be modified in substandard practices.
- On the importance of "high potential incidents," Bird writes: "An incident with high potential for harm (HIPO) should be investigated as thoroughly as an accident. In this context, then, an incident is an undesired event which under slightly different circumstances, could have resulted in harm to people, damage to property of loss of process."
- Frank Bird used the ideas of Heinrich, developed them further and proposed a systematic comprehensive approach that is called a "modern loss control management system." He moved the focus of attention from the acts of individuals to the

management system as the best way to influence behaviors in a lasting way. He integrated safety into all aspects of the job so that it became a part of "how we do things" instead of safety as an "add on." He applied risk management concepts to focus on the critical few aspects with the greatest potential for loss. He emphasized the study of near misses, especially those with high potential.

3.9.2 Herbert William Heinrich

Lots of details on the life and work of Heinrich can be found in the master thesis of Busch (2018).

Heinrich started to work for The Travelers Insurance as a boiler and industrial plant inspector in 1913. He worked for the Engineering and Inspection Division that had 339 professional staff (Busch, 2018).

In the 1920s, Heinrich published a number of articles on the hidden costs of accidents. Heinrich also gave presentations and speeches on this theme.

In 1928, he presented the results of a study of 75,000 accidents and came up with the postulate that 98% (88% attributable to unsafe acts and 10% attributable to unsafe conditions) of all accidents were preventable. This postulate was a real breakthrough in safety! For the first time, an authority on safety stood up and declared that "all accidents" are preventable. This is remarkable because for a long time, different researchers tried to quantify the number of accidents that were preventable.

Law and William (1909), for instance, wrote in his introduction in 1909:

> The prevention of accidents absolutely is of course impossible. . . . The United States not having progressed so far, it is probable that more than 57.95% of the accidents that occur are preventable.

Beyer (1916) analyzed causes of accidents and summarized:

> To summarize – we may conclude from the results of this and the previous Chapter, that 50% of our present industrial accidents are preventable, – 25% by Mechanical Safeguards and 25% by Education.

The importance of the postulate that all accidents are preventable cannot be underestimated. As long as people believe that a large part of occupational accidents (e.g., 50%) are not preventable, they will have a hard time to be fully engaged in accident prevention activities. People will justify their lack of effort by "Whatever I do, accidents will continue to happen." Heinrich introduced a complete change by postulating that "all accidents are preventable."

Heinrich presented in 1929, during the Annual Meeting of the Connecticut Society of Civil Engineers, his triangle 1:29:300.

> A major injury is defined as any case that is reported to insurance carriers or to the state com-
> pensation commissioner. A minor injury is a scratch, bruise, or laceration such as is commonly
> termed as a first-aid. The great majority of reported or major injuries are not fatalities or frac-
> tures or dismemberments, they are not all lost-time cases, and even those that are do not nec-
> essarily involve payment of compensation.   (Heinrich and Granniss, 1959, page 30)

What Heinrich calls a "major injury" is what nowadays would be named "a recordable accident" (= fatalities + lost time + medical treatment + restricted work case).

The main point that Heinrich pointed out is that people should focus on the causes
why incidents occurred and they should not focus on the injury. He emphasized that
most of the injuries (i.e., 29 first aids) are not the "major injury" and that even in the
"major injuries" the prevailing were not the fatalities or permanent disabilities. Heinrich understood very well that not all substandard conditions or substandard acts
could lead to a major injury. He wrote (Heinrich and Granniss, 1959, page 28):

> There are certain type of accidents, of course, where the probability of serious injury may vary
> in accordance with circumstances. For example, a type of accident such as a fall, if occurring
> on a level field of soft earth or on a rug-covered floor in the home may not be potentially as
> serious as the fall of a steel erector from the top of a skyscraper.

When nowadays people pretend that one should focus on the so-called SIF contrib-
utors (serious injuries and fatalities), then they simply confirm what Heinrich wrote
(Figure 3.12).

3.9.3 Frank E. Bird Jr.

A more comprehensive tribute to Frank E. Bird can be found on the website of DNV
(Det Norske Veritas): https://www.dnvgl.com/oilgas/international-sustainability-rat
ing-system-isrs/tribute-to-frank-bird.html.

In 1951, at the age of 30, Frank went to work at Lukens Steel Co. in Coatesville,
PA. At that time, Lukens was the largest independent specialty steel company in the
USA, with nearly 5,000 employees. Starting as a second helper at the open hearth,
he soon was asked to become a safety trainee and rose rapidly to become director of
safety and security.

While at Lukens, he implemented leading-edge operational safety activities for
all levels of management, as well as for hourly and salaried employees. He led research studies and implemented seminal concepts regarding safety, security and
loss control. Frank reduced the complexity of traditional steel company safety programs by focusing greater attention on the higher risk aspects of the work. He went
on to serve for 18 years as Supervisor of Safety for the Lukens Steel Company (Bird
and George, 1985).

In partnership with George Germain, who was in charge of training and development for Lukens, Frank coauthored the book *Damage Control, A New Horizon in*

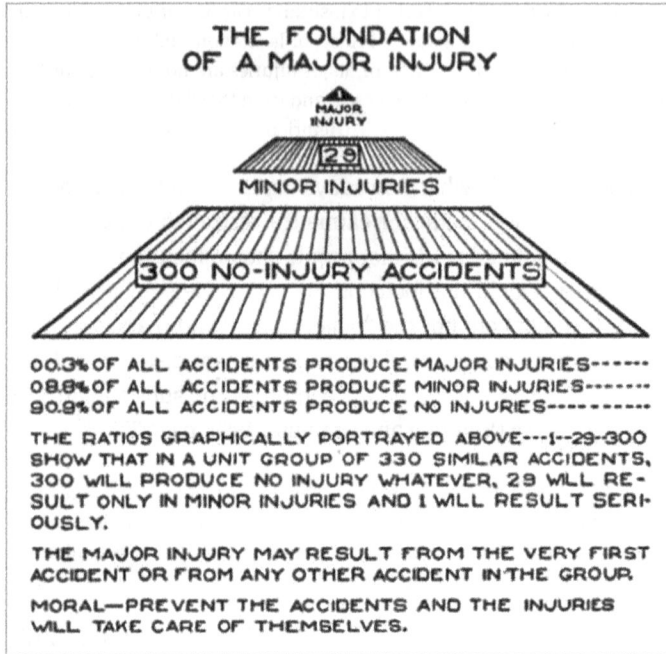

THE FOUNDATION OF A MAJOR INJURY

MAJOR INJURY

29

MINOR INJURIES

300 NO-INJURY ACCIDENTS

00.3% OF ALL ACCIDENTS PRODUCE MAJOR INJURIES------
08.8% OF ALL ACCIDENTS PRODUCE MINOR INJURIES-------
90.9% OF ALL ACCIDENTS PRODUCE NO INJURIES----------

THE RATIOS GRAPHICALLY PORTRAYED ABOVE---1--29-300 SHOW THAT IN A UNIT GROUP OF 330 SIMILAR ACCIDENTS, 300 WILL PRODUCE NO INJURY WHATEVER, 29 WILL RESULT ONLY IN MINOR INJURIES AND 1 WILL RESULT SERIOUSLY.

THE MAJOR INJURY MAY RESULT FROM THE VERY FIRST ACCIDENT OR FROM ANY OTHER ACCIDENT IN THE GROUP.

MORAL—PREVENT THE ACCIDENTS AND THE INJURIES WILL TAKE CARE OF THEMSELVES.

Figure 3.12: Heinrich triangle.

Accident Prevention and Cost Improvement, which was published by the American Management Association in 1966.

At Lukens, he headed a 7-year study (1959–65) of 90,000 incidents. It revealed a ratio of 1:100:500, that is, for every one disabling injury there were 100 minor injuries and 500 property damage accidents.

In July 1968, Frank's growing influence led to his becoming director of engineering services for Insurance Co. of North America (ESTA). There, he headed a study of industrial accidents, including analysis of 1,753,498 accidents reported by 297 companies. These companies represented 21 different industrial groups, employing 1,750,000 employees who worked more than 3 billion man-hours during the exposure period analyzed. The study resulted in the 1:10:30:600 ratio.

Bird defines a serious or major injury as "resulting in death, disability, lost time or medical treatment" (Bird and George, 1985, page 21) and minor injury as "requiring first aid" (Figure 3.13). Bird also mentions that the ratio within the major injuries was "one lost time injury per 15 medical treatments."

Bird developed a "loss causation model" on the idea for the Heinrich domino model.

The domino concept of Heinrich (Figure 3.14), which explain the factors in an accident sequence, was used to develop an integrated management system. It reflects his genius for concepts and models that led to practical applications; in this case,

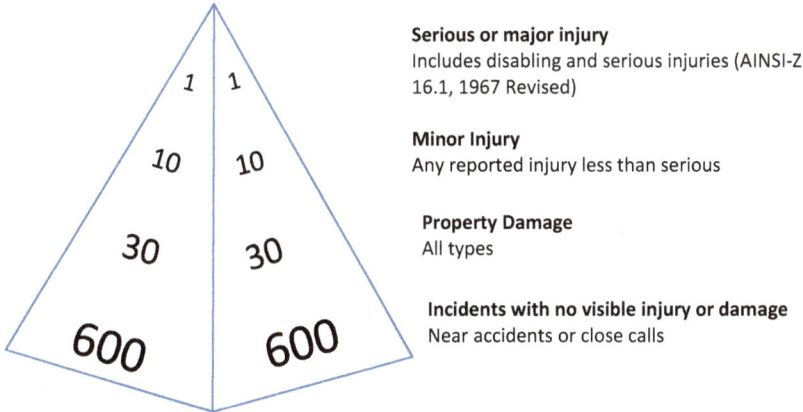

Serious or major injury
Includes disabling and serious injuries (AINSI-Z 16.1, 1967 Revised)

Minor Injury
Any reported injury less than serious

Property Damage
All types

Incidents with no visible injury or damage
Near accidents or close calls

Figure 3.13: Bird pyramid.

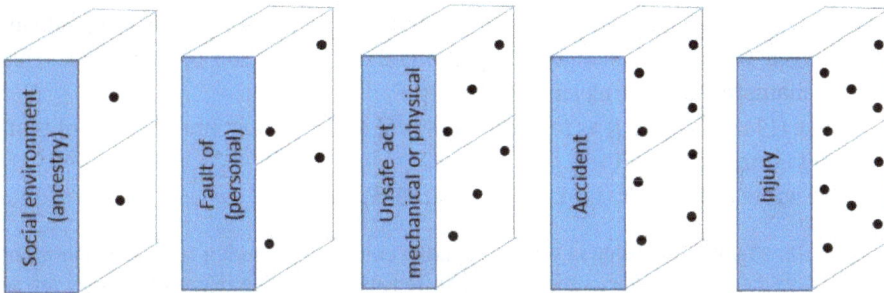

Figure 3.14: The five factors in an accident sequence (based on Heinrich).

improvements in the management system as the best way to control incidents. Read from right to left, it is a problem-solving model widely used in incident investigation.

Immediate causes are "unsafe conditions" or "unsafe acts." These terms were not new. They have been used since the 1920s but Bird spelt them out so that they could be used in the field. The same holds for the basic causes. Personal factors such as mental state, mental stress, behavior and skill level have been listed by many authors (e.g., Blake, 1943) but Bird put them in a management model that can directly be used in the field (Figure 3.15). Examples of job factors are training, knowledge, management supervision, leadership, contractor selection, engineering/design, work planning and purchasing. The approach is now a "management loss control approach" that includes safety and production.

Lack of Control		Basic Causes		Immediate Cause		Incident		Loss
1. Inadequate Programme 2. Inadequate Programme Standard 3. Inadequate Compliance to Standard	⇨ ⇨ ⇨	Personal Factors Job Factors	⇨ ⇨ ⇨	Substandard Practices And Conditions	⇨ ⇨ ⇨	Contact With Energy Or Substance	⇨ ⇨ ⇨	People Property Process Environment Quality

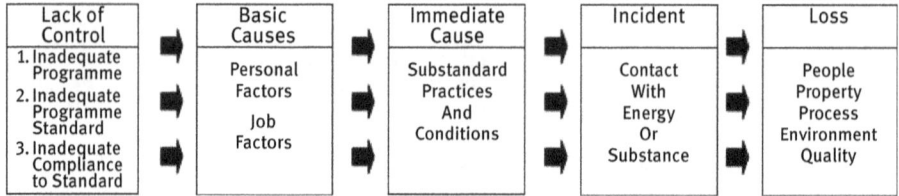

Figure 3.15: The DNV loss causation model (Bird).

3.9.4 Discussion on the 1–29–300 and the 1–10–30–600 rule

The idea behind the 1–29–300 rule (Heinrich) and behind the 1–10–30–600 rule (Bird) is very well explained in the original books of, respectively, Heinrich and Bird.

They first try to explain the difference between an incident or accident and the outcome (the loss or the damage). The incident (which is a sudden, unexpected event that causes or can cause damage) is the result of a number of factors that can be managed while the outcome or damage is more a question of (bad) luck. This was the mainstream thinking since the 1920s.

Blake (1943) resumes it as follows "causes of accidents are more important than causes of injuries."

Bird wrote in his book (Bird and George, 1985, page 21):

> The 1-10-30-600 relationship in the ratio indicate clearly how foolish it is to direct our major effort at the relatively few evens resulting in serious or disabling injury when there are so many significant opportunities that provide a much larger basis for more effective control of total accident losses. It is worth emphasizing at this point that the ratio study was of a certain large group of organizations at a given point in time. It does not necessarily follow that the ratio will be identical for any particular occupational group or organization. That is not its intent. The significant point is that major injuries are rare events and that many opportunities are afforded by the more frequent, less serious events to take actions to prevent the major losses from occurring. Safety leaders have also emphasized that these actions are most effective when directed at incidents and minor accidents with a high loss potential.

Bird also writes

> An incident with high potential for harm (HIPO) should be investigated as thoroughly as an accident. In this context, then, an incident is an undesired event which under slightly different circumstances, could have resulted in harm to people, damage to property of loss of process.

Heinrich and Bird were very careful not to make correlations within the group of major injuries, that is, correlations between accidents that result in fatalities and accidents that result in nonfatal injuries. The reason is very simple: it was well known that there *is no causal link between these accidents*.

The recommendations of all prominent authors (Beyer, Heinrich, Blake, Vernon, Bird, etc.) are always the same: "Accidents are caused but the degree of injury is a question of luck or bad luck." This is the underlying logic for the 1–10–30–600 metaphor: "focus on causes of accidents and not on the magnitude of the injury."

It is strange that the metaphors from Heinrich and Bird are since about 20 years considered as wrong. ICSI (Descazeaux et al., 2019) which refers to the "Bird pyramid" explains that the "The proportionality between minor incidents and serious accidents would only hold true if the mechanisms leading to the accidents were the same. Yet, all the evidence seems to indicate that is not the case. We therefore give too much importance to minor accidents (injury free), by devoting a lot of time and energy to them."

In their publication in 2019, ICSI first give an incorrect representation of the Bird pyramid from the 1960s. According to ICSI, the Bird pyramid is:
– 1 fatality;
– 10 lost time incidents;
– 30 incidents without lost time; and
– 600 injury free incidents.

ICSI uses this incorrect representation to pretend that this relationship cannot be correct. ICSI finally comes to the same conclusion as Blake, Heinrich and Bird already did more than 50 years ago. And ICSI is not alone in making this error. Gilbert et al. (2018) writes:

> The famous assertion of a constant ratio between unsafe behavior, minor injuries, and fatal accidents (Heinrich, Bird) is a first example. This belief has been used worldwide throughout the industry for decades to prevent severe accidents through chasing daily noncompliance and minor accidents. Yet it has been refuted by many researchers (Hopkins, Hovden, Abrechtsen & Herrera) and characterized as an urban myth. A recent study conducted by BST on occupational accidents in seven global companies (ExxonMobil, Potash Corp, Shell, BHP Billiton, Cargill, Archer Daniels Midland Company and Maersk) shows a de-correlation between the evolution of fatal and non-fatal accident rates over a given period.

Some authors first create a myth by wrongly presenting the metaphors of Heinrich and Bird and next they explain why the myth is wrong. Examples are Saloniemi and Oksanen (1998), Barnett and Wang (2000), Bellamy et al. (2008), Gallivan et al. (2008) and Yorio and Moore (2018). Some conclude that the safety pyramid is not as obvious and straightforward as many assume it to be or that the ratio is wrong. Other criticize the ratio between the number of fatal accidents and the number of recordable accidents while this is not discussed in the work of Bird and Heinrich. Dekker (2019) discussed the Bird pyramid in the context of major industrial disasters (e.g., the Macondo in 2010).

The origin of the confusion is not clear. A scholar study of the books of Bird and Heinrich makes very clear that the ratio 1–10–30–600 has nothing to do with major disasters. It seems that many authors invent a myth to explain afterward why the myth is wrong. It remains an interesting debate whether the theory of Bird, Heinrich and others are applicable to major accidents but there is no doubt that this is not at all what Heinrich and Bird wrote. The whole idea behind the pyramid of Heinrich and Bird is to draw the attention on the causes of accidents instead of putting too much focus on the injury and to pay more attention to HiPo events rather than to all events. This is very clearly explained in the work of Heinrich and in the work of Bird and in addition this is the mainstream thinking since the 1920s.

Some situations will very seldom lead to a fatality (e.g., slippery floor) while other situations (e.g., fall from height) often result in bad injuries or even in a fatality. It is therefore important that each company has a dedicated strategy toward those situations or tasks that have a high probability of permanent disability or even a fatality.

Heinrich and Granniss (1959) formulate it as follows: "In reality, the terms 'accident' and 'injury' are so merged, it is assumed that no accident is of serious importance unless it produces a serious injury." He adds:

> An injury is merely the result of an accident. The accident itself is controllable. The severity or cost of an injury that results when an accident occurs is difficult to control. It depends upon many uncertain and unregulated factors – such as the physical or mental conditions of the injured person, the weight, size, shape or material of the object causing the injury, etc.

The postulate that a company should mainly focus on situations with SIF potential is attractive but should be done with care. First of all, the notion of "minor injury" is only meaningful to those who are not the victims of the injury. And second, and more important, it is very subjective to assume that certain incidents can hardly cause a major injury.

A large oil company had 155 fatalities in a period of 10 years (period 2005–2015). An in-depth analysis of these accidents showed that about 90% of these fatalities were related to five working situations:
- accidents with a truck during transport of goods (53%);
- while working on energized systems (17%);
- fall from height (9%);
- during lifting and hoisting operations (6%); and
- during circulation onsite (4%).

Hence, at first sight, it is obvious that these are the five areas to focus on if the company wants to avoid occupational accidents with fatalities. That is what the company did and the number of fatal accidents decreased drastically.

But there are also examples where the consequence was a fatality while the typology of the accident was not such that a fatal outcome would have been expected.

In 2019, in one of the refineries, a person fell from a ladder from a height of 1.5 m. His head hit a metal plate and he died a couple of weeks later in hospital. In many similar situations (fall from 1.5 m), the outcome would most of the time be a minor injury. However, in this particular case, the person unfortunately died.

In the same company in the same year, a young man, in another site, fell from a five-step ladder from less than 1.5 m and he died.

Fatal accidents always cause shocks through all levels of the organization. In-depth investigations are started, and additional measures to be implemented in the whole corporation are studied. The reaction would have been completely different if the same incident (fall from 1.5 m) would have resulted in a first aid.

On 4 December 2013, a contractor was removing the rust from a pump drain pipe 1.5″ before repainting. He used a power brush. The task is considered as a low-risk task. Housekeeping at the workplace is good and there is enough space to perform the job without difficulty. There is no time constraint to perform the job. The person was wearing a neck warmer. The power brush came close to his neck warmer which was gripped by the brush and the worker was not able to switch off the power. The neck muff he was wearing was rolled into the power brush and it fastened tight his respiratory tract and killed the worker (Figure 3.16).

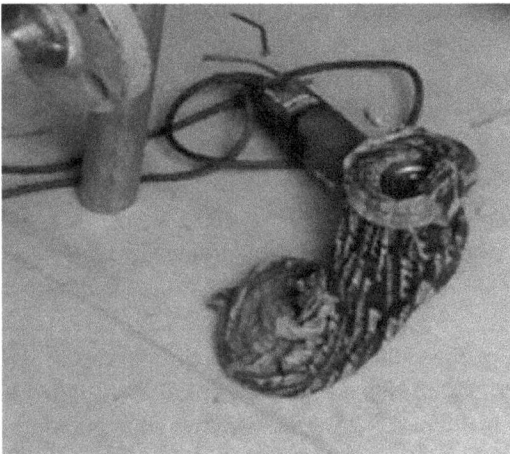

Figure 3.16: Neck warmer gripped in power brush.

These three examples of real accidents show the difficulty to identify HiPo events. These three examples among many others demonstrate that it is not easy to identify substandard situations (practices and conditions) that can result in a fatality. The examples also show the pertinence to focus on all types of substandard conditions and all substandard acts.

3.9.5 Is there a correlation between fatal accidents and nonfatal accidents?

Since the 1920s, it is obvious that the focus has to be on the causes of accidents and not on the type of injury. This explains why safety experts (Heinrich, Beyer, Vernon, etc.) are not interested in the ratio of fatalities to smaller injuries. It is the basis of their theory.

In this paragraph, it will be demonstrated that a mathematical correlation between the number of accidents with fatalities and the number accidents with nonfatal accidents is very divergent, which in fact confirms the theory of Heinrich and Bird (and many others) that it is not useful to focus on injuries but that the focus has to be on the causes of incidents and accidents.

Calder (1899) analyzed accident in the UK (1896) in 168,503 registered factories for a global workforce of 4,398,983 workers. The overall ratio was *1 fatality per 79 accidents that result in nonfatal injuries*. The ratio for males was *1 fatality per 75 accidents that result in nonfatal injuries while for female workers it was 1 fatality per 349 accidents that result in nonfatal injuries*.

Beyer (1916) mentions a comprehensive study of industrial accidents in 1908 carried out by the Department of Commerce and Labor in the USA. The study estimates that in 1908 there were 30,000–35,000 fatal accidents and 2,000,000 nonfatal accidents. This gives an overall ratio of *1 fatality per 57 accidents that result in nonfatal injuries to 1 fatality per 67 accidents that result in nonfatal injuries*.

Beyer gives a breakdown of 90,000 injuries that occurred in Massachusetts during one year, ending 30 June 1915. Table 3.17 gives the ratio of fatal accidents per number of nonfatal accidents for different causes.

Table 3.1 shows *fatality per 12 accidents that result in nonfatal injuries to 1 fatality per 1,625 accidents that result in nonfatal injuries. There were 4,331 eye injuries but none of them resulted in a fatality*.

Figure 3.17 shows the evolution of the ratio of the number of fatalities to the number of injuries between 1910 and 1940 (based on the tables of Aldrich, 1997). It is obvious from this graph that there is no relationship between the number of fatal accidents and the number of injuries.

It was also obvious for safety engineers that there is no correlation between the fatality rate and the injury rate. Companies like DuPont had implemented a rigorous recording scheme as early as 1910.

Figure 3.18 shows the evolution of the fatality rate and the injury rate per million hour within DuPont for the period 1912–1937. These figures were well known by the safety professionals and hence there was no doubt that there was no correlation between the number of fatalities and the number of injuries.

Vernon (1936) gives an overview of the number of fatal and nonfatal accidents paid for in 1931 under the Workmen's Compensation Act for Northern Ireland and Great Britain. The data concern 6,913,974 persons employed in different occupations

Table 3.17: Ratio of nonfatal injuries to fatal injuries for a number of situations (Beyer, 1916).

Cause of the accident	Number of accidents		Ratio
	Fatal injury	Nonfatal injury	
Machinery	7	11,375	1625
Burns	14	3,316	237
Hoists	4	630	158
Falls	66	8,417	128
Occupational diseases	2	104	52
Boilers and steam explosions	1	36	36
Elevators	33	1,036	31
Cranes	11	306	28
Electricity	25	495	20
Excavating	14	164	12
Eye injuries	0	4,331	

(mines, docks, quarries, construction work, shipping, railways and factories). The overall ratio was *1 fatal accident for every 164 nonfatal accidents*. The ratio observed in 1930, according to Vernon was *1 fatal accident for every 169 nonfatal accidents*. The ratio varies considerably in the different occupations, the extremes being *1–40* in shipping accidents and *1–217* for factory accidents.

Blake (1943) gives the statistics from the US Bureau of Labor Statistics for 1941. He reports that 19,200 workers lost their lives, 100,600 others suffered various permanent disabilities and some 2,060,400 more were injured. This gives an overall ratio of *1 fatal accident for every 107 nonfatal accidents*.

The author studied detailed (confidential) data from a number of companies in the oil and gas industry. The period 2010–2019 covers about 25.6 billion man-hours. In that period, the ratio was *1 fatality for every 30 lost time injuries and 1 fatality for 110 recordable injuries*.

The discussion above shows that it is not possible to give a useful indication of the ratio between accidents with a fatality and accidents with nonfatality. This has been reported since more than 100 years by several authors. Hence, it is very strange that some articles continue to open this discussion. As explained in many textbooks since more than 100 years, incidents are caused while the outcome (first aid, recordable accident or even a fatality) is a question of (bad) luck. And by definition it is not useful to make statistical correlations between the different degrees of injuries.

It makes sense that companies focus in particular on the situations or jobs that have a greater potential for causing a fatality but at the same time it is important to keep in mind that the removal of the causes (substandard practices and substandard conditions) are the key for improving safety at the workplace.

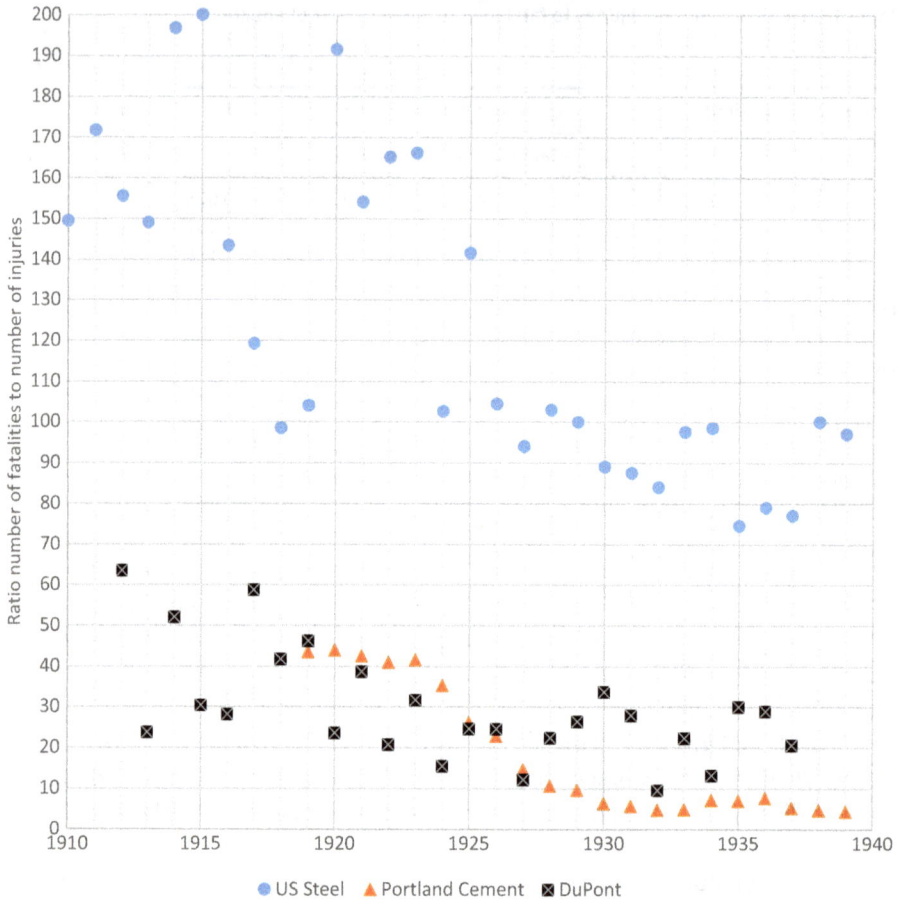

Figure 3.17: Ratio of number of fatalities to the number of injuries in the period 1910–1940 for three US companies.

3.9.6 Safety management systems

All ingredients for the concept of an "integrated safety management system" were developed in the period 1900–1950:
- The thought that safety should be managed in the same way as any other business was mainstream.
- Establishing a safe environment was a combination of technical measures, organizational issues and last but not least the motivation and engagement of each individual.
- Accidents had to be prevented by eliminating the causes.

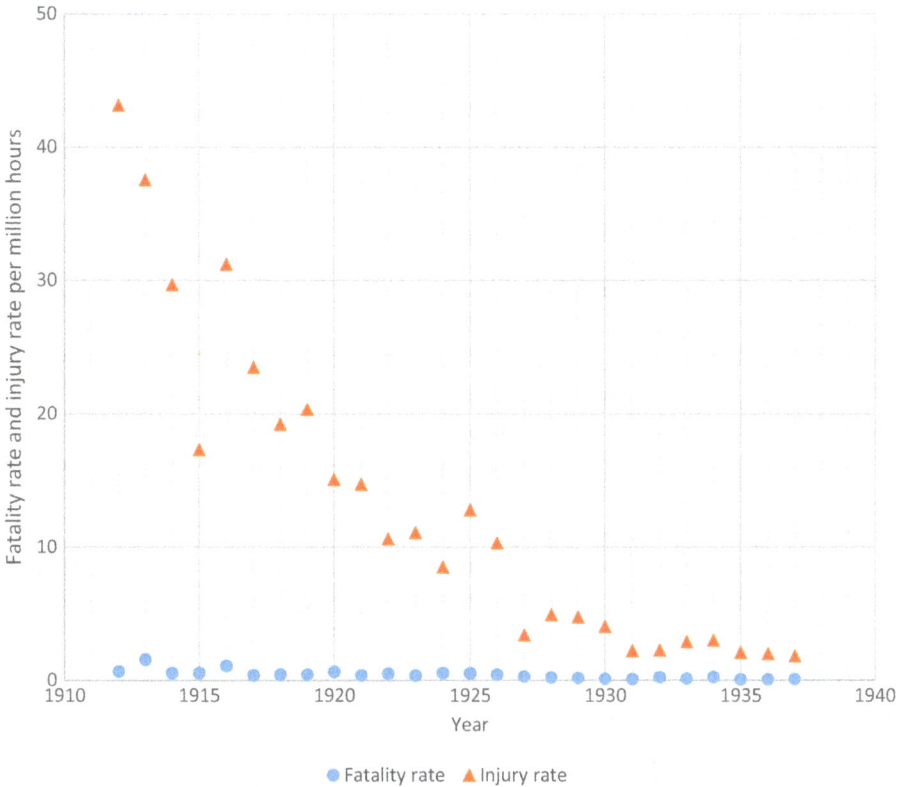

Figure 3.18: Evolution of the fatality rate and the injury rate at DuPont between 1912 and 1937.

- All type of activities (safety committees, inspections, JSA, selection and hiring, education, etc.) were in place in (large) companies.
- The role of the top management was recognized as crucial.

In the 1950s, people started to combine the theories and approaches of management of business with safety topics. Simonds and Grimaldi (1956) were to my knowledge the first to publish a book dedicated to *Safety Management*.

Factory Mutual (1957 and 1967 for the 2nd edition) uses the term "Property Conservation and a Property Conservation Program" and lists the following fundamental goals of any property conservation program:

1. to set a policy and establish a plan;
2. to create and sustain employee interest;
3. to plan safe buildings, equipment processes;
4. to eliminate the causes of fire, explosion and other losses through proper education, supervision, housekeeping and maintenance;

5. to provide automatic sprinklers and other protective equipment when needed;
6. to maintain the protective equipment in readiness; and
7. to organize and train employees for emergency action.

The property-conservation program of Factory Mutual is oriented to fire and explosion but the proposed approach demonstrates that a system approach was mainstream in the 1950s and 1960s.

The evolution toward integration of safety into a professional management system experienced many significant advances in the 1940s, 1950s and 1960s of the twentieth century. Bird and George (1985) list a number of books considered as landmark publications and their list start with the book of Blake (1943). This shows that Bird recognizes that the book of Blake from 1943 was a milestone for his insights into "modern safety management."

The evolution from various parallel safety initiatives to a fully integrated loss control system accelerated in the 1960s and has not disappeared since.

Bird developed the so-called Modern Safety Management Systems which are characterized by three acronyms. The role of these acronyms as fundamentals of a safety management system is given in Table 3.18 and explained below.

Table 3.18: Three acronyms that form the fundamentals of a safety management system.

Acronym	Meaning
IEDIM	This acronym refers to the goal of a safety management system. It is the reason why a safety management system is implemented.
PEMEP	This acronym helps to understand the different and multiple safety activities that are needed to compose a safety management system.
ISMEC	This acronym refers to the method to implement a safety activity in a safety management system.

The acronym IEDIM stands for:
- **I**dentification of all hazards
- **E**valuation of associated risks
- **D**evelopment of a plan
- **I**mplementation of the plan
- **M**onitoring of the plan

The acronym sums up the purpose of the whole management system. Every activity in a safety management system serves the purpose of IEDIM. If all hazards are identified and all associated risks have been evaluated and for each of them a plan is developed, put in place and monitored, then there is full control of the situation.

Unfortunately, the acronym of IEDIM has been replaced by PDCA (plan, do, check and act) and the result is that many people do not fully understand the final goal of a safety management system.

> In the 1980s there was a growing popularity for the PDCA management method. The PDCA (Plan-Do-Check-Act) approach is used in business to explain the continuous improvement of processes and products. It is also known as the Deming circle or the Shewhart cycle.

> In 1987 the International Organization for Standardization published the first version of ISO 9000 standards on Quality Management Systems. In March 1992, BSI Group published the world's first environmental management systems standard, BS 7750, as part of a response to growing concerns about protecting the environment. BS 7750 supplied the template for the development of the ISO 14000 series in 1996.

> In 2018, ISO 45001 was published, an ISO standard for management systems of occupational health and safety (OH&S). The goal of ISO 45001 is the reduction of occupational injuries and diseases, including promoting and protecting physical and mental health. The layout of the ISO standards on managements systems has since then be following the PDCA cycle in the process based approach. Since about 10 years the ISO standards encourage risk-based thinking.

Many people in industry, however, confuse the PDCA for quality purposes with the "PDCA" for OHS. PDCA for quality is logical: there is a plan to start with and from that plan we strive for continuous improvement.

In OHS, the most important element is the identification of the hazards and the evaluation of the risks before even starting to think about a plan. One of the reasons why accidents occur is that people have a plan (e.g., taking a sample, dismantling a pump and working at height) and that they want to execute the plan as quickly as possible. That is why IEDIM should be preferred in safety rather than PDCA. It draws attention to identification of the hazards and evaluation of the associated risks as the starting point.

The acronym PEMEP stands for:

People Personnel
 Third parties
 Public
Equipment Machines
 Installations
 Tools
Materials Raw materials
 End products
 Waste
 Utilities

Environment   Workers environment
                  Environment (broad sense)
Processes   Organizational processes
                  Production processes

Accidents happen at the interface of the different PEMEP elements. Hence, if we want to avoid accidents we need to manage the interrelationship between the different elements of the PEMEP matrix. The following cross-table shows how it works. For each combination the question to be asked is "how can the PEMEP element in the first column be influenced in order to avoid accidents with the PEMEP element in the upper row" (Table 3.19).

The answers to these questions lead to the different elements of a safety management system. The matrix (Table 3.20) is just an example which can be further completed.

The acronym PEMEP clarifies why a safety management system is very broad. It covers all subjects that need to be addressed in a safety management system. The PEMEP principles are not in the original work of Bird. They were developed by the author of this book with my colleagues, Ron Rhodes and Dirk Roosendans, while working for Atofina in the early 2000s.

The third acronym from ISMEC is used to decide which safety activities need to be developed. ISMEC stands for:
– **I**dentification of the work (activities) to be done
– Establishing **S**tandards for each of the activities (i.e., who will do what, when and how)
– **M**easure the performance of the work done
– **E**valuate the efficiency and effectiveness on a regular basis
– **C**orrect if needed

Figure 3.19 illustrates ISMEC as published in the trainings of Det Norske Veritas and the book of Frank Bird.

Building on the activities of the management function of control, "ISMEC" became widely known and used. The loss causation model showed how lack of control led to accident losses. ISMEC illustrated the steps toward gaining the desired control.

In 1974, Bird founded the International Loss Control Institute (ILCI) and became its executive director. The institute's development was sponsored by the Department of Insurance, School of Business Administration, Georgia State University in Atlanta, where he also became an adjunct professor. ILCI's headquarters was established in Loganville, GA.

Table 3.19: PEMEP principle.

	People	Equipment	Materials	Environment	Processes
P	How can *people* be influenced in order to avoid accidents that result from contact with other *people*?	How can *people* be influenced in order to avoid accidents that result from contact with *equipment*?	How can *people* be influenced in order to avoid accidents that result from contact with *materials*?	How can *people* be influenced in order to avoid accidents that result from contact with their *environment*?	How can *people* be influenced in order to avoid accidents that result from contact with *processes*?
E	How can *equipment* be influenced in order to avoid accidents that result from contact with *people*?	How can *equipment* be influenced in order to avoid accidents that result from contact with *equipment*?	How can *equipment* be influenced in order to avoid accidents that result from contact with *materials*?	How can *equipment* be influenced in order to avoid accidents that result from contact with the *environment*?	How can *equipment* be influenced in order to avoid accidents that result from contact with *processes*?
M	How can *materials* be influenced in order to avoid accidents that result from contact with *people*?	How can *materials* be influenced in order to avoid accidents that result from contact with *equipment*?	How can *materials* be influenced in order to avoid accidents that result from contact with *materials*?	How can *materials* be influenced in order to avoid accidents that result from contact with their *environment*?	How can *materials* be influenced in order to avoid accidents that result from contact with *processes*?
E	How can the *environment* be influenced in order to avoid accidents that result from contact with *people*?	How can the *environment* be influenced in order to avoid accidents that result from contact with *equipment*?	How can the *environment* be influenced in order to avoid accidents that result from contact with *materials*?	How can the *environment* be influenced in order to avoid accidents that result from contact with the *environment*?	How can the *environment* be influenced in order to avoid accidents that result from contact with *processes*?
P	How can *processes* be influenced in order to avoid accidents that result from contact with other *processes*?	How can *processes* be influenced in order to avoid accidents that result from contact with *equipment*?	How can *processes* be influenced in order to avoid accidents that result from contact with *materials*?	How can *processes* be influenced in order to avoid accidents that result from contact with their *environment*?	How can *processes* be influenced in order to avoid accidents that result from contact with *processes*?

Table 3.20: Examples of barriers according to PEMEP principle.

People	Equipment	Materials	Materials	Processes
P				
– Individual and group communication – Promotion campaigns	– Training on use of equipment – Work instruction – Safety rules – Training	– Product information – Personal protective equipment	– ISO 14001 – Awareness program – Information about hazards related to the work environment	– Return of experience – Hazard analysis – Training – Emergency Preparedness – Safe work procedures
E				
– Purchase rules – Inspection of equipment – Ergonomics – Equipment safeguards	– Spacing – Standardization of technologies – Vibration monitoring – Design	– Material selection – Corrosion monitoring – Inspection – Maintenance	– Inherent safe design – Best available technology – Emission monitoring	– Risk analysis – Reliability analysis – Process control system – Interlock
M				
– Product management – Labeling – Reduction of quantities – Replacement	– Selection of less aggressive products – Containment – Moderate (inhibitors)	– Separation (avoidance of accidental mixing) – Sampling and analysis	– Containment – Product stewardship – Waste management – Sustainable development	– Compatibility analysis (material vs. equipment) – Moderate process conditions
E				
– Industrial hygiene – Noise reduction – Emission reduction – Stress management	– Temperature control – Humidity control – Housekeeping	– Management of storage conditions	– Facility layout – Electrical area classification	– Housekeeping – Passive equipment protection (e.g., coating)
P				
– Inherent safe design – Spacing – Automatization	– Safe margins (design vs operational conditions)	– Engineering and design – Limit quantities	– Best available technology – Emission reduction program	– Reliability studies – Technical standards – Preventive maintenance

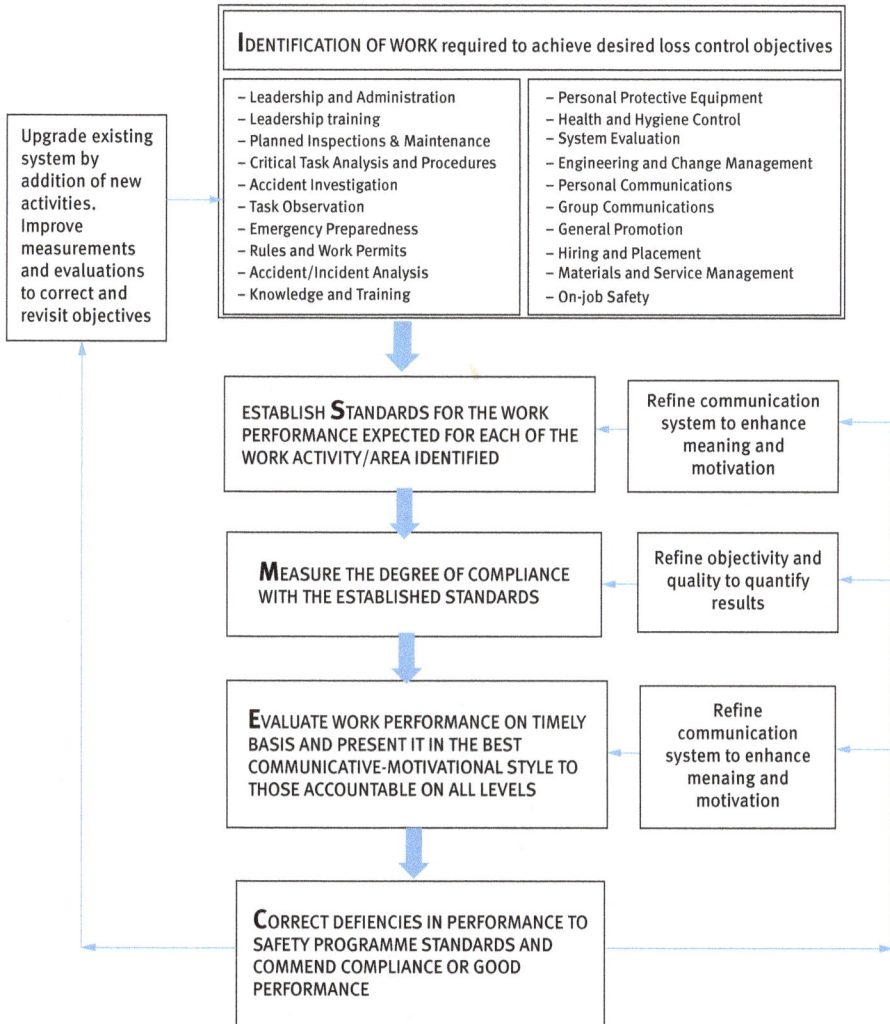

	IDENTIFICATION OF WORK required to achieve desired loss control objectives
Upgrade existing system by addition of new activities. Improve measurements and evaluations to correct and revisit objectives	– Leadership and Administration · – Personal Protective Equipment – Leadership training · – Health and Hygiene Control – Planned Inspections & Maintenance · – System Evaluation – Critical Task Analysis and Procedures · – Engineering and Change Management – Accident Investigation · – Personal Communications – Task Observation · – Group Communications – Emergency Preparedness · – General Promotion – Rules and Work Permits · – Hiring and Placement – Accident/Incident Analysis · – Materials and Service Management – Knowledge and Training · – On-job Safety

ESTABLISH **S**TANDARDS FOR THE WORK PERFORMANCE EXPECTED FOR EACH OF THE WORK ACTIVITY/AREA IDENTIFIED — Refine communication system to enhance meaning and motivation

MEASURE THE DEGREE OF COMPLIANCE WITH THE ESTABLISHED STANDARDS — Refine objectivity and quality to quantify results

EVALUATE WORK PERFORMANCE ON TIMELY BASIS AND PRESENT IT IN THE BEST COMMUNICATIVE-MOTIVATIONAL STYLE TO THOSE ACCOUNTABLE ON ALL LEVELS — Refine communication system to enhance menaing and motivation

CORRECT DEFIENCIES IN PERFORMANCE TO SAFETY PROGRAMME STANDARDS AND COMMEND COMPLIANCE OR GOOD PERFORMANCE

Figure 3.19: ISMEC principle to implement management control (DNV ISRS & Bird, 1985).

ILCI professionals continued to improve the old and create the new in terms of practical tools and techniques for controlling risks and managing loss control. Under Frank's direction and in collaboration with the Industrial Accident Prevention Association of Toronto, several influential texts emerged such as the *Professional Accident Investigation* series by Ray Kuhlman and the *Effective Series of Supervisory Training Manuals* by George Germain.

ILCI produced a great variety of printed and audiovisual materials to facilitate the implementation of risk management and loss control in organizations of all sizes, including manuals, books, training materials, leader's guides and performance measurement tools. The enormously influential practical loss control leadership first appeared in 1985 and quickly became a standard text in the field. In whole or in part, it has been translated into more than a dozen languages. These materials helped untold numbers of organizations to train and guide their employees in continual improvement of safety and loss control.

While working with the South African Chamber of Mines, Bird was intrigued by the potential of the group's basic Five Star Program for measuring and recognizing the level of safety performance of gold mines and related organizations. He expanded the conference material for the course he was conducting (The Consultant's Guide to Loss Control) and sent it back to Loganville for M. Douglas Clark to edit and develop into a generic safety audit system.

The resultant ISRS has been the basis of more safety management systems worldwide than any other document or strategy. It has been developed into specialized versions for industries ranging from maritime to mining and petrochemicals and customized to fit specific corporations.

3.9.7 Safety culture

The term "Safety Culture" which is commonly used in industry refers to "the company culture that delivers a certain safety performance". Hence "an excellent safety culture" means "a company culture that results in an excellent safety performance".

Before the 1980s, there was a lot of interest in "organizational culture" which was perceived as an instrument for improving companies. The term "safety culture" emerged in the wake of the Chernobyl disaster in 1986, and has been used ever since by numerous industries to describe the "security" status of a company (Tomei and Russo 2019).

On Sunday 26 April 1986 at 01:23, an accident occurred at unit 4 of the Chernobyl Nuclear Power Plant, near the city of Pripyat in the north of the Ukraine, which resulted in the destruction of the reactor core and part of the building in which it was housed. Large amounts of radioactive materials were released. The International Atomic Energy Agency (IAEA) and the Soviet Union agreed to hold a post-accident review meeting in Vienna. The meeting was held over the period 25–29 August 1986.

The International Nuclear Safety Advisory Group (INSAG) was requested by the IAEA Director General "to prepare on the basis of the information presented and the discussion, a summary report of the Meeting." This report was to be available for his consideration and transmission to the Board of Governors of the IAEA before its September 1986 meeting.

The INSAG report (IAEA, 1986) refers a couple of times to "nuclear safety culture." The general observations and conclusions in the executive summary state:

As described by the Soviet experts and discussed in detail among the experts the accident was caused by a remarkable range of human errors and violations of operating rules in combination with specific reactor features which compounded and amplified the effects of the errors and led to the reactivity excursion. A vital conclusion drawn from this behavior is the importance of placing complete authority and responsibility for the safety of the plant on a senior member of the operations staff of the plant. Of equal importance formal procedures must be properly reviewed and approved and must be supplemented by the creation and maintenance of a "nuclear safety culture". This is a reinforcement process which should be used in conjunction with the necessary disciplinary measures.

On page 69 of the reports, INSAG writes:

Since the immediate cause of the accident was successive violations of operating rules, disabling safety systems, it is necessary to understand how such violations were possible in practice. This requires an understanding of the existing plant procedures, organization and responsibilities, as well as the means used to violate the rules, especially the bypassing of safety systems. Discussion of this issue might lead to exchange of information on the various methods used:
a. For the establishment and testing of procedures; and
b. For enlisting the plant operator's support to established procedures

These experiences would be useful for all of us in understanding the human need for positive reinforcement of good performance in addition to the acknowledged need for a disciplinary regime. In each national culture the most successful methods might be quite different from the general norms.

INSAG observes that "there is a need for a nuclear safety culture in all operating nuclear power plants" (page 76 of INSAG-1 report) and recommendation 3 of Section VII of the report proposes that IAEA should devote special effort to create a "nuclear safety culture" in nuclear power plant operation.

Since 1986, the experts of INSAG continued to explore the topic of "safety culture":

– INSAG-4, *Safety Culture* published in 1991, was one of the first attempts to define what is meant by safety culture and to turn the concept into practical language.
– INSAG-13, *Management of Operational Safety in Nuclear Power Plants*, published in 1999, built on this by considering the organizational issues that underpin an excellent safety culture.
– INSAG-15, *Key Practical issues in strengthening safety culture*, published in 2002 further extends the INSAG-13.

Other publications from IAEA on safety culture include:

– IAEA put in place a framework for the Assessment of the Safety Culture in Organizations Team that issued guidelines in 1996 with key indicators for different areas that need to be considered when assessing safety culture.
– IAEA-TEC-DOC-1329, Safety Culture in Nuclear Installations: Guidance for use in the enhancement of Safety Culture in 2000?

Safety culture is INSAG-4 is defined as:

> The assembly of characteristics and attitudes in organizations and individuals which establishes that, as an overriding priority, nuclear plant safety issues receive the attention warranted by their significance.

There are in literature many definitions of safety culture. The Human Factors Study Group of Great Britain's Advisory Committee on the Safety of Nuclear Installations (ACSNI) in their third report defines safety culture as a subset of the overall company culture that influences the safety outcome of the safety management system (ACSNI, 1993):

> The safety culture of an organization is the product of individual and group values, attitudes, perceptions, competencies, and patterns of behavior that determine the commitment to, and the style and proficiency of, an organization's health and safety management. Organizations with a positive safety culture are characterized by communications founded on mutual trust, by shared perceptions of the importance of safety and by confidence in the efficacy of preventive measures.

The Human Factors Study Group of the ACSNI was set up to look at the part played by human factors in nuclear risk and its reduction. The third report of the study group ACSNI considers the role played by organizational factors and management in promoting nuclear safety. Organizational failures are recognized as being as important as mechanical failures or individual human errors in causing major accidents.

Over the last 30 years, an overwhelming number of articles and books have been published about the importance of the right safety culture. But despite all the efforts, people continue to struggle to agree on a common definition of "safety culture."

Human engineering at the request of HSE UK made a review of the literature surrounding safety culture published between 1986 and 2005 (Human Engineering, 2005). They conclude:

> The research has highlighted some confusion and inconsistency in the literature over the use of the terms "safety climate" and "safety culture". The review has provided a useful framework for approaching these terms. The term safety culture can be used to refer to the behavioral aspects (i.e. "what people do"), and the situational aspects of the company (i.e. "what the organization has").

> The term safety climate should be used to refer to psychological characteristics of employees (i.e. "how people feel"), corresponding to the values, attitudes, and perceptions of employees with regard to safety within an organization.

Human engineering added a new notion – "safety climate."

In the view of the author of this book, the notions of "safety culture" and "safety climate" complete the notion of "a safety management system" by measuring "how the people (management, the staff and the contractors) experience the usefulness of the safety management system."

Two companies can have exactly the same safety management system but in one company it is considered as an administrative burden to do the work while in the other company people adhere to it and they believe that the safety management system, including continuous improvement, is deemed useful in their daily work as well as in their strategy to avoid accidents. Both companies have the same safety management system but a different safety culture.

This also means that ALL companies have a safety culture irrespective of how we define it. Hence, the more important question is how a company can let its safety culture evolve so that it becomes more important. The Human Factors Study Group of Great Britain's ACSNI (1993) consider that the symptoms of poor cultural factor can include:

- widespread, routine procedural violations;
- failure to comply with the company's own SMS (although either of these can also be due to poor procedure design); and
- management decisions that appear consistently to put production or cost before safety.

ACSNI considers the following key indicators for a good safety culture: management commitment to health and safety, visible management (leading by example when it comes to health and safety), good communications between levels of employee and active employee participation. They developed a list of open questions to help organizations to provide a helpful picture of the overall style of the company (see Table 3.21).

Table 3.21: Open questions that help to measure the company's safety culture according to ACSNI (1993).

Key indicators	Some questions
Management commitment and visible management	– Where is safety perceived to be in management's priorities (senior/middle/first line)? – How do they show this? – How often are they seen in the workplace? – Do they talk about safety when in the workplace and is this visible to the workforce? – Do they "walk the talk"? – Do they deal quickly and effectively with safety issues raised? – What balance do their actions show between safety and production? – Are management trusted over safety?

Good communication	– Is there effective two-way communication about safety?
	– How often are safety issues discussed
	  – with line manager/subordinate?
	  – with colleagues?
	– What is communicated about the safety program of the company?
	– How open are people about safety?

Table 3.21 (continued)

Key indicators	Some questions
Employee involvement	– How are people (all levels, especially operators) involved in safety? – How often are individual employees asked for their input safety issues? – How often do operators report unsafe conditions or near misses etc.? – Is there active, structured operator involvement, for example, workshops, projects, safety circles? – Is there a continuous improvement/total quality approach? – Whose responsibility is safety regarded to be? – Is there genuine cooperation over safety – a joint effort between all in the company?
Training/ information	– Do employees feel confident that they have all the training that they need? – How accurate are employees' perceptions of hazards and risks? – How effective is safety training in meeting needs (including managers!)? – How are needs identified? – How easily available is safety information?
Motivation	– Do managers give feedback on safety performance (and how)? – Are they likely to notice unsafe acts? – Do managers (all levels – S/M/first) always confront unsafe acts? – How do they deal with them? – Do employees feel they can report unsafe acts? – How is discipline applied to safety? – What do people believe are the expectations of managers? – Do people feel that this is a good place to work (why/why not)? – Are they proud of their company?
Compliance with procedures	– What are written procedures used for? – What decides whether a particular task will be captured in a written procedure? – Are they read? – Are they helpful? – What other rules are there? – Are there too many procedures and rules? – How well are people trained in them? – Are they audited effectively? – Are they written by users? – Are they linked to risks?
Learning organization	– Does the company really learn from accident history, incident reporting and so on? – Do employees feel confident in reporting incidents or unsafe conditions? – Do they report them? – Do reports get acted upon? – Do they get feedback?

Pidgeon and O'Leary (2000) argue "a 'good' safety culture might reflect and be promoted by four factors":
- Senior management commitment to safety
- Realistic and flexible customs and practices for handling both well-defined and ill-defined hazards
- Continuous organizational learning through practices such as feedback systems, monitoring and analysis
- Care and concern for hazards shared across the workforce

Human Engineer (2005) who studied the evolution of safety culture between 1986 and 2005 consider that the five key indicators are:
- Leadership
- Two-way communication
- Employee involvement
- Learning culture
- Attitude toward blame

It is obvious that the same components (leadership, commitment, learning, etc.) are ever present. Some companies use the Dupont Bradley curve to estimate their safety culture as presented in Figure 3.20 (Hewitt, 2011).

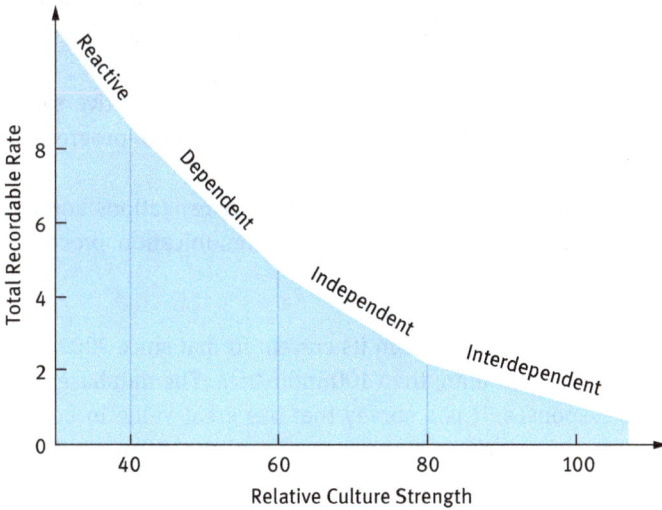

Figure 3.20: DuPont Bradley curve (Hewitt, 2011).

The chart shows four categories of companies in terms of recordable injury rate versus the relative culture strength (RCS) measured by DuPont's survey (the text below is copied from the article from Hewitt, Vice President, Global Workplace Safety Practices at DuPont Safety Resources, 2011):

1. **Reactive:** These companies handle safety issues by natural instinct, focusing on compliance instead of a solid safety culture. Responsibility is delegated to the Safety Manager, and there is generally lack of management involvement in safety issues
2. **Dependent:** While there is some management commitment, supervisors are generally responsible for safety control, emphasis, and goals. Attention to safety is made a condition of employment, but with an emphasis on fear and discipline, rules and procedures. Such companies do value all their people and will provide safety training
3. **Independent:** These companies stress personal knowledge of safety issues and methods as well as commitment, and standards. Safety management is internalized and stresses personal value and care of the individual. These companies engage in active safety practices and habits and recognize individual safety achievements.
4. **Interdependent:** These companies actively help others conform to safety initiatives – they become "other's keepers" in a sense. They contribute to a safety network and have strong sense of organizational pride in their safety endeavors

The Safety Perception Survey was first published in 1999 by Dr. James M. Stewart, safety management consultant and a former executive at DuPont, Canada. DuPont acquired the survey in 2000, modified it, incorporated it into the company's consulting methodology and turned it into a powerful tool that can be used for benchmarking and action planning. The survey is based on a number of questions related to:
- **Leadership.** The commitment and visibility of the management.
- **The structure.** The accountability of the line management, a supportive safety staff, an integrated committee, performance measurement and progressive motivation.
- **Processes and actions.** This element requires thorough investigations and follow-up, effective audits and reevaluation, effective communication processes and safety management skills.

The DuPont™ Safety Perception Survey exists in its current format since 2002, and it covers more than 50 countries and more than 100 industries. The database contains more than a million responses. It is a survey that has great value in benchmarking. It is a well-known safety culture measurement tool but there are many more available. Tomei and Russo (2019) give an overview of 20 Safety Culture Measurement Tools among which is the DuPont™ Safety Perception Survey.

The Bradley curve and the explanation that is used by the DuPont consultants make sense and have great value. However, it is important to realize that the curve is a construction and not a mathematical causal correlation between the measurement done via the method and the recordable injury rate. Bradley made a graph by setting out the measurement of the RCS with the OSHA Total Recordable Injury

Rates provided by the companies that were surveyed. He used an average of rates from the 3 consecutive years prior to the survey to represent the safety performance of an organization. When a survey was launched in the last 6 months of a year, that year's rate was included in the 3-year average. When a survey was launched in the first 6 months of a year, he used the prior 3 years of rate data (Hewitt, 2011). He obtained a graph shown in Figure 3.21.

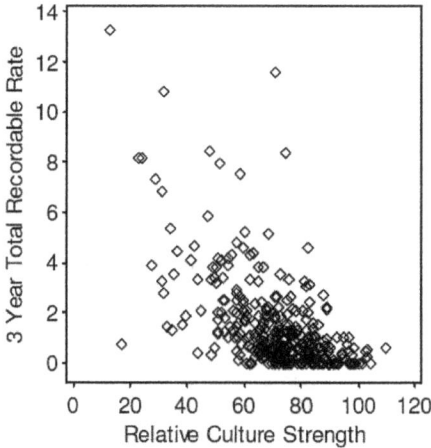

Figure 3.21: Relative culture strength versus OSHA 3-year total recordable rate (Hewitt, 2011).

The graph shows that for a RCS of less than 60 (categories of "reactive" or "dependent"), all types of TRIR performance are possible (from excellent to very bad). Between 60 and 80 (category of "independent") the center of gravity shifts to low TRIR numbers but there is no guarantee at all; some companies have a TRIR which is far above the companies which are in the "reactive" category. When the culture strength is above 80 (category of "interdependent"), the TRIR seems to correlate pretty well.

Westrum and Adamski (1999) considered that there are three types of safety culture that can be distinguished: pathological, bureaucratic and generative (see Table 3.22).

Professor Patrick Hudson from the Centre for Safety Research at the University in Leiden in conjunction with Shell extended the model of Westrum to a five-step evolutionary model of safety culture maturity model as shown in Figure 3.22.

A description of the levels is given in Table 3.23.

The determination of the level is done by asking all employees from all levels in the organization to fill in a questionnaire. Examples of the use of the Hudson model can be found in the open literature. Aleksandrova, for instance, describes how a survey was conducted in the mining industry facilities in the Irkutsk region (Aleksandrova and Timofeeva, 2020).

Table 3.22: Three types of safety cultures according to Westrum and Adamski (1999).

Subject	Safety culture		
	Pathological	**Bureaucratic**	**Generative**
Information	Hidden	Ignored	Actively sought
Messengers	Are "shot"	Are tolerated	Are trained
Responsibilities	Are shirked	Compartmented	shared
Bridging	Discouraged	Allowed	Shared
Failures	Covered up	Just and merciful	Causes enquiry

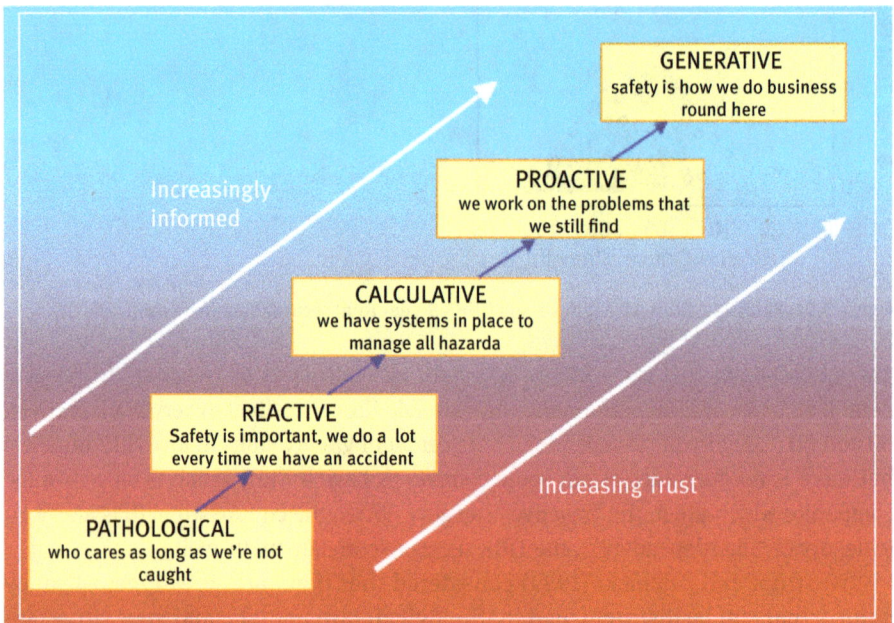

Figure 3.22: The evolutionary model of safety culture according to Hudson (2001).

Aleksandrova remarks, however, that:

It is necessary to understand that "safety culture" cannot be detached from "technological culture" and in general from "corporate culture". The systemic approach also includes the selection, training, adaptation, development of competencies, as well as the development of internal motivation of employees to meet safety requirements. In its turn, a safe workplace implies technologies, collective protection equipment, organization of safe work, preventive medical measures (examinations, vaccinations, screenings), encouraging employees to work safely, and only a full set of protection against risks will allow creating safety culture at an enterprise.

Table 3.23: Maturity levels in Hudson's safety culture model.

Pathological (nascent)	Labor Protection and Industrial Safety is a procedural decision; not a business risk. The safety department is primarily responsible for ensuring safety. Accidents are considered an inevitable part of the job.
Reactive (reactive, passive)	The accident rate of the organization corresponds to the industry average. Safety is a business risk, but is viewed in terms of rules and regulations. Management makes efforts to prevent accidents and is convinced that the cause of the majority of accidents is the unsafe behavior of line personnel. Managers of the organization are involved in the work on Labor Protection and Industrial Safety reacting to incidents and fines.
Calculative (system)	The accident rate is low, but significant progress is not observed. Management is convinced that the involvement of line personnel in Labor Protection is crucial for future improvements, but not for ongoing operations. Most of the personnel assume responsibility for their own safety and are willing to work with management on the improvement of Labor Protection and Industrial Safety. Management recognizes that many of the root causes of accidents often stem from its decisions. Mechanical systems are used to control hazards.
Proactive (interacting)	The majority of the personnel is convinced of the importance of Labor Protection and Industrial Safety. All employees are aware that accidents are caused by many factors, and their root causes often stem from wrong management decisions. Line personnel take responsibility for their own safety and the safety of others. Considerable efforts are made to prevent accidents by means of preventive measures. Labor Protection and Industrial Safety indicators are continuously monitored using all available data.
Generative (productive, continuously improving)	Preventing damage to the health of employees is a core corporate value. Accidents have not occurred for a long time, but employees are aware that they can happen at any time. The organization is constantly striving for improvements and is looking for optimal ways to improve the mechanisms for controlling hazards. Efforts are made to ensure safety of workers outside the workplace.

3.10 Some final thoughts about the evolution of OHS since 1900

3.10.1 Short resume of the OHS approach since 1900

"OHS" has been considered of utmost importance since the beginning of the industrial revolution. It has been a concern from all points of view: ethical, business, economics, costs and so on.

Since the beginning of the industrial revolution (end of eighteenth century), the "human element" was central to the debate about performance in general and OHS performance in particular. It was obvious that "human performance" was crucial to achieve whatever objective (production, availability, safety). It took, however, a long time during the nineteenth century to gain insight into the underlying reasons for occupational accidents. In the first half of the nineteenth century, liberal ideology with great individual freedom was the predominant social train of thought, preventing the intervention of the state in the employer–worker relationship. All types of theories emerged (accident proneness, free agent doctrine, motivation of people, just culture, etc.). These are still under study and discussion.

Safeguarding machines to avoid accidents was also part of mainstream thought since the middle of the nineteenth century.

Extreme poverty among the working class sparked a socialist movement in the second half of the nineteenth century, so that much more attention was paid to the conditions of the workers.

From the 1900s onward, we have seen more and more studies on the causes of accidents and the measures to be taken to make the workplace safe. Very quickly it became clear that prevention pays off politically and that it is based on three main pillars: human performance, technical measures and organizational measures.

It is tragic and cynical, but WWI provided a great deal of insight into accident prevention. The weapon factories during WWI had to run at full regime and a lot of study was done on the prevention of incidents and the maximization of the output. As a result of the insights gained in the factories during WWI, the organization of labor was considered as a major topic by the Allied Powers when they established the Treaty of Peace of Versailles in 1919. Part XIII of the Treaty is dedicated to the topic and puts in place "The International Labour Office" at the seat of the League of Nations as part of the organization of the League.

By the 1920s, there was a general consensus on the importance of management and that OHS had to be managed in a similar manner as all other business topics such as production. In the meantime, different activities were developed to improve safety at the workplace (safety committees, inspections, JSA, etc.). The term "Inherent safe design" was not used but the philosophy was in place in the 1920s. Technical standardization organizations emerged all over the world in the 1920s.

Until the 1950s, the safety performance was achieved via technical measures, organizational measures and studies about the human performance.

From the 1950s onward, the notion of safety management and safety management systems appeared. The new element was that all safety activities were imbedded in a common management approach.

From 1986 (following the Chernobyl accident), the notion of safety culture was introduced.

Over the past 20 years, old recipes have been given a new look and too much attention has been devoted to making existing recipes complicated.

3.10.2 Quantitative decrease in accidents in the twentieth century

There is no doubt that the number of work-related accidents decreased significantly over the twentieth century and in particular in the second half of the twentieth century. Table 3.24 gives an idea on the number of fatalities per 100,000 workers.

Table 3.24: Evolution of the number of fatalities per 100,000 workers.

Year	Fatalities per 100,000 workers	Reference
1898	16	Calder (1898) based on a population of about 4.4 million workers. The number of accidents in this period is without doubt underreported.
1931	30	Vernon (1936) reported accidents compensated in 1931 in the UK under the Workmen's Compensation Act. There were 2,295 fatalities and 377,376 nonfatal injuries in a working population of 6,913,974 workers. The number of fatalities per 1,000 workers varied between 0.1 (in factories) and 1.2 (in mines) depending on the occupation.
1941	36	Blake (1943) reports 18,000 deaths for a total average employment of 49,640,000.
1974	2.9	For modern times, the UK will often start with 1974 as reference. In 1974, the Health and Safety at Work etc. Act 1974 came into force and it marked a departure from the framework of prescribed and detailed regulations that was in place at the time. The act introduced a new system based on less prescriptive and more goal-based regulations, supported by guidance and codes of practice. For the first time, employers and employees were to be consulted and engaged in the process of designing a modern health and safety system.
1990	1.9	UK Industry. https://www.hse.gov.uk/statistics/fatals.htm
2000	1.1	UK Industry. https://www.hse.gov.uk/statistics/fatals.htm
2010	0.6	UK Industry. https://www.hse.gov.uk/statistics/fatals.htm
2018	0.5	UK Industry. https://www.hse.gov.uk/statistics/fatals.htm

The number of fatalities per 100,000 workers in the UK industry decreased by a factor of more than 50 over the last 80 years. Major progress was made in the second half of the twentieth century.

Progress has also been observed in other Western Industrialized countries. However, this progress is the same for all countries and it varies between industries. The number of occupational fatal accidents in the UK is among the lowest in the world.

Hämäläinen et al. (2006) studied occupational accidents all over the world for the year 1998. The number of workers (total employment equivalent) in 1998 was estimated as 2,164,739,590. The number of occupational fatalities was 345,719, which corresponds to 16 fatalities per 100,000 workers. Table 3.25 compares a number of regions of the world.

Table 3.25: Comparison on occupational fatalities between different regions in the world in 1998 (Hämäläinen et al., 2006).

Region/country	No. of workers	Percentage of global workforce	No. of occupational fatalities	Percentage of total no. of occupational fatalities	No. of occupational fatalities per 100,000 workers
China	699,771,000	32.3	73,615	21.3	10.5
India	419,560,000	19.4	48,176	13.9	11.5
Other Asia and Islands	328,673,800	15.2	83,084	24.0	25.3
EU	203,877,354	9.4	11,671	3.4	5.7
USA	131,463,000	6.1	6,821	2.0	5.2
Japan	65,140,000	3.0	2,077	0.6	3.2

Large multinational companies have a global approach to OHS. These corporations are operating in all parts of the world and they establish a common safety policy. It seems that company culture is much stronger than country culture in OHS. The author studied detailed data of a number of Western major oil companies. Over the last 10 years (2008–2018), these companies had a total of 8,962 lost time accidents for an equivalent of about 13 million workers. The number of fatalities per 100,000 workers was 2.6.

Within one region, there is a difference between different countries. Figure 3.23 compares the 28 member states of the European Union.

Another way of demonstrating the improvement in safety performance is by comparing, for instance, the no-injury records.

Table 3.26 compares the best in class (expressed in number of man-hours with no injury) in American industry for a number of sectors between 1942 and 1968.
The record of no-injury man-hours in the American Chemical Industry increased from about 11.4 million hours in 1943 to about 45.8 million hours in 1968.

3.10.3 New victims but same accidents

A number of studies conducted and books published since the 1900s (e.g., Calder, 1899; Pittsburgh survey, 1910; Beyer, 1916; Blake, 1943; Heinrich and Granniss, 1959; Bird 1960) laid the basis for what is called today modern safety management. In addition to the efforts of individual companies, much progress has also been made under regulatory pressure.

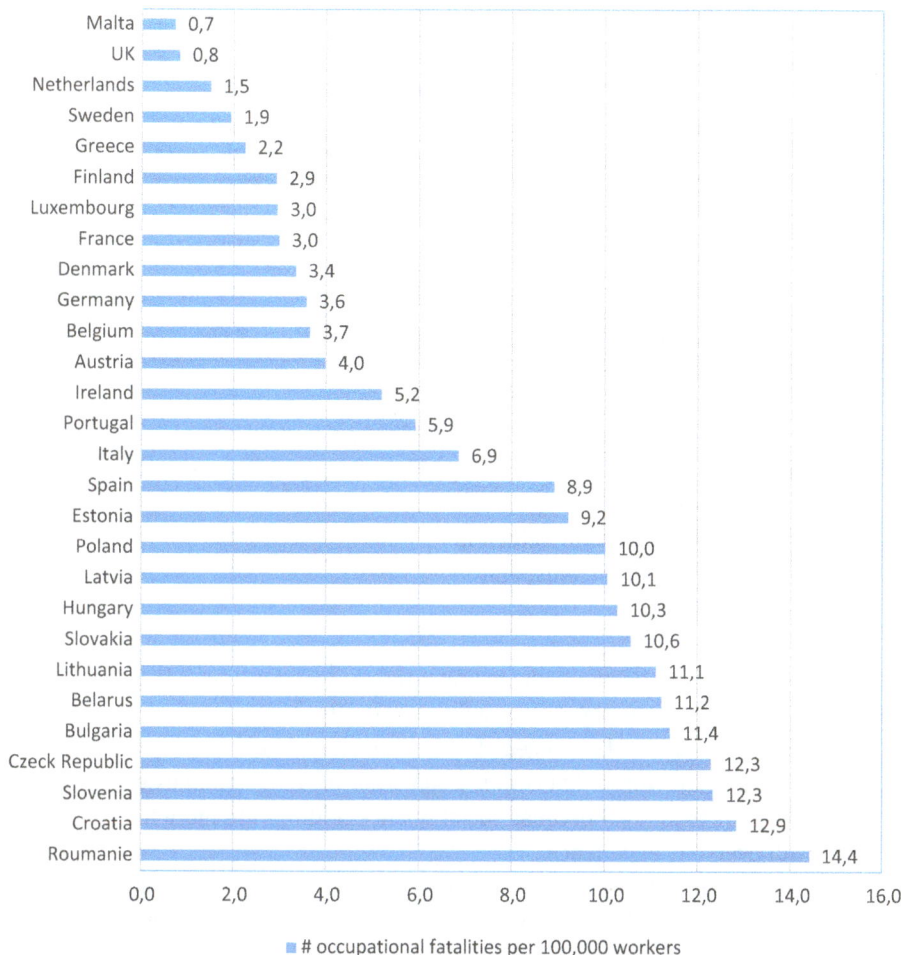

Figure 3.23: Number of occupational fatalities per 100,000 workers in the member states of the EU in 1998.

Table 3.26: Evolution of no-injury record in man-hours in American industry between 1943 and 1968.

Industry	No-injury record in man-hours in American industry		
	Blake (1943)	**Heinrich and Granniss (1959)**	**Hearings health subcommittee 1968**
Chemical	11,361,846	28,743,768	45,808,775
Machinery	11,114,600	17,604,263	17,604,263
Automobile	5,808,921	15,071,411	17,496,659

Table 3.26 (continued)

Industry	No-injury record in man-hours in American industry		
	Blake (1943)	Heinrich and Granniss (1959)	Hearings health subcommittee 1968
Rubber	5,688,369	10,250,040	15,966,227
Steel	5,325,144	8,241,906	17,133,243
Foundry	2,034,419	7,407,010	9,634,319
Mining	1,243,854	2,875,085	8,664,118

But time and time again, there are casualties in nearly identical accidents with the same causes. The main causes of occupational fatal accidents are still
- falls (mainly from height);
- struck by a moving object (or vehicle);
- contact with energized systems; and
- caught in machinery.

All these accidents can be avoided. One may rightly wonder how it is possible that so many accidents still happen. Some main reasons are:
1. Most of the workers and managers do not have enough scholar knowledge about OHS. They are convinced that OHS is a *soft science*.[1] To a certain extent, this conviction can be explained because social science findings are more likely to intersect with everyday experience.
2. Professional OHS people are hired but the line management who rightly has the final and full responsibility, will ultimately decide considering other constraints. The advice of the professional OHS people will be one of the many topics that will be considered in his or her decision. Given the previous point, this can be problematic in situations when the advice of the OHS professional is part of the "hard science" part of the OHS knowledge domain. Take the following example: since more than 100 years there is no doubt that the quality of a JSA prior to starting the job is an indispensable and essential action to ensure that the job can be done in a safe way. A JSA, therefore, has to be considered as hard science in the field of OHS. In practice, however, workers will consider the time they spent in a

---

1 Hard science and soft science are colloquial terms that were used to compare scientific fields on the basis of perceived methodological rigor, exactitude and objectivity. Roughly speaking, the natural sciences (physics, chemistry, biology, astronomy and geoscience) are considered "hard," whereas the social sciences (e.g., archaeology, cultural anthropology, psychology, sociology and economics) are usually described as "soft."

JSA a waste. From their own experience, they have done the job many times and they do not always feel the need for a JSA.

3. The lack of scholar knowledge at the level of middle and upper managers in the field of OHS is used by consultants who recycle old ideas and present them as a sort of *Deus Ex Machina*, that is, a seemingly unsolvable problem of the recurrence of accidents will suddenly and abruptly be resolved by an unexpected and unlikely occurrence. The result is that some managers are not aware of the full extent of OHS strive for simple solutions (e.g., "let's implement a just culture and all problems will be solved" and "let's give special training for supervisors"). Unfortunately, OHS problems, in general, are *simple problems within a complex organization.*

3.10.4 Fundamentals for occupational health and safety

Most, if not all, of the theory needed to avoid OHS accidents have been developed in the first half of the twentieth century. Additional interesting insights were developed in the period 1950–2000.

Since then, the industry is flooded by consultants and academics who produce (sometimes complicated) publications and services, which in fact are nothing more than a re-explanation of these existing old theories. This consultancy activity is valuable because it continues to put the focus on the importance of OHS. I can only encourage consultants to continue to push industry.

However, the main challenge is that consultants as well as industry should know the fundamentals and should avoid getting lost in theoretical considerations.

Blake (1943) lists a number of fundamentals that should be in the mind of all workers:

1. Accidents can be prevented.
2. Prevention is not difficult. It is primarily a matter of the unremitting application of common sense.
3. Accidents are caused, they do not simply happen.
4. It is important to discover the cause (or causes) of each accident, then to eliminate those causes so that similar accidents will not recur.
5. Causes of accidents are more important than causes of injuries.
6. In general, the great majority of accidents in all firms are due to similar causes regardless of size of the establishment or type of industry.
7. Safety in any firm is not a one-man job. Accidents cannot be eliminated unless every man in the organization from the top executive down to the most lowly laborer accepts his/her share of the responsibility.
8. Many of the methods used to control quantity, quality and cost of production can be used with equal effectiveness to control accidents.

There is one very important fundamental that needs to be added in order to avoid accidents:

9. OHS is a combination of hard science and soft science. Leadership and management skills are important but the experience and know-how to safely execute a job is fundamental.

The following example illustrates the point: a company had a number of fatalities with working at height. The management of the company decided that for a large number of activities people would be obliged to wear a safety harness and that when working at height people had to be hooked up. At first sight, the decision seemed "easy" to implement, namely:

(1) defining at different locations anchor points;
(2) installing lifelines (anchor lines);
(3) purchasing antifall harness; and
(4) obliging the workers to wear the harness and to be hooked up when working at height.

When the company started to roll out the program, managers only started to realize how much knowledge (hard science) was needed with regard to working at height. Just a few examples to clarify:

(1) The general characteristics and assembly of personal fall protection systems is given in the European Standard EN 363. This standard exists since 1993 but the latest revision is dated December 2018. The previous revisions (1993, 2002 and 2008) have been aborted.
(2) The technical requirements placed on anchor devices are defined in standards such as EN 795. This standard sets out the loads that horizontal fall arrest solutions must withstand and how they are to be tested and certified. EN 795 distinguishes five types of anchorage device, from A to E but does not consider multiuser applications. For anchor devices that allow more than one user to be attached at the same time, the standard CEN/TS16415 (2013) can be used.
(3) Standard EN 353 relates to guided-type fall arresters including a rigid anchor line commonly found on fixed ladders and other vertical structures. In 2004, however, the UK's HSE issued a safety warning regarding EN 353–1:2002. Information received by the HSE from various committees, manufacturers and end users raised concerns with the use of such products following reported cases of serious injuries and some fatalities. These resulted from the user falling backward, thus not engaging the locking device due to a force perpendicular to the line. Since then, EN 353–1:2002 has been withdrawn on safety grounds and replaced by EN 353–1:2014. This introduces various locking tests, including a fall-back test as well as other function tests to take into account any foreseeable similar occurrences.

(4) The full body harnesses (personal protective equipment) have to comply with EN 361:2002. This standard specifies dynamic performance, static strength and corrosion resistance. For static strength, for instance, harnesses are subjected to a 15-kN tensile force applied upward followed by a 10-kN force applied downward. Lanyards are subjected to either 22 or 15 kN applied between the attachment points, depending on the material used. The application of an increased force tests the effects of aging (e.g., abrasion, wear and tear) on the lanyard's protective capacity. Tensile forces are usually applied and held for at least 3 min to ensure the breaking strength of the product exceeds the force specified by the standard. In preparation for dynamic performance testing, the harness is fitted to a 100-kg solid torso dummy and connected by the harness attachment to a 2-m length of 11 mm mountaineering rope, chosen specifically to generate a known force in the harness in the event of a fall. This rope is attached to a secure anchor point and the dummy is released over a distance of 4 m. This test is conducted twice on each harness attachment, once from a head-up position and once from a head-down position (i.e., with the dummy upside down on release). To pass the test, the harness must hold the dummy following both drops, with the dummy held in a position that does not exceed 50° from the upright position. These tests are conducted on each of the harness' attachment points.

(5) The technical requirements for the energy absorbers are specified in EN 355.

The example demonstrates that the hard science is an important part of OHS. Too many people in industry still ignore the hard science that is behind every single detail needed to ensure zero accidents.

On 27 September 2019, workers were preparing a roof of a warehouse in a petrochemical plant in South Korea. A worker fell through a hole in the roof and he was hit by a stacker crane. The worker was wearing a safety harness and he was hooked up to an anchor point via a rope. The worker was killed because he was hit by a stacker crane during his fall. However, the investigation revealed that the pendulum effect was not taken into account and that the different fall factors (fall clearance calculations, length of the connector, etc.) were not well assessed. So even if the worker would not have been hit by the passing stacker crane, he still would have been badly injured in case of a fall.

If zero accidents is really the objective, then it will be necessary to ensure that (1) the hard science of OHS is available in the company (e.g., in technical standards), (2) the workforce is trained and experienced in the use of the hard science and (3) leadership and supervision guarantee that the rules are implemented and followed.

3.10.5 Occupational health and safety is a result-oriented activity

"OHS" is a part of labor law which regulates the relationship between workers, employing entities, trade unions and the government. They work together in order to create the conditions so that the work can be done in safe conditions. The approach is result oriented, which means performing tasks without accidents. In case of an uncertainty about the safe conditions, the work has to be stopped or even not started ("zero risk concept"). Rules have been developed over more than 200 years and are internationally driven via the United Nations. The rules are important from an ethical point of view but they are also necessary to ensure welfare, social justice and fair international competition.

In the USA, for instance, CFR (United States Code of Federal Regulations) Title 29, labor is 1 of 50 titles comprising CFR, containing the principal set of rules and regulations issued by federal agencies regarding labor. The regulations relating to OSH are in Chapter XVII of this Title 29.

OHS is a result-oriented obligation for which the only acceptable result is zero harm. Many companies monitor their performance in the field of OHS with lagging indicators (recordable accidents, first aid cases, etc.). There is nothing wrong with monitoring lagging statistics as long as everybody understands that *injuries, no matter how minor, are simply unacceptable*. A "minor" injury is in most cases "minor" for the people who were not injured. For the injured person the perception of "minor" is often different.

The legislation in all countries of the world (except the UK who accept the notion of risk in OHS) is very clear on that. The legislation in France, for instance, stipulates:

> L'employeur est tenu par la loi de prendre toutes les mesures nécessaires pour assurer la sécurité et protéger la santé physique et mentale de ses salariés (article L. 4121–1 du Code du travail). L'employeur ne doit pas seulement diminuer le risque, mais l'empêcher. Cette obligation est une obligation de résultat (Cour de cassation, chambre sociale, 22 février 2002, pourvoi nr. 99–18389), c'est-à-dire qu'en cas d'accident ou de maladie liée aux conditions de travail, la responsabilité de l'employeur pourra être engagée.

> Translation: The employer is required by law to take all necessary measures to ensure the safety and protect the physical and mental health of his employees (article L. 4121–1 of the Labor Code). The employer must not only reduce the risk, but prevent it. This obligation is an obligation of result (Cour de cassation, social chamber, February 22, 2002, appeal nr. 99–18389), that is to say in the event of an accident or illness linked to working conditions, the responsibility of the employer may be engaged.

The Court of Cassation in France stated in 2002 that the law obliges the "employer to prevent the risk" which is not the same as "preventing the accident or preventing the harm in case of an accident."

The European Council Directive 89/391/EEC of 12 June 1989 on the introduction of measures to encourage improvements in the safety and health of workers at work states in its article 6 point 1 that "the employer shall take the measures necessary for the safety and health protection of workers, including prevention of occupational risks and provision of information and training, as well as provision of the necessary organization and means." The directive clearly mentions "the prevention of occupational risks" which is not the same as "the prevention of occupational accidents." In article 6.2, it mentions that risks that cannot be avoided shall be evaluated. This evaluation, however, will be used "to assure an improvement in the level of protection afforded to workers with regard to safety and health" (article 6 point 3).

It is obvious that the interpretation of the notion of risk is important. When risk is defined as "a probability of occurrence of an event" it will be difficult for the employer to ensure that the probability can be prevented. If the meaning is about "uncertainties," then the interpretation is that employer should define the measures to be taken to ensure that there are no uncertainties associated with the execution of the task.

The notion of "risk" which will be very important in the chapters dealing with technological risks or process safety is less relevant in the field of OHS. The agreement between an employer and a worker as part of labor law is that the job has to be done without any harm to the worker. The employer and the employee have to prepare the task and they have to put in place the measures and conditions so that the task can be executed with no negative impact on the health of the worker. In case of uncertainty that the work can be done safely, the work should be stopped or even not started. "Uncertainty" should not be confused in this context with "unknown."

3.10.6 Occupational health and safety in the twenty-first century: mental health and well-being

The main challenges in the field of OH S in the Western world for the twenty-first century will probably be physical fitness of the workers and mental health of the workers.

The World Health Organization (WHO) defines mental health as a state of well-being in which every individual realizes his or her own potential, can cope with the normal stresses of life, can work productively and fruitfully and is able to make a contribution to her or his community.

Anja Baumann and Matt Muijen of the WHO Regional Office for Europe write it as follows:

> Good health contributes to quality and productivity at work, which in turn promotes economic growth and employment and the ability to invest in good employment practices. Numerous measures have been shown to be effective in promoting mental health and well-being, preventing and managing mental illness and helping reintegrate people into work – including effective programs to tackle stigma and discrimination. (WHO, 2010)

Work contributes to the personal fulfilment, the financial and social prosperity, and is the most effective way to improve the well-being of individuals, their families and their communities (Waddell and Burton, 2006). Returning to work help people to recover from mental health problems. There is now substantial scientific evidence that a "bad job" or "work-related factors" (psychological hazards or organizational stressors) can negatively affect both mental and physical health with associated negative effect on the human and business performance. Promoting mental health at work has become a vital response to these challenges (European Framework for Action on Mental Health and Well-being, 2016).

The manager will have to become a manager-coach and invest his/her time in the factors that influence the mental health of the workers. Companies will need to set up programs to help the worker in improving his/her physical health.

4 Technological risks

4.1 Introduction of new technologies

The industrial revolution was accompanied by a rapid scientific evolution and the introduction of new technologies. These new technologies were not always fully understood and a new type of accidents occurred. These accidents were the result of lack of knowledge in combination with an increasing complexity of the technology.

New technologies introduced new type of accidents. The environment was confronted with high consequences accidents that were comparable to natural disasters. These accidents were different from the known occupational health and safety accidents for a number of reasons:
- The frequency of occurrence was relatively low.
- The consequences were beyond imagination.
- It was difficult to blame the organization (at least in the beginning) because the accident was considered as unforeseeable.
- The events were "complex" in a way that they included multiple interrelated conditions, events and actions, including latent failures.

In the next paragraph, a number of technologies and associated major risks will be discussed. The notion "risk," as will be explained later, refers to "uncertainty."

4.2 Discovery of major risks related to new technologies or products

4.2.1 What are major risks?

The notion of "risk" is used by people in different contexts. Some examples:

Expression	Discussion
"I would not buy these shares because the *risk* that you will lose your money is very high"	The word risk is mainly used here to refer to the likelihood that the person will lose his money. The maximum potential consequence is known but it is not sure whether you will lose the money or not.
"The nuclear industry is a high *risk* industry"	The word risk is mainly referring to the potential consequences. The nuclear industry is considered by some people as a "high risk industry" irrespective of the likelihood of an accident. There is uncertainty whether an accident will or will not happen but the term "high risk" in this phrase is only referring to the potential consequence.

https://doi.org/10.1515/9783110632132-004

(continued)

Expression	Discussion
"giving up your job before you have signed a new contract is a *risk*"	The word risk is referring to the uncertainty of having a job. It is possible that after having resigned, the situation change and you don't get the new contract.

The term "risk" is a concept invented by human beings to discuss about "uncertainties." When there is no uncertainty, then there is no risk. The degree of risk (minor, major) will, however, refer to the potential outcome and not to the degree of uncertainty. Hence, when people talk about a "major risk," they normally refer to the uncertainty about an event with a potential major impact.

When the "possibility of a major undesired event" cannot be excluded, humans will consider that there is a major risk.

If the distinction between reality and possibility is accepted, the term risk denotes the possibility that an undesirable state of reality (adverse effects) may occur as a result of natural events or human activities (Hunter et al. 1994, NRC 1983, Luhmann 1990).

The distinction between reality and possibility is fundamental in risk analysis. In a fatalistic view of nature and society (*accidents are considered as inevitable*), there is no need for risk analysis other than to please one's curiosity (Renn, 1992).

Three situations can be distinguished:

A The undesired event (fire, explosion, etc.) is impossible. In this case, risk doesn't exist.

B There is no doubt that the undesired event (fire, explosion, etc.) will happen. We don't know exactly when it will happen but there is absolute certainty it will happen. In this case also, risk doesn't exist. The activities need to be stopped.

C The undesired event or accident cannot be excluded but, it is also possible that the accident will never occur.

Risk only exists in the third situation. If we decide to continue the activity, it is possible that we will have all the benefits and that the undesired events don't occur. Hence, there is a good argument to continue the activity.

But, whenever the undesired event or accident occurs, it is for sure that we will regret our decision that we continued the activity.

The most relaxing situations are situation A or B, because the decision ("to stop the activity") is straightforward and simple.

It is important to understand that the notion of risk is characteristic for situation C, which is a situation where *nobody knows whether it is better to continue the activity for its benefits or whether it is better to stop the activity and thus avoid the accident*. Once the accident occurs, the concept of risk doesn't exist anymore.

4.2.2 Discovery of major risks with steam boilers

The idea to use steam as working fluid to perform mechanical work was known since the first century. An aeolipile, a simple, bladeless radial steam turbine which spins when the central water container is heated, was mentioned by Marcus Vitruvius Pollio (80–70 BC – after c. 15 BC), in his work De Architecture.

From the late seventeenth century onwards, a number of pioneers sought to harness the power of steam in engines capable of pumping out water from (coal) mines. Denis Papin (1647–1713), a French physicist and mathematician built in 1679, the steam digester, a closed vessel with a safety valve that can be tightly closed by a screw and a lid. The steam is confined until a high pressure is generated. The internal temperature and pressure will increase due to the heating until the pressure exceeds a defined limit, which will force the safety valve to open.

The first steam-powered vehicle was supposedly built in 1679 by Ferdinand Verbiest, a Flemish Jesuit in China. The vehicle was a toy for the Chinese Emperor. While not intended to carry passengers and therefore not exactly a car but a carriage, Verbiest's device is likely to be the first ever engine powered vehicle. Also it seems that the Belgian vehicle served as an inspiration for the Italian Grimaldi (early 1700) and the French Nolet (1748) steam carriage successor.

Thomas Savery (1650–1715), an English inventor and engineer, patented in 1698, what he called the "Miner's Friend" and invented the first commercial-steam powered device.

Thomas Newcomen (1664–1729) combined the practice of Savery with the theory of Papin and developed in 1712, the first commercially successful engine that could transmit continuous power to a machine (Forty, 2019). The system worked as follows:
1. Water is heated and transformed in steam.
2. A steam inlet valve opens and lets the steam in a cylinder after which the valve is closed.
3. Water is injected in the cylinder, which cools the steam that transforms in water again.
4. The pressure in the cylinder drops close to vacuum.
5. The piston is pushed downward by the atmospheric pressure into the cylinder.
6. The steam inlet valve opens again, which pushed the piston upward (not clear without a drawing) and that takes us back to step 3 above.

The importance of Newcomen's development is reflected in some 100 machines that he installed.

James Watt (1736–1819) was a Scottish instrument maker. He was asked to repair a Newcomen machine when he realized that a lot of energy was lost because of the thermal cycle of the piston (heating and cooling). He developed, between 1763 and 1775, a solution:

- A combined furnace and boiler that produces the steam.
- The steam is sent to the cylinder via a steam inlet pipe.
- The steam enters at each end of the cylinder alternatively forcing the piston up or down.
- The steam is sent to a cold water tank in which it condenses and from which it is sent back to the boiler.

In 1769, he patented this innovative idea of a double action engine that produced power with both the up and down strokes. One problem was, however, that James Watt could not find anyone to accurately bore the cylinders for his engine designs because it was impossible to seal the pistons effectively. In 1774, when John Wilkinson (1728–1808) invented a boring machine that was capable of working to far closer tolerances, it became possible for James Watt in partnership with Matthew Boulton to build for Samuel Whitbread in 1784, the Whitbread Engine, the first rotative beam engine in which reciprocating motion of the beam is converted in rotary motion to produce a continuous power source.

By the nineteenth century, stationary steam engines powered the factories, ships and locomotives.

The hazards associated with this new technology, however, were not known and the society discovered them via a number of major accidents. Some of these major accidents are:

- On July 31, 1815, the Philadelphia train, in Philadelphia, County Durham, England, suffered a boiler explosion. This first recorded boiler explosion on a train caused major loss of life, as 16 people were killed.
- In 1837, the packet steamer Union exploded in the Humber Dock basin (Hull, UK), resulting in the death of over 20 people including bystanders on the dock side, and a large number of injuries, the vessel itself being sunk by the explosion.
- On April 25, 1838, the Moselle, one of the fastest river boats suffered a boiler explosion just east of Cincinnati, killing 160 of the estimated 280–300 passengers. The Moselle was constructed between December 1, 1837 and March 31, 1838. The boat was brand new.
- On June 14, 1838, a boiler exploded on the Steamship Pulaski, causing massive damage. The ship sank in 45 minutes 30 miles off the coast of North Carolina, resulting in the loss of life of 128.
- On June 13, 1858, the Pennsylvania was steaming near Ship Island, just below Memphis Tennessee when its boiler exploded. Estimates at the time put the passenger manifest at 450 with an initial loss of life of 250.
- Sultana was a Mississippi River side-wheel steamboat constructed in 1863 and intended for the lower Mississippi cotton trade. On April 27, 1865, three of the boat's four boilers exploded. The ship was carrying 2,137 passengers and 1,800 of them died. It was the worst maritime disaster in the US history.

Sir William Fairbairn, first Baronet of Ardwick (February 19, 1789–August 18, 1874) was a Scottish civil engineer, structural engineer and shipbuilder. Fairbairn was one of the first engineers to conduct systematic investigations of failures of structures, including the collapse of textile mills and boiler explosions. He became the third president of the institution of mechanical engineers (established in 1847 by George Stephenson, the "father of railways").

He also conducted experiments on pressurized cylinders of glass and was able to show that the highest stress in the wall occurs around the diameter. It is known as the hoop stress and is twice the value of the longitudinal stress, which occurs along the length of the cylinder. The precise value depends only on the wall thickness and the internal pressure. Fracture is governed by the hoop stress in the absence of other external loads since it is the largest principal stress. The first theoretical analysis of the stress in cylinders was developed by the mid-nineteenth-century engineer William Fairbairn, assisted by his mathematical analyst Eaton Hodgkinson. Their first interest was in studying the design and failures of steam boilers. Fairbairn realized that the hoop stress was twice the longitudinal stress, an important factor in the assembly of boiler shells from rolled sheets joined by riveting.

It is important to understand the difference between accidents with steam boilers and work place accidents such as falls from height or slip and trip accidents. In the latter type of accidents, it is not difficult to identify measures, which make the occurrence of the accident simply impossible to occur. In the case of steam boilers, it took a while to understand why accidents happened. It took more than one hundred years between the first idea of Papin (the steam digester in 1679) and the first Whitbread Engine (Watt Boulton engine in 1784). It took another 100 years and several disasters before Sir Fairbairn in 1861, at the request of the UK Parliament, conducted research into metal fatigue. The knowledge of how hoop stress increased with diameter and how stresses were independent of drum length was not known before and led to the Fairbairn-Beeley five tube stationary boilers.

The underlying cause of these accidents is the lack of knowledge.

4.2.3 Discovery of major risks with nuclear energy

The text below is based on Wikipedia, the report of The President's Commission on The Accident at Three Mile Island (Kemeny, 1979) and report of the Nuclear Regulatory Commission Special Inquiry Group (Rogovin, 1980)

On 28th of March 1979, there was a partial meltdown of reactor number 2 of the Three Mile Island Nuclear Generating Station (TMI-2) near Harrisburg in Pennsylvania (USA) (see Figure 4.1 for the principle of TMI-2). The accident was rated as 5 on the 7-point International Nuclear and Radiological Event Scale (INES) because of the limited release of radioactive material. This accident can however, be seen as the worst that ever happened with a commercial nuclear reactor because it was a

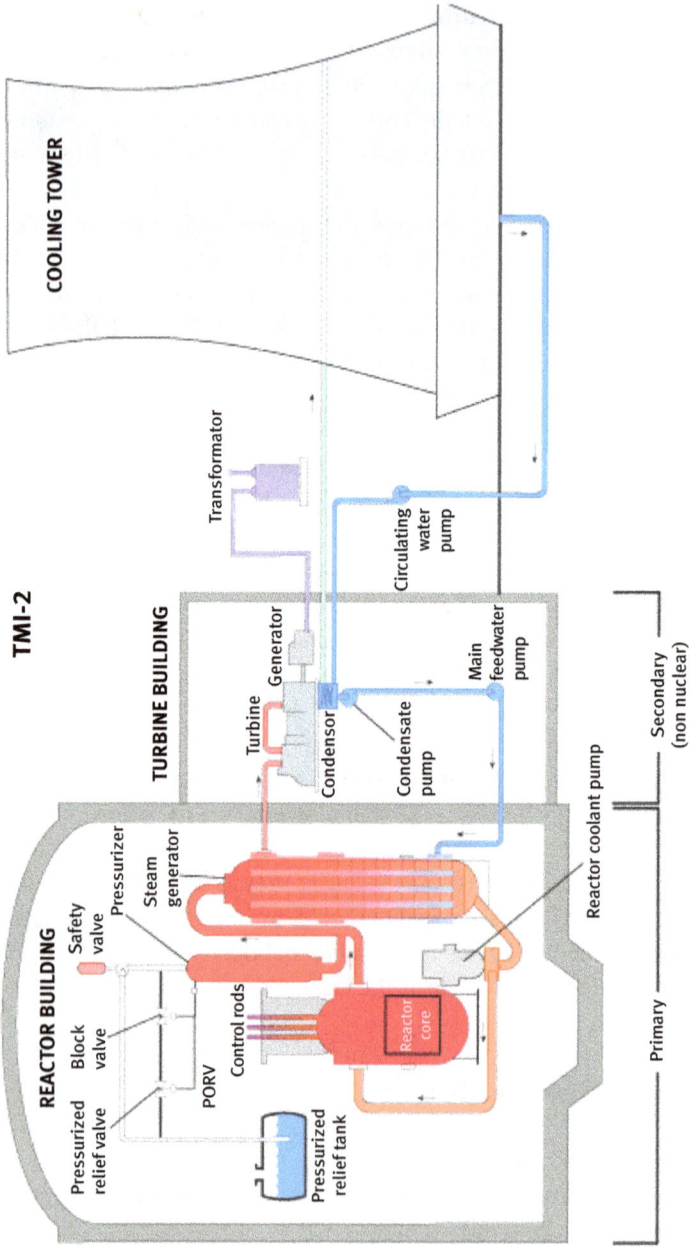

Figure 4.1: Principle of TMI-2 nuclear reactor (Three Mile Island).

meltdown caused by the operators due to lack of understanding of the system. From a consequence point of view, the accidents in Chernobyl (1985) and Fukushima (2010) were worse but main difference is that in Three Mile Island, the operators were taking actions to "create the disaster" without even realizing what was going on.

28 March 1979 at about 4:00 On March 28, 1979, when the reactor of unit 2 was operating at 97% power.

Operators decided to clean resin filters (used in the secondary loop water) with water and compressed air. During this operation (normally only compressed air was to be used), small amounts of water forced its way through a check valve into an instrument airline. This would eventually cause the boiler feed-water pumps, condensate booster pumps and condensate pumps to turn off and cause an emergency shutdown of a power generation turbine. Heat and pressure in the reactor coolant system increased, causing the reactor to perform an emergency shutdown. Within 8 seconds, the control rods ("bars with a special composition that are used to control the fission rate of uranium") were inserted into the core to halt the nuclear chain reaction. Shut down took about 1 second. At this point, a relief valve failed to close, but instrumentation did not reveal the fact.

But, because the steam turbine also tripped, heat was no longer being removed from the reactor's primary water loop and the reactor continued to generate heat as a result of radioactive decay.

Three auxiliary pumps were activated automatically but the system was unable to pump any water because valves had been closed for routine maintenance. This was a violation of a NRC rule (NRC = Nuclear Regulatory Commission). The pressure in the pressurizer increased and its pilot-operated pressure relief valve (PRV) opened to relieve the pressure.

Due to a mechanical failure, the pilot-operated safety valve remained stuck open, after that the excess pressure had been released. On the control panel in the control room, however, the position of the pilot-operated valve (open or closed) was indicated by the status of the solenoid being powered or not. The operators interpreted the unlighted lamp as evidence for the PRV to be in closed position.

Hence, at this moment, the situation is as follows:
1. Reactor coolant continued to flow to the reactor.
2. The open pilot-operated safety valve allows the pressure in the pressurizer to continue to decrease.
3. The coolant water is boiling in the coolant channels. Due to the significant temperature difference between the boiling temperature of coolant water and the

wall, the regime of boiling went from nucleate boiling to film boiling, which in turn decreased the heat transfer from the core to the water.

4. The operators are convinced that the pilot-operated safety valve is closed and since the water level in the pressurizer was high, the operators judged that the core was properly covered with water and hence they failed to recognize the accident as a loss-of-coolant accident.

28 March 1979 at about 5:20 The primary loop's four main reactor coolant pumps began to cavitate as a steam bubble/water mixture, rather than water, passed through them. The pumps were shut down, and it was believed that natural circulation would continue the water movement. Steam in the system prevented flow through the core, and as the water stopped circulating, it was converted to steam in increasing amounts.

28 March 1979 at about 6:00 The top of the reactor core was exposed and the intense heat caused a reaction to occur between the steam forming in the reactor core and the zircaloy nuclear fuel rod cladding, yielding zirconium dioxide, hydrogen and additional heat. This reaction melted the nuclear fuel rod cladding and damaged the fuel pellets, which released radioactive isotopes to the reactor coolant, and produced hydrogen gas that is believed to have caused a small explosion in the containment building later that afternoon.

28 March 1979 at about 6:45 At 6 am, there was a shift change in the control room. A new arrival noticed that the temperature in the pilot-operated relief valve tail pipe and the holding tanks was excessive. They closed a "block valve" – to shut off the coolant venting via the pilot-operated relief valve, but around 120,000 l of coolant had already leaked from the primary loop. It was not until 6:45 am, that radiation levels in the primary coolant water were around 300 times expected levels, and the general containment building was seriously contaminated.

It was still not clear to the control room staff that the primary loop water levels were low and that over half of the core was exposed.

28 March 1979 at about 6:45 A group of workers took manual readings from the thermocouples and obtained a sample of primary loop water. Seven hours into the emergency, new water was pumped into the primary loop and the backup relief valve was opened to reduce pressure so that the loop could be filled with water. After 16 hours, the primary loop pumps were turned on once again, and the core temperature began to fall. A large part of the core had melted, and the system was still dangerously radioactive.

On the third day following the accident, a hydrogen bubble was discovered in the dome of the pressure vessel, and became the focus of concern. A hydrogen explosion might not only breach the pressure vessel, but, depending on its magnitude, might compromise the integrity of the containment vessel leading to large-scale release of radioactive material. However, it was determined that there was no oxygen present in the pressure vessel, a prerequisite for hydrogen to burn or explode. Immediate steps were taken to reduce the hydrogen bubble, and by the following day, it was significantly smaller. Over the next week, steam and hydrogen were removed from the reactor using a catalytic recombiner and, controversially, by venting straight to the atmosphere.

The Three Mile Island accident inspired Charles Perrow's Normal Accident Theory, in which an accident occurs, resulting from an unanticipated interaction of multiple failures in a complex system. TMI was an example of this type of accident, because it was "unexpected, incomprehensible, uncontrollable and unavoidable."

Perrow concluded that the failure at Three Mile Island was a consequence of the system's immense complexity. Such modern high-risk systems, he realized, were prone to failures, however well they were managed. It was inevitable that they would eventually suffer what he termed a "normal accident." Therefore, he suggested, we might do better to contemplate a radical redesign, or if that was not possible, to abandon such technology entirely.

"Normal" accidents, or system accidents, are so-called by Perrow because such accidents are inevitable in extremely complex systems. Given the characteristic of the system involved, multiple failures which interact with each other will occur, despite efforts to avoid them. Events which appear trivial initially cascade and multiply unpredictably, creating a much larger catastrophic event.

Normal Accidents contributed key concepts to a set of intellectual developments in the 1980s that revolutionized the conception of safety and risk. It made the case for examining technological failures as the product of highly interacting systems, and highlighted organizational and management factors as the main causes of failures. Technological disasters could no longer be ascribed to isolated equipment malfunction, operator error or acts of God.

The underlying cause of this accident is the complexity of the process.

4.2.4 Discovery of major risks with new products

One of the most talked-about examples in the twentieth century is the introduction of thalidomide on the market. Thalidomide is a molecule that was discovered by the German pharmaceutical company Chemie Grünenthal (now Grünenthal GmbH) around c. 1953. It was first marketed in 1957 in West Germany, where it was sold without prescription under the trade name Contergon. The component thalidomide was however present in many medicines: Dai-Nippon Seiyaku/Dai Nippon Pharmaceutical laboratory marketed in Japan since 1958, the sleeping pill Isomin and in 1959, a specialty for stomach aches, Proban M., Pharmacolor distributed it since 1958 in Switzerland, Sweden Astra AB (en) launched the Neurosedyn in February 1959, in Belgium, it was marketed under the name from Softenon. In 1959, Two companies shared the Canadian market: Frank W. Horner Ltd. (Talimol) and the Canadian subsidiary of Merrell (Kevadon) since 1959. In Brazil, Denmark, Finland, Ireland, the Netherlands, Norway, Sweden, it was marketed in 1959.

When first released, thalidomide was promoted for anxiety, trouble sleeping, "tension" and morning sickness.

While initially considered safe, the drug was responsible for teratogenic deformities in children born after their mothers used it during pregnancies, prior to the third trimester. In November 1961, thalidomide was taken off the market due to massive pressure from the press and public.

Experts estimate that the drug thalidomide led to the death of approximately 2,000 children and serious birth defects in more than 10,000 children, about 5,000 of them in West Germany. About 40% died around the time of birth. Those who survived had limb, eye, urinary tract and heart problems.

The precise mechanism of action for thalidomide is not known, although efforts to identify thalidomide's teratogenic action generated 2,000 research papers and the proposal of 15 or 16 plausible mechanisms by the year 2000.

The underlying cause of this disaster is lack of knowledge.

In some cases, the hazards are known but it takes time before society reacts. A typical example is the use of X-rays.

On November 8, 1895, the German scientist Wilhelm Röntgen discovered an unknown type of electromagnetic radiation that he named X-radiation to signify that it was an unknown type. It is also called Röntgen radiation. He discovered that so that it could have medical applications as he could make a picture of the bones of his wife's hand. The radiation penetrated the skin. The many applications of X-rays immediately generated enormous interest. The hazards of the use of X-rays were very soon reported by scientists who came with all types of stories of burns, hair loss and worse in technical journals as early as 1896.

From the early 1920s through the 1950s, X-ray machines were developed for shoe fitting in commercial shoe stores (see Figure 4.2). Concerns about the impact of poorly controlled use made an end to that practice.

4.2.5 Discovery of major risks in the process industries

Principles of chemistry, physics and mathematics have been used by early civilizations, for instance, for fermentation and evaporation. Manufacture of chemicals till the end of the eighteenth century consisted of modest craft operations.

Increase in demand, public concern at the emission of noxious effluents and competition between rival processes provided the incentives for greater efficiency.

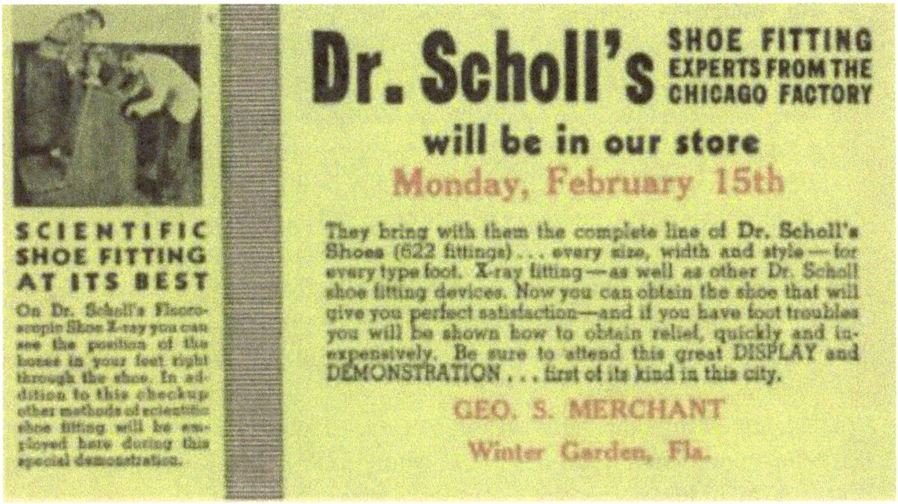

Figure 4.2: Publicity of the use of X-ray fitting for shoes.

This led to the emergence of combines with resources for larger operations and caused the transition from a craft to a science-based industry. The emergence of the chemical industry as an independent branch is associated with the industrial revolution.

The first sulfuric acid plants were built in Great Britain in 1740 (Richmond), France in 1766 (Rouen), Russia in 1805 (Moscow Province) and Germany in 1810 (near Leipzig). The development of the textile and glass industries prompted the initiation of soda production. The first soda plants were built in France in 1793 (near Paris), Great Britain in 1823 (Liverpool), Germany in 1843 (Schönebeck) and Russia in 1864 (Barnaul). In the mid-nineteenth century, artificial fertilizer plants appeared in Britain (1842), Germany (1867) and Russia (1892).

At about the same time, the modern history[2] of the oil and gas industry started:

- In 1847, the Scottish chemist James Young observed natural petroleum seepage in the Riddings coal mine, and from this seepage distilled both a light thin oil suitable for lamps and a thicker oil suitable for lubrication.
- In 1852, the Polish engineer Ignacy Łukasiewicz improved Gesner's method to more easily distil kerosene and petroleum, opening the first "rock oil" mine in Bóbrka, Poland, in 1854.

---

2 Oil and gas had already been used in some capacity, such as in lamps or as a material for construction, for thousands of years before the modern era, with the earliest known oil wells being drilled in China in 347 AD.

- In 1745, under the Empress Elizabeth of Russia, the first oil well and refinery were built in Ukhta by Fiodor Priadunov. The world's first commercial oil well was drilled in Poland in 1853, and the second in nearby Romania in 1857, and in 1857 the American Merrimac Company drilled the first oil well in the town of La Brea, Trinidad.
- In 1859, the first modern oil well in America was drilled by Edwin Drake in Titusville, Pennsylvania. The discovery of petroleum in Titusville led to the Pennsylvania "oil rush," making oil one of the most valuable commodities in America.
- John D. Rockefeller founded the Standard Oil Company in 1865, becoming the world's first oil baron. Standard Oil quickly became the most profitable in Ohio, controlling about 90% of America's refining capacity and a number of its gathering systems and pipelines.
- The late nineteenth and early twentieth centuries marked the creation of major oil companies (Standard Oil in 1870, Chevron in 1879, SHELL in 1890, Exxon in 1911, TOTAL in 1924).

By the end of the twentieth century, industry leadership had shifted from Great Britain to Germany. The rapid process of concentration in the chemical industry, the high level of scientific and technological development, the strengthening of the monopoly on patents, and commercial politics led to Germany's conquest of the world market. Until World War I, it retained a monopoly on the production of organic dyes and intermediates. The chemical industry in the USA began developing appreciably later than in the European countries, but as early as 1913, the USA led the world in volume of chemical production as a result of the country's extremely rich mineral resources, well-developed transportation systems, large domestic market and its exploitation of the experience of other countries.

By World War II, petrochemicals were being used in the USA to produce plastics and fibers. Petrochemicals are chemical products derived from oil or natural gas. Polymer science and chemical engineering were becoming the driving forces of the industry. Polymer science uses petrochemicals to make products such as plastics, resins, paints and adhesives. Chemical engineering made the production of such products possible, and at a low enough cost to be profitable. After World War II, the industry experienced a shift in production from organic chemicals, such as coal, to petrochemicals.

Today, the chemical industry converts raw materials (oil, natural gas, air, water, metals and minerals) into more than 70,000 different products. The plastics industry contains some overlap, as most chemical companies produce plastic as well as other chemicals.

The worldwide global GDP is estimated to be about $86 trillion (2019). The total revenues from the oil and gas drilling sector is estimated to approximately 3.8% ($3.3 trillion) of the global economy (market research by IBISWorld) and the chemical industry established its importance in national economies and grew to about $3.7 trillion USD in sales.

On a regular base, the development of the hydrocarbon industry (oil, gas and chemicals) was confronted with major accidents. Examples of major accidents are given in many textbooks (e.g., Lees, 1980; Marshall, 1987; Atherton and Gil, 2008). Some well-known examples are:

- Multiple BLEVE accident in Feyzin on January 4, 1966, killed 18 people and injuring another 81 people (BLEVE = Boiling Liquid Expanding Vapor Explosion).
- VCE accident in Flixborough on June 4, 1974, killed 28 people and injured another 36 people (VCE = Vapor Cloud Explosion).
- On June 23, 1969, a barrel of an insecticide, endosulfan, fell in the Rhine downstream of Bingen, polluted this river for 600 kilometers (to its mouth), killing more than 20 million fish according to some estimates.
- Leak of MIC (methyl isocyanate) in a pesticide plant in Bhopal, Madya Pradesh, India killed thousands of people and injured several hundreds of thousands of people.
- . . .

Typical characteristics for these accidents are:
- The Physical and chemical phenomena that were (and sometimes still are) not well understood in particular, by the operating companies
- The consequences go far beyond the fence of the site and are beyond imagination
- The complicated course of the accident, which requires in-depth knowledge and understanding of the processes

Patrick Lagadec (1981) writes about "the discovery of major accidents." The start of the European Union was in 1957 when Belgium, France, Italy, Luxembourg, the Netherlands and West Germany signed the Treaty of Rome, which created the European Economic Community (EEC) and established a customs union. Occupational Health and Safety was dealt with since the start but European regulation of major accidents only started after the set-up in 1981 of the Environment Directorate-General (via the European Directive 82/502/EEC of June 24, 1982 on major-accident hazards of certain industrial activities.

4.2.6 Discovery of major risks in renewable energy: BESS

Climate change, coupled with the continuing fall in the costs of some renewable energy equipment, such as wind turbines and solar panels, are factors leading to increased use of renewables.

Over the course of the twentieth century grid, electrical power was largely generated by burning fossil fuel. When less power was required, less fuel was burned.

Wind power is uncontrolled and may be generating at a time when no additional power is needed. Solar power varies with cloud cover and at best, is available only during daylight hours, while demand often peaks after sunset.

Interest in storing energy from these intermittent sources grows as the renewable energy industry begins to generate a larger fraction of overall energy consumption. Essentially, all Energy Storage Systems capture energy and store it for use at a later time or date. There are various types of energy storage systems and the common way of classifying an ESS (Energy Storage System) is by the type of energy used, such as hydroelectric dams, where water is stored in the reservoir during periods of low demand and released when demand is high.

Battery energy storage systems (BESS) are an emerging technology that is becoming a popular component to a resilient and efficient electric strategy. BESS use electro-chemical solutions and include different types of batteries such as Lithium-ion, Lead-acid, Sodium-Sulphur and Zinc-Bromine.

Arizona Public Service (APS), a subsidiary of Pinnacle West Capital that serves about two-thirds of the Phoenix metropolitan area (State of Arizona) with electricity, constructed a 2 MW battery energy storage system ("the McMicken Energy Storage Facility") to power the neighborhood in Surprise, a city in Maricopa County, in the US state of Arizona. Surprise is, the second-fastest-expanding municipality in the greater Phoenix metropolitan area and was the sixth-fastest-expanding place among all cities and towns in Arizona. The population grew from 30,848 at the 2000 census to 141,664 (estimated) in 2019.

The battery systems consist of 27 racks of 14 modules each, for a total of 378 modules of lithium-ion batteries. The BESS was assembled with Lithium ion (Li-ion) batteries manufactured by LG Chem. Korea's LG Chem is the world's number 1 Li-ion battery manufacturer by capacity (Rawles, 2019). Construction began on the facilities in 2016 and they came online in 2017.

All types of BESS offer pros and cons in terms of capacity, discharge duration, energy density, safety, environmental risk and overall cost. However, BESS utilizing Li-ion batteries are by far, the most widely used system today. This is primarily due to their high energy density and steady decrease in cost. Lithium-ion offer good energy storage for their size and can be charged/discharged many times in their lifetime. Several types of Li-ion batteries exist: Lithium Cobalt Oxide ($LiCoO2$), Lithium Manganese Oxide ($LiMnO2$), Lithium Iron Phosphate (LFP – $LiFePO4$), Lithium Nickel Cobalt Aluminum Oxide ($LiNiCoAlO2$, NCA) and Lithium Titanate.

The installation base of grid-connected energy storage was negligible before 2000 and utility industry experience with it has been poor.

Between 2003 and 2017, 734 MW of large-scale battery storage power capacity was installed in the USA, two-thirds of which was installed in the past three years (The International Association of Engineering Insurers, 2019). In their publication, IMIA writes:

However, these systems are considered relatively new technology and could in many ways be seen as prototypical. As with most developing technologies there are often some challenges to tackle before the technology can be seen as proven.

In 2012, a 1.5-MW APS battery facility near Flagstaff, Arizona, caught fire. The utility took lessons from the investigation of that incident and has continued with advancements in battery design and safety standards to expand its stationary storage for renewable energy resources.

Since 2018, at least 26 accidents with Li-Ion BESS have been reported from which 16 were in 2018, 7 in 2019 and 1 in 2020 (VRB, 2020)

On April 19, 2019, a suspected fire was reported at the McMicken Energy Storage Facility. At 17:48 local time, first responders arrived to investigate. Several hours later, at approximately 20:04, an explosion occurred from inside the BESS. The explosion injured several firefighters and essentially destroyed the BESS and its container. The cascading thermal runaway that happened on April 19, 2019 is a typical catastrophic failure of a lithium-ion battery system, where multiple cells in a battery fail due to a failure starting at one individual cell. Thermal runaway can occur due to exposure to excessive temperatures, external short circuits due to faulty wiring or internal shorts due to cell defects. Thermal runaway events result in the venting of toxic and highly flammable gases and the release of significant energy in the form of heat. If ignited, these gases can cause enclosed areas to overpressurize, and if unmitigated, this overpressure can result in an explosion and severe damage to the battery and surrounding equipment or people. An explosion scenario can be even more severe for a large battery pack, where the heat generated by one failed cell can heat up neighboring cells and lead to a thermal cascade throughout the battery pack.

There is still some debate on the cause of the cascading thermal runaway, but one of the contributing factors of the accident on which all investigations agree is that the fire suppression system was incapable of stopping thermal runaway (Hill, 2020; LG Chem, 2020). LG Chem writes in its report to the Arizona Corporation Commission:

> LG Chem agrees that, today, there is an industry consensus that NOVEC 1230 clean agent cannot stop module to module propagation during a thermal runaway event. This was not known at the time the third parties responsible for the McMicken system design selected NOVEC 1230. Moreover, the relevant safety standards in place then did not require prevention of module to module propagation in the event of thermal runaway. Today, new standards are in place that require mitigation measures to control module to module spread, and LG Chem is a leader in the development of technologies that allow new BESS installations to meet or exceed these new standards.

One of the underlying cause of the escalation of the accident is lack of knowledge.

4.3 Technological accidents are complex and complicated because of uncertainties

Organizations, in general and large organizations in particular, are complex systems. A complex system, broadly defined, is a set of entities that, through their interactions, relationships or dependencies, form a unified whole.

Hence, an organization with all its own people, contractors, assets and activities is a social complex system because of the many components which may interact with each other.

Complex systems are characterized by the inherent difficulty experienced by an observer in explaining and describing the behavior of the system at a macro level in terms of its constituent parts.

Complex systems are typically made up of a large number of constituent entities that interact with each other and also with its environment. Complex systems theories identify different types of complex systems. In reality, most large organizations are so-called complex adaptive systems (CAS) that exhibit fundamental CAS principles like self-organization, complexity, emergence, interdependence, space of possibilities, co-evolution, chaos and self-similarity.

CAS are contrasted with ordered and chaotic systems by the relationship that exists between the system and the agents, which act within it. In a CAS, the system and the agents co-evolve. The system lightly constrains agent behavior, but the agents modify the system by their interaction with it. This self-organizing nature is an important characteristic of CAS; and its ability to learn to adapt.

As explained above, occupational health and safety refers to the relationship between the employer and the worker. The employer and the worker have to create the conditions so that the work can be done in safe conditions. The approach requested is a result-oriented result, which means performing the task without accident. In case of an uncertainty about the safe conditions, the work has to be stopped or even not started ("zero risk concept").

Occupational health and safety problems in general, are simple problems within a complex organization. Some characteristics of a simple problem are:
- The recipe is essential
- Recipes are tested to assure easy replication
- Experience increases success rate
- A good recipe produces good results every time
- There is no uncertainty

Hence, the legal obligation of "prevention of risks" which is synonym of "prevention of uncertainties" is possible for a large number of tasks. For instance, working at height risks (= uncertainties) can be prevented.

Complexity theory makes a difference between (see, for instance, Westley et al., 2006)
- Simple refers to "known knowns"
- Complicated refers to "known unknowns"
- Complex refers to "unknown unknowns"

When new technologies are used, however, the problem for making it safe becomes complicated and sometimes complex because uncertainties are inherent to many new technologies.

These uncertainties originate from (1) lack of knowledge, (2) lack of understanding or (3) the variability of parameters.

While the legal demand for occupational health and safety is very clear ("zero accidents and hence zero uncertainty before starting a task"), the use of new technologies will oblige more than ever before to accept uncertainties and to deal with uncertainties.

4.4 Why do we have to deal with risks?

Risks are inherent to business (see Figure 4.3) and therefore there is no other option than to deal with the risks. Given the fact that uncertainties are inherent to the use of new technologies, one could ask to apply the "precautionary principle" which may be defined as "caution practiced in the context of uncertainty." There are many interpretations of the precautionary principles. One of these interpretations is:

> The Precautionary Principle says that if some course of action carries even a remote chance of irreparable damage to the ecology, then you shouldn't do it, no matter how great the possible advantages of the action may be. You are not allowed to balance costs against benefits when deciding what to do.   — Freeman Dyson, Report from 2001 World Economic Forum

But, meeting the basic needs of human beings (Food, Water, Air and Shelter) in our current society is not possible without the hydrocarbon industry in at least the coming decades. It is obvious that our society needs a sustainable development but for at least the coming decades, the hydrocarbon industry will be an integral part of this sustainable development.

In response to the "precautionary principle," one can argue that "accepting the risk" doesn't mean that the accident is acceptable. When we accept the risk, it means that we believe that the accident will not happen and hence that it is worthwhile to continue the activity.

Industry can however, not "impose" its conviction toward the stakeholders. The moto "don't worry, just trust us because we know what we are doing" is not a responsible attitude.

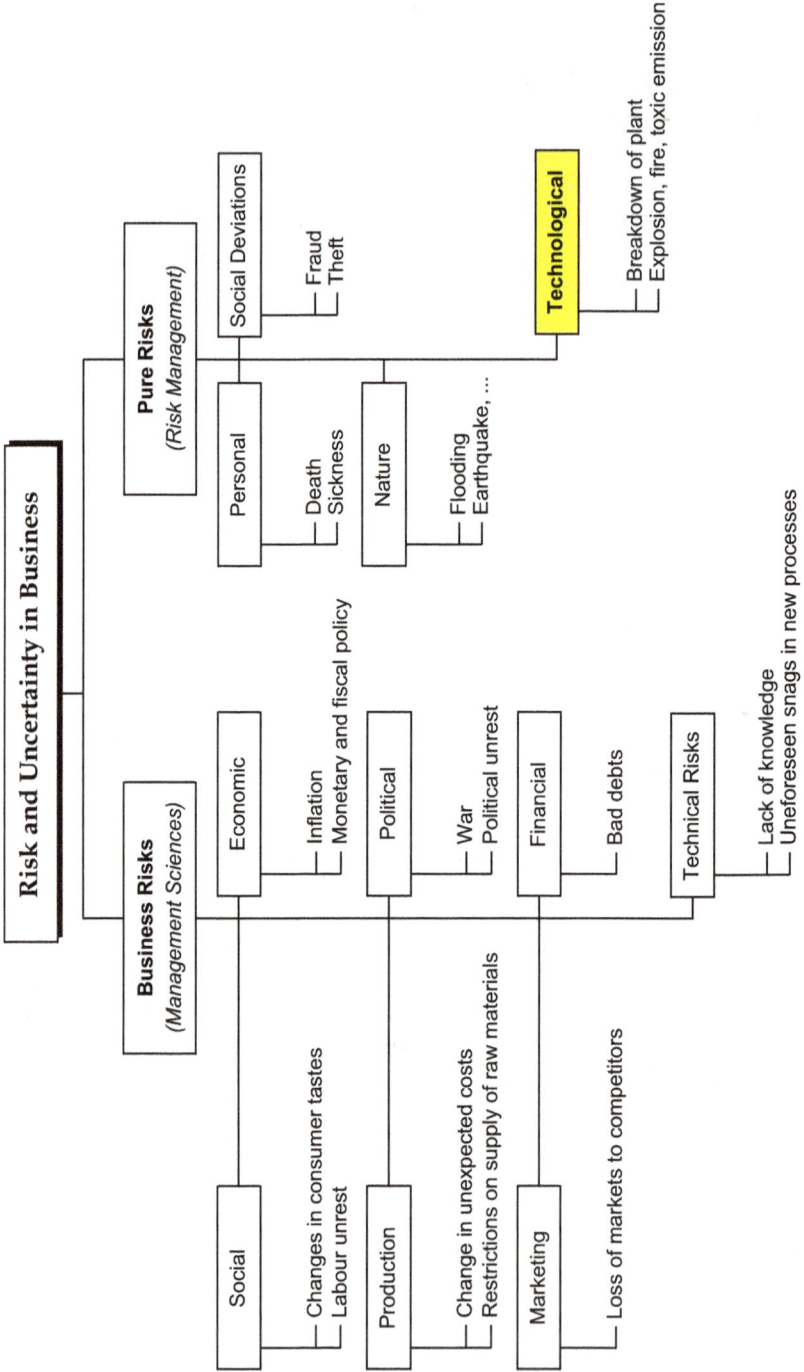

Figure 4.3: Risk and uncertainty in business.

It is important that all stakeholders (concerned population, authorities, workers, etc.) have a common understanding of the risks that are taken so that the society (including industry) can take the decision to continue or to stop the activity.

4.5 How to deal with risks?

4.5.1 Different risk approaches

There are many ways to answer the question, "How much risk is socially, politically, economically and ethically acceptable?" The study of risk, and people's reactions to it, has been a topic of empirical investigation and philosophical debate for several decades. However, there remain significant controversies regarding basic issues, such as how should risk be defined and the range of variables which should be taken into account when evaluating it (Weyman and Kelly, 1999).

Bradbury (1989) in her PhD research identifies two concepts of risk:
1. A concept of risk as an objective physically given property of a hazard, and
2. A concept of risk as a socially constructed attribute.

She advances that the social construction of risk provides a firm, theoretical basis for the design of policy. The discussion links the perception, management and communication of risk to the more fundamental issue of the nature and role of science and technology.

"Risk" defined as an objective property of an event or activity and measured as the probability of well-defined adverse effects, is very easy to use in discussions about the ranking of the risks. If, on the other hand, "risk" is seen as a cultural or social construction, then the discussion about the ranking will depend on social values and lifestyle preferences. Society is not only concerned about risk minimization. People are willing to suffer harm if they feel it is justified or if it serves other goals (Renn, 1992).

The literature provides an assortment of classifications (Starr, 1969; Lowrance, 1976; Vlek and Stallen, 1979). The classification I prefer is the classification by Renn, in which he discusses seven risk possible approaches (Renn, 1992):
–   Multidimensional descriptive theories:
    –   Psychological approach
    –   Sociological studies
    –   Cultural-historical studies
–   One-dimensional pragmatic approaches:
    –   Actuarial approach (insurance approach)
    –   Toxicological and epidemiological approach
    –   Economical approach
    –   Engineering approach including probabilistic assessments

In general, one can say that one-dimensional pragmatic approaches are very effi-cient tools to perform risk management while multidimensional descriptive theories are very powerful to understand why an individual or a group of people show a certain behavior toward risk.

Multidimensional theories have demonstrated that the intuitive understanding of risk is a multidimensional concept and that defining risk to the product of proba-bilities and consequences is an oversimplification of reality.

Each of these approaches will be discussed briefly. The most common way of dealing with risks in industry is to use an "engineering approach" which will be ex-plained further in more detail.

4.5.2 Psychological risk approach

A "personally experienced risk" might be defined by describing the whole complex of measurable reactions to a risky decision of a situation (Vlek and Stallen, 1980). This includes physiological measurements such as heart rate, behavioral actions like avoidance behavior and cognitive reactions such as verbal descriptions.

One of the most important assumptions in this approach is that risk is inherently subjective. Risk does not exist. Human beings have invented the concept "risk" to help them understand and cope with the uncertainties of life.

There has been a sizeable amount of research into personal risk-taking tendencies (e.g., Kogan and Wallach, 1967). It seems that perceived risk is situationally deter-mined and less personally.

Personal risk acceptance tendencies are influenced by a large number of factors, such as voluntariness of exposure (e.g., availability of options, personal influence on the decision, controllability of consequences, distribution of consequences in time and distribution of consequences in social–geographical space, context of probability assessment, context of accident evaluation, combinations/interactions of accident probability and seriousness (Vlek and Stallen, 1980).

Some authors (e.g., Otway et al, 1978) grouped together different aspects (involun-tary exposure, uncontrollable consequences) in "psychological risk factors." Slovic, (1987) grouped together characteristics that refer to the dread of the risk ("Factor 1") and other characteristics that refer to the knowledge of the risk ("Factor 2") (Table 4.1.).

Based on a psychometric study of a variety of hazards from different thematic fields, a graph is then produced like in Figure 4.4. The location of the potential event is based on a number of characteristics, which are related to respectively the dread of the event (combined in factor 1) and to the knowledge about the event (combined in factor 2). The higher the risk is judged on the factor dread, the higher is the perceived risk and the more people want to see its risks reduced and regulated.

The different location on the graph helps to explain why people have the intui-tive feeling that different potential undesired events cannot be compared with each

Table 4.1: Grouping of 15 risk characteristics in 2 factors according to Slovic (1987).

Factor 1: dread of risk		Factor 2: knowledge of risk	
Controllable	Uncontrollable	Not observable	Observable
Not dread	Dread	Unknown for those	Known to those
Not global catastrophic	Global catastrophic	exposed	exposed
Consequence not fatal	Consequence fatal	Effect delayed	Effect immediate
Equitable	Not equitable	New risk	Old risk
Individual	Catastrophic		
Low risk for future	High risk future		
generations	generations		
Easily reduced	Not easily reduced		
Risk decreasing	Risk increasing		
Voluntary	Involuntary		

other. The risks of dying because of "traffic accidents" are perceived as completely different than risks associated with, for instance, "nuclear reactor accidents."

This list of factors demonstrates that the intuitive understanding of risk is a multidimensional concept and cannot be reduced to the product of probabilities and consequences. It appears to be a common characteristic in almost all countries in which perception studies have been performed that most people perceive risk as a multidimensional phenomenon and integrate their beliefs with respect to the nature of risk, the cause of risk, the associated benefits and the circumstances of risk (Renn, 1992; Slovic, 2010).

Exploration of psychometric methods have identified a number of contextual variables of risk that affect the perceived seriousness of risks (Slovic et al 1981, Vlek et al 1981, Renn 1990 and Covello 1983):

- The catastrophic potential (e.g., the expected number of fatalities in one accident). This means that "high-consequence-risks" are usually perceived as more threatening that "low-consequence-risks" irrespective of the probability of occurrence;
- Certain risk characteristics such as: the perception of dread, the conviction of having personal control over the magnitude or probability of the risk, the familiarity with the risk, the potential to blame a person or institution responsible for the creation of the risky situation;
- The beliefs associated with the cause of risk. Attitudes encompass a series of beliefs about the nature, consequences, history and justifiability of a risk cause.

A more comprehensive discussion on the psychology of risk can be found in books dedicated to the subject (e.g., Kouabenan et al., 2006; Breakwell, 2007; Asailly, 2010 in the collection of professor Franck Guarnieri).

Factor 2
Unknown risk

Laetrile
Microwave Ovens

DNA Technology

Water Fluoridation
Saccharin
Water Chlorination
Coal Tar Hairdyes
Oral Contraceptives

Nitrates
Hexachlorophene
Polyvinyl
Chloride
Diagnostic
X-Rays

Electric Fields
DES
Nitrogen Fertilizers

SST

Valium
IUD
Darvon

Antibiotics
Rubber
Mfg.

Cadmium Usage
Mirex
Trichloroethylene
Pesticides
Asbestos
Insulation

2,4,5-T
Uranium Mining
PCBs

Radioactive Waste

Nuclear Reactor
Accidents
Nuclear Weapons
Fallout

Auto Lead
Lead Paint

DDT
Mercury
Fossil Fuels
Coal Burning (Pollution)

Satellite Crashes

Caffeine
Aspirin

Vaccines

Factor 1
Dread risk

Skateboards

Auto Exhaust (CO)
D-CON

LNG Storage &
Transport
Coal Mining (Disease)

Nerve Gas Accidents

Smoking (Disease)
Power Mowers   Snowmobiles

Large Dams
SkyScraper Fires

Trampolines
Tractors

Nuclear Weapons (War)

Alcohol
Chainsaws

Home Swimming Pools
Downhill Skiing
Recreational Boating
Electric Wir & Appl (Shock)
Bicycles

Elevators
Electric Wir & Appl (Fires)
Smoking
Motorcycles
Bridges
Fireworks

Underwater
Construction
Sport Parachutes
General Aviation
High Construction
Railroad Collisions
Commercial Aviation
Alcohol
Accidents
Auto Racing

Coal Mining Accidents

Auto Accidents

Handguns
Dynamite

Factor 2

Not Observable
Unknown to those Exposed
Effect Delayed
New Risk
Risk Unknown to Science

Controllable	Uncontrollable
Not Dread	Dread
Not Global Catastrophic	Global Catastrophic
Consequences Not Fatal	Consequences Fatal	Factor 1
Equitable	Not Equitable
Individual	Catastrophic
Low Risk to Future Generations	High Risk to Future Generations
Easily Reduced	Not Easily Reduced
Risk Decreasing	Risk Increasing
Voluntary	Involuntary

Observable
Known to those Exposed
Effect Immediate
Old Risk
Risks Known to Science

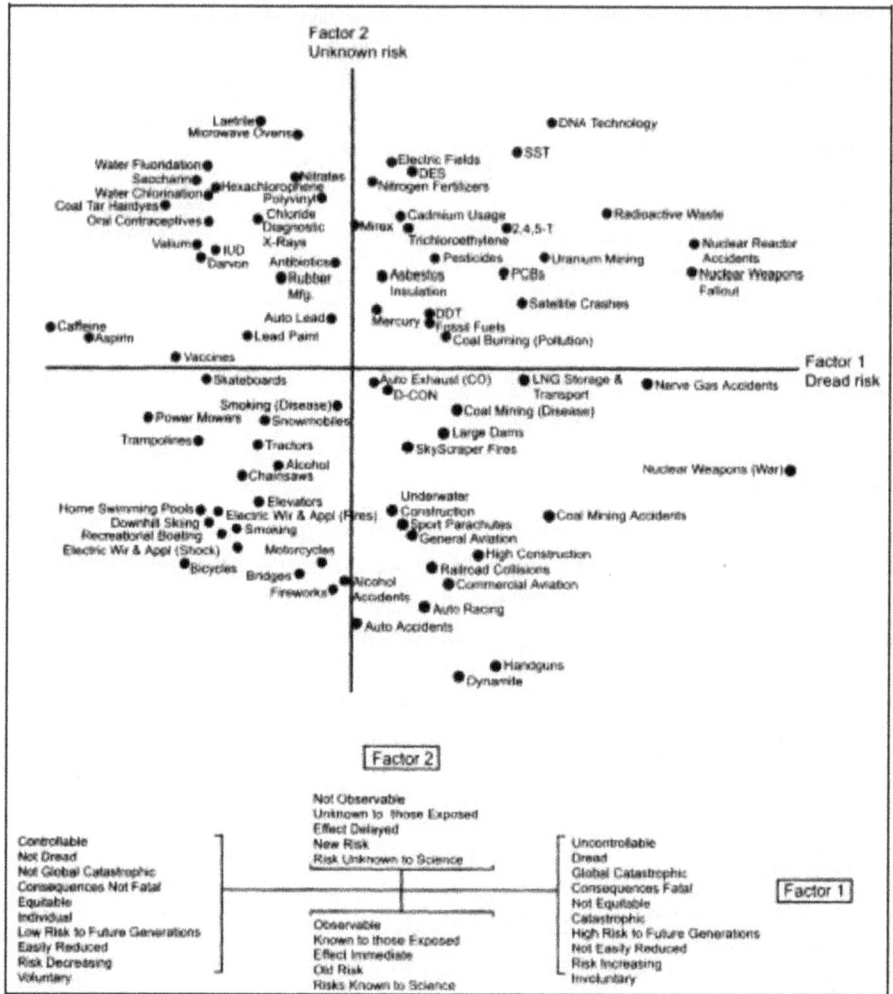

Figure 4.4: Location of 81 risks in a psychometric measurement (taken from Schmidt, 2004 – original source: Slovic, 1987).

4.5.3 Sociological risk approach

The social sciences are a set of academic disciplines having in common the study of the human society, and social interactions between individuals, groups and their environments. It encompasses a wide array of academic disciplines.

Within sociology, risk and uncertainty have become a major interest, but there is a variety of different points of view available. While most approaches can agree on the idea that risk is a possible threat in the future, they conceptualize risk in connection to at least three different core ideas: risk in (rational) decision-making, risk in calculative-probabilistic calculation and risk as part of a modern worldview (Zin, 2008). There are as many perspectives within sociology as there are sociologists (Renn, 1992).

The rational actor concept considers that social actions are the result of deliberate intentions by individuals or groups to promote their interests. Rationality is widely used since the beginning of the twentieth century as an assumption of the behavior of individuals in microeconomic models. The so-called rational choice theory appears in almost all economics textbook treatments of human decision-making. By applying the principles of neoclassical economics and behaviorism to the analysis of social facts, the American sociologist George C. Homans developed in the early 1960s, a "theory of social exchange" which made him one of the early propagators of rational choice theory in the social sciences. A more in-depth study of the rational actor concept in risk theories is given by Jaeger et al. (2013).

The social mobilization theory explains that people join social movements such as Youth for Climate without any rational beneficial payoff. Gustave Le Bon was amongst the first to write about the subject in 1895 (Le Bon, 1895). Gustave Le Bon shows in this work that the behavior of individuals together is not the same as when individuals reason in isolation – he thus explains the irrational behavior of crowds. The book of Le Bon and its theses have been widely criticized – both methodologically and theoretically – and are now considered outdated but the idea of social mobilization without any rational action is still observed today. The explosion in hall 12 on 4 August 2020 in Beirut was the final trigger to the social mobilization that resulted in resignation of the government.

Organizational risk theories emphasize the structural aspects of large corporations and institutions. Organizational factors (complacency of people, organization of control, movement of people, training of people, etc.) influence the level of technological risks and not seldom in a way that is not visible. Routinization of tasks and diffusion of responsibilities are both factors that influence the risk. As a result, the objective mathematical calculation of a risk is not possible. It is only when a major disaster happens that people will be able to understand how organizational factors influenced the way the organization dealt with the risk. Risk management of complex technologies requires institutional operation and control activities (Clarke, 1989).

Systems theory regards risks as an element of a larger social or institutional unit. Risk issues evolve in an evolutionary process in which groups and institutions organize their knowledge with other social systems through communication (Luhman, 1990). Various systems of knowledge compete in a society. Organizations are complex, interconnected, soft systems themselves, so systematically establishing this awareness requires multiple factors including leadership, accountability and performance management systems (Barber and Burns, 2002). According to Barber, current risk management approaches fail to address the unpredictability and interconnectedness of risks. The weaknesses of the traditional approaches are the result of the underlying assumptions with the biggest limitation being the assumption that the complex response of a human/technical system can be reduced to a series of discrete risks each with its own separate cause and effect. Risk databases assume that risks are discrete and can be treated separately.

4.5.4 Cultural risk approach

Whereas the sociological analysis of risk links social judgements about risks to individual or social interest values, the cultural theory assumes that cultural patterns structure the mind-set of individuals and social organizations to adopt certain values and reject others.

Each culture selects its own portfolio of risks. Cultural analysis implies that the definition of undesirable events, the generation and estimation of possibilities as well as the constructions of reality depend on the cultural affiliation of the respective social group (Renn, 1992).

Cultural theory argues that risks are defied, perceived and managed according to principles that inhere in particular forms of social organization. According to cultural theory, institutional structure is the ultimate cause of risk perception.

Anthropologist Mary Douglas (Douglas et al, 1983) argues that the role of cultural ways of life determine what risks individuals see as worthy of taking. Rayner (1990) advocates that the key to understanding perception, communication and management of technological risk lies in understanding how various institutional or organizational cultures lead human agents actively to select different issues for attention as risks, while ignoring other candidates for concerns that actuarially could be as dangerous to life or limb.

4.5.5 Actuarial approach of risk

Actuaries are skilled professionals whose comprehensive training includes the use of statistical analysis to understand risks and uncertainties. When doing business, one is exposed to a full range of risks. Business risks are unexpected events that can result in a gain or a loss. Business risks are speculative risks. Pure risks are unexpected events that can result in a loss. When the unexpected event does not materialize, then there is no gain but also no loss.

Speculative risks are almost never insured by insurance companies, unlike pure risks. Insurance companies require policyholders to submit proof of loss (often via bills) before they will agree to pay for damages. Losses that occur more frequently or have a higher required benefit normally have a higher premium. Key principles adopted in an actuarial approach to risk management focus on the identification, quantification, mitigation and control of risks rather than the governance arrangements that might be placed around a risk management framework.

4.5.6 Toxicological and epidemiological approach of risk

Toxicology is a scientific discipline, overlapping with biology, chemistry, pharmacology and medicine that involve the study of the adverse effects of chemical substances on living organisms. Epidemiology is the study and analysis of the distribution (who, when and where), patterns and determinants of health and disease conditions in defined populations.

Toxicological and epidemiological risk analysis comprises a process of hazard identification, dose-response assessment and exposure assessment, providing as an output, the estimation of the incidence and severity of adverse effects likely to occur in a human population in relation to actual or eventual exposure to hazardous compounds.

Cancer is a major cause of mortality. Let's assume that we want to manage the risk of colorectal cancer.[3] Colorectal cancers are the fourth most commonly diagnosed cancers and rank second among cancer deaths in the United Sates.

The multi-dimensional approaches, as explained in the previous chapter can help us to get insights in the way people feel about the risk, but they will not allow to manage this cancer-risk. To enable management of this type of risk, we have to understand the risk factors. First, we will try to identify the factors that influence the (probability of) occurrence of the type of cancer.

Suspected risk factors for colorectal cancer are age, diet, polyps ("polyps are benign growths on the inner wall of the colon and rectum"), personal medical history, family medical history, ulcerative colitis ("a condition in which the lining of the colon becomes inflamed").

Next, researches will study the different risk factors and they will try to identify correlations between the cancer incidence and the risk factor. Research has shown that cooking certain eats at high temperature creates chemicals (heterocyclic amines) that may increase cancer risk. Incidence rates show wide divergence by racial/ethnic group, with rates in the Alaska Native population that are four times as high as rates in the American Indian population in New Mexico (figures for the latter population are $1.86\ 10^{-4}$ in men and $1.53\ 10^{-4}$ in women).

---

3 The body is made up of many types of cells. Normally, cells grow, divide and produce more cells as they are needed to keep the body healthy and to function properly. Sometimes, however, the process goes astray – cells keep dividing when new cells are not needed. The mass of extra cells forms a growth or tumor. Tumors can be either benign or malignant. Malignant tumors are cancer. Cells in malignant tumors are abnormal and divide without control or order. These cancer cells can invade and destroy the tissue around them. The colon and rectum are parts of the body's digestive system, which removes nutrients from food and stores waste until it passes out of the body. Together, the colon and rectum form a long, muscular tube called the large intestine (also called the large bowel). The colon is the first 6 feet of the large intestine, and the rectum is the last 8 to 10 inches. Cancer that begins in the colon is called colon cancer, and cancer that begins in the rectum is called rectal cancer. Cancers affecting either of these organs may also be called colorectal cancer.

Epidemiological studies try to get insight in the correlation between, for instance, meat consumption and cancer incidence. An example of such a correlation is given in Figure 4.5.

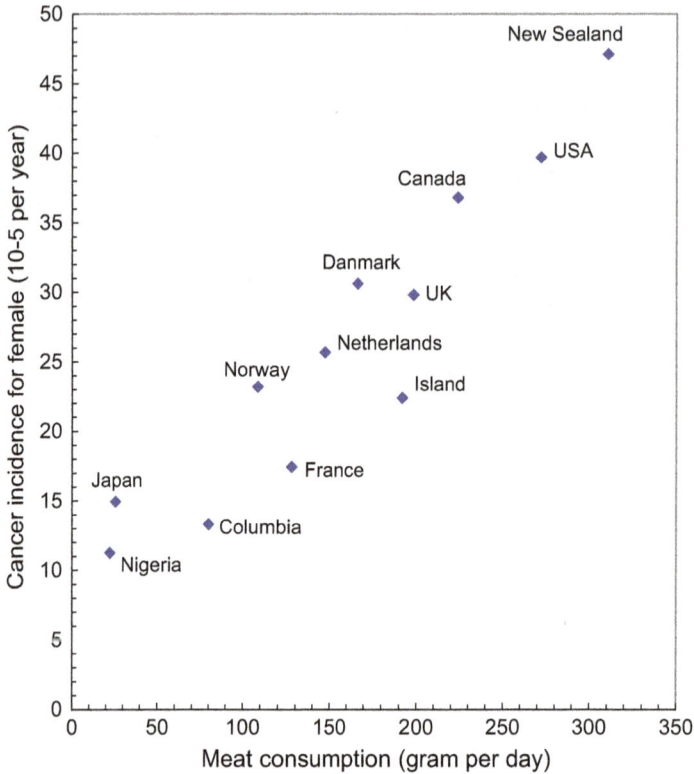

Figure 4.5: Example of epidemiological correlation between meat consumption and cancer incidence.

The main challenge in a toxicological or epidemiological risk approach is first to identify the possible risk factors and next to establish whether these risk factors are correlational and or causal.

Figure 4.6 shows some results of a risk study of the influence of noise and air pollution on dementia (Alzheimer). The study was done for the period 2005–2013 on a population of 130, 978 adults aged 50–79 years with no recorded history of dementia or care home residence (Carey et al, 2018). The risks are expressed by "HRs (Hazard Ratio) from Cox models that were adjusted for age, sex, ethnicity, smoking and body mass index." Hazard ratios are used in survival analysis, a branch of statistics for analyzing the expected duration of time until an event (in this case dementia) happens. A hazard ratio of 2 for instance, would indicate a two times higher risk of dementia.

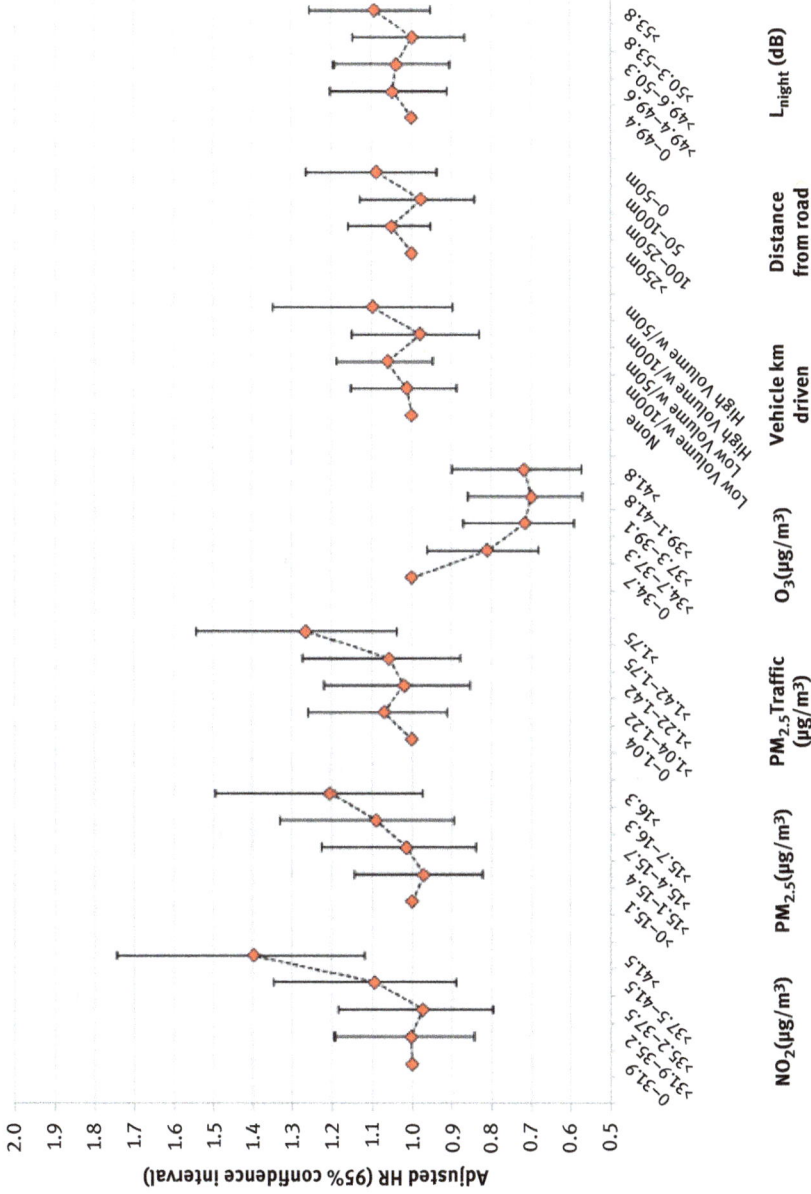

Figure 4.6: Example of an epidemiological study on the influence of noise and air pollution on dementia (Carey et al, 2018).

Some of the conclusions of the study:
- There was a positive exposure response relationship between dementia and all measures of air pollution except O_3, which was not readily explained by further adjustment
- Adults living in areas with the highest fifth of NO_2 concentration (>41.5 µg/m^3) versus the lowest fifth (<31.9 µg/m^3) were at a higher risk of dementia (HR = 1.40).

The examples given above show how epidemiologists use the notion of risk in their domain. These types of approaches are specific for this discipline.

4.5.7 Economic approach of risk

The economic approach consists in weighing the costs against the possible damage. This approach is closely related to the actuarial approach (see, for instance, Hibbert and Turnbull, 2003).

In an actuarial approach, a premium is determined on the basis of a large group of insured persons. In an economic approach, it is rather an individual consideration of the possibility of paying for the damage.

The basic idea of the economic approach: "when the maximum potential loss is much smaller than your financial possibilities, then from an economic point of view there is no problem at all to take the risk." A typical example is, participating with a small amount in a lottery. The chances of winning a lottery jackpot vary widely depending on the lottery design but, in all cases will be low.

In a simple 6-from-45 lotto, a player chooses six numbers from 1 to 45 (no duplicates are allowed). If all six numbers on the player's ticket match those produced in the official drawing (regardless of the order in which the numbers are drawn), then the player is a jackpot winner. For such a lottery, the chance of being a jackpot winner is 1 in 8, 145,060 $\left(\frac{45!}{6!(45-6)!} \right)$

But from an economic approach, the probability for winning is not important. The bet is very small (e.g., 5 US$) while the potential gain is very significant (e.g., 1,000,000 US$).

Some economic studies have estimated a value of statistical life (VSL) which is an economic value used to quantify the benefit of avoiding a fatality. The VSL is more of an estimate of willingness to pay for small reductions in mortality risks rather than how much a human life is worth.

While there is much controversy whether monetizing human life is ethical, it remains a common practice in some domains. Kirigia et al. (2020) for instance, estimates the fiscal value of human lives lost from coronavirus disease (COVID-19) in China. Jonkman et al. (2003) distinguishes four approaches to make a monetary valuation of human life. His view is interesting but primarily academic. In industry, there is a great reluctance to use monetary values to human life.

4.5.8 Engineering approach of risk

The engineering approach considers that risk is an objective property of a potential undesired event. This property is defined by the combination of the potential consequence and the likelihood that the event and its potential consequence will manifest. Because it is a property, it can be calculated and it is by definition, objective.

The engineering approach evolved after World War II and is widely used in the industry. The high technology industry employs many technical people for whom a pragmatic way of calculating risk objectively and mathematically is very welcome, much to the annoyance of sociologists and psychologists.

The confrontation with major accidents was a real challenge for engineers and the fact that the risks (uncertainties) could be calculated is very attractive.

Since World War II, more and more methods have been developed and the engineering approach has become a discipline within engineering.

The hydrocarbon industry is an industry that is managed by technically (scientifically) oriented managers who are used to manage and to take decisions based on metrics. These managers don't feel confortable when a subject can not be measured.

The engineering approach enables to calculate risk, to communicate in a very simple way about risks and to assess the calculated risks against reference values. The discussion and decision about acceptable risks is relieved of difficult sociological, psychological, philosophical or cultural considerations.

Kaplan and Garrick (1981) have discussed a number of alternative definitions of risk. These include:

- Risk is a combination of uncertainty and damage.
- Risk is a ratio of hazards to safeguards.
- Risk is a triplet combination of event, probability and consequences.

The work of Kaplan was amongst the most influential scientific works in the field of risk engineering (Aven, 2020). Aven discusses the legacy of the work of Kaplan and the influence of some other articles in the 80s and 90s.

The reality is that the engineering risk approach has circumvented but not solved the problem. Technical risk analysis is not superior to any other construct of risk. They are also based on group conventions, specific interests of elites and implicit value judgements. In the next chapter, we will go more in detail on the engineering approach and the fundamental difficulties will be highlighted.

5 Process safety: an engineering discipline

5.1 Engineering

A common accepted description of engineering is (based on The American Engineers' Council for Professional Development):

> The creative application of scientific principles to design or develop structures, machines, apparatus, or manufacturing processes, or works utilizing them singly or in combination; or to construct or operate the same with full cognizance of their design; or to forecast their behavior under specific operating conditions; all as respects an intended function, economics of operation and safety to life and property.

The Industrial Revolution created a demand for machinery with metal parts, which led to the development of several machine tools. The invention and development of steam engine would not be possible without the understanding of the principles of science and applied science behind steam power. With the increasing demand for machinery in the eighteenth century, engineering became a profession.

The "Smeatonian Society of Civil Engineers" was founded in 1771, and was originally known as the "Society of Civil Engineers," being renamed following its founder's death, John Smeaton (1724–1792). Smeaton was an English civil engineer responsible for the design of bridges, canals, harbors and lighthouses. It was the first engineering society to be formed anywhere in the world, and remains the oldest.

In France, the "Ecole des Ponts et Chaussées," founded in 1747 is considered as the first school where engineers graduated.

In the 1820s and 1830s, in order to overcome the old system, each large and medium German land established technical schools at the level of trade schools. Most of today's prominent German Universities of Technology (Technische Universitäten) were founded in these decades.

The Belgian State established in 1834–1836 the first schools that delivered certified engineers: "École du Génie civil" in Gent, "École des Arts et Manufactures et des Mines" in Liège and the "École Militaire" in Brussels (Linssen et al., 2013).

The US census of 1850 listed the occupation of "engineer" for the first time with a count of 2,000. In 1870, there were a dozen US mechanical engineering graduates, with that number increasing to 43 per year in 1875. In 1890, there were 6,000 engineers in civil, mining, mechanical and electrical.

The first PhD in engineering (technically, applied science and engineering) awarded in the USA went to Josiah Willard Gibbs at Yale University in 1863.

Classical branches in engineering are mining engineering, chemical engineering, electrical engineering and civil engineering.

https://doi.org/10.1515/9783110632132-005

Engineers use mathematics and they apply all type of sciences (physics, hydraulics, etc.) to identify, understand and interpret the constraints on a design in order to yield a successful result.

Process safety is about applying engineering methods to design, build and operate a process plant in a safe way. Uncertainties are inherent to all engineering disciplines and hence uncertainty is embedded in process safety.

5.2 Process installations

The "hydrocarbon industry" refers to the different industrial companies that are involved in the transport, manufacture and conversion of raw materials (minerals, oil and natural gas) into all types of different products. The main industries are part of one of the following industries:
– The mining industry
– The oil industry, or the oil patch, includes the global processes of exploration and extraction
– The refining industry with refineries, storage terminals and marketing of petroleum products
– The petrochemical industry
– The chemical industry
– Transport of hazardous products mainly by sea-going vessels, inland water ways, pipelines, rail or road

All these industries make extensive use of industrial processes which take place in process units or industrial installations. These units consist of mechanical equipment (pressure vessels, columns, tanks), rotating machines (pumps, compressors, turbo expanders), heating devices (heat exchangers, furnaces), process piping, valves, safety devices (bursting discs, pressure safety valves, flares), computer control systems, instrumentation and so on.

A process installation will have different phases during its life cycle such as:
– Conceptual phase
– Feasibility study
– Design phase
– Operational phase
– Demolition

The possibility for an undesired event exists in all phases of the life cycle and an accident during the operational phase can sometimes be caused by decisions taken during the feasibility study.

5.3 The framework to perform risk management of process installations

Four questions have to be addressed at all time (i.e., during the whole life of a process installation) (CCPS, 1998):

1. What can go wrong? What failure scenarios can we realistically expect with this process?
2. What impact can those failures have? Can we live with such consequences?
3. Do we need to worry about these potential failure scenarios actually happening? How likely is it?
4. Do we need to take additional actions? What is the risk? Can we tolerate the potential consequences at the estimated likelihood?

A professional jargon and a framework (Figure 5.1) have been developed within the field of process safety:

- The first question will be answered by "Hazard Identification." Different techniques have been developed to perform hazard identification.
- The second question will be answered by "Consequence and Impact analysis."
- The third question will be addressed by statistics. This will be called "a probabilistic approach."
- The first three questions will enable to quantify the risks as a combination of the potential consequence and the associated likelihood of occurrence.
- The final question will then be addressed by comparing the calculated risks with some reference values. This step is called risk assessment. Based on the assessment, it can be decided to avoid the risk (i.e., the activity is stopped), to transfer the risk (e.g., to transfer it to a company which is more experienced in dealing with the risk) or to implement additional reduction measures can be decided. Risk reduction measures can be technical or organizational.

The different steps will be discussed in the next chapters.

5.4 Application of industry standards and best practices

5.4.1 Engineering standards

As explained above, uncertainties are inherent to the use of complex and new technologies. Uncertainties are however not necessarily eternal. Uncertainties can be removed by scientific progress or via return from experience from previous accidents. This progress and return of experience (REX; also called experience feedback system) will be incorporated into new engineering standards or into revisions of existing engineering standards. These engineering standards evolve over time. The following examples illustrate this:

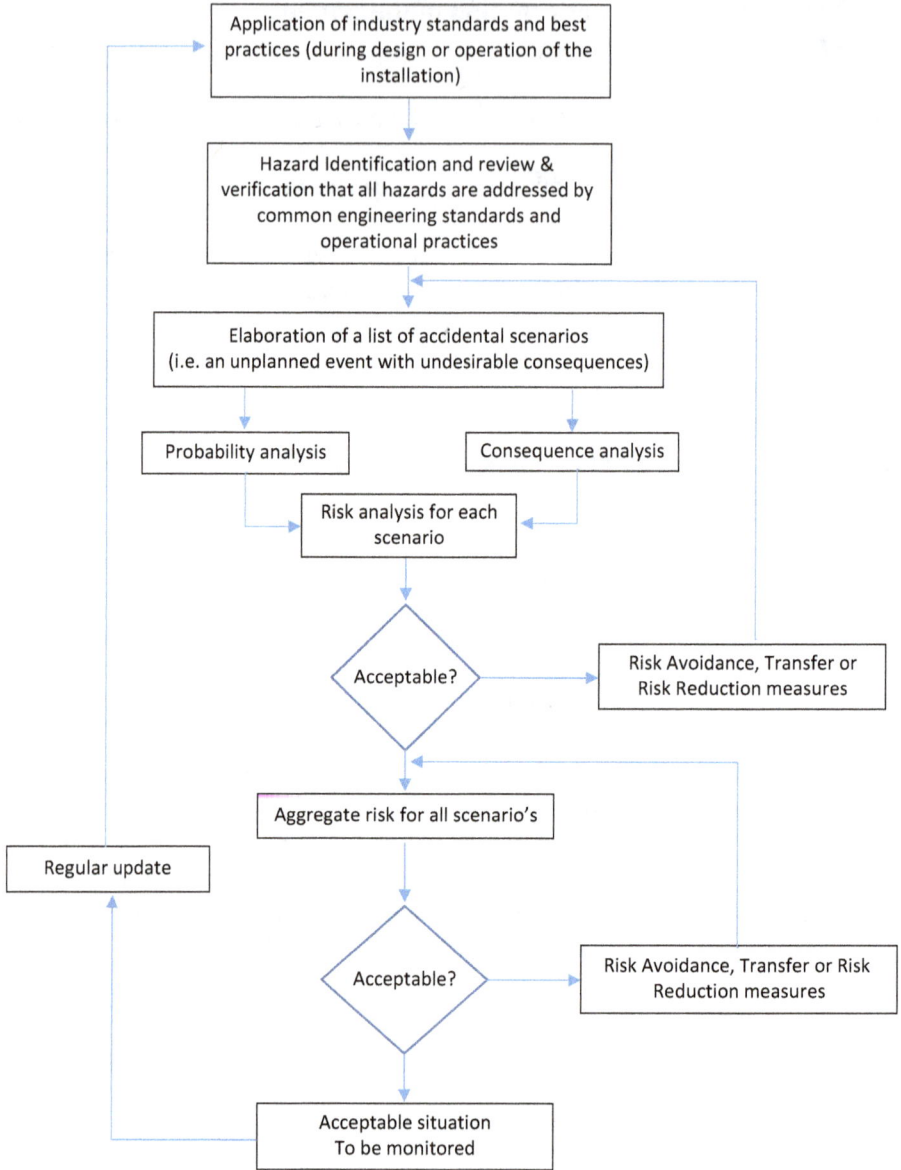

Figure 5.1: Framework to perform risk management of a process installation.

Disastrous accidents with boilers and pressure vessels (in particular the boiler explosion in 1905 in a shoe factory in Brockton, Massachusetts that killed 58 people and injured 117 others) resulted in the first edition of the ASME Rules for construction of Stationary Boilers and For Allowable Working Pressures, that was adopted in the spring of 1915.

Pressure Vessels were designed based on experience and on the principle that the hoop stress had to be kept low with respect to yield and to use ductile material to accommodate local peak stresses. With the development of the nuclear technology in the 1950s, pressure vessel design requirements needed to be improved. Advances in mechanics theory and analysis methods provided new and more scientific methods for the pressure vessel design.

In 1963, ASME published the B&PV Code Section III: Nuclear Vessels based on the principles of limit analysis (Shakedown analysis, Fatigue Analysis) and Stress Analysis was used to determine higher allowable loads and more consistent margins of safety.

Scientific knowledge and experience continued to evolve and the latest new ASME Boiler & Pressure Vessel Code, Section VIII, Division 2 was published on July 1, 2007. This 2007 Edition became mandatory for vessel design in July, 2009. Familiarity with the new organization and the application of these rules is essential to both the new and experienced vessel engineer to ensure proper vessel design.

Similar examples can be found in all technologies and all domains of engineering. Hence, it is obvious that the basis for process safety is the knowledge of the engineering standards and engineering best practices. Companies will actively participate in trade associations where they share their experience and knowledge and develop industrial engineering standards. A well-known trade association is, for instance, American Petroleum Institute (API). It was formed in 1919 as a standards-setting organization. API counts about 600 member companies and developed more than 700 standards to enhance operational and environmental safety, efficiency and sustainability. The standards are kept up to date via API Standards Committees that are made up of subcommittees and task groups that work and maintain these standards.

API is just one example of a trade association that develops engineering standards. Associations with similar objectives exist in all types of industries. The recent explosion on 4 August 2020 reminds us about the potential accidents that can occur with the unsafe storage of ammonium nitrate. The International Fertilizer Association (IFA) was founded in 1927 and has a membership of 480 companies in 68 countries. IFA has a technical and HSE committee with the mission to "promote the efficient, safe and secure production, storage and transportation of plant nutrients in a sustainable manner."

5.4.2 The role of the process safety engineer

Knowing and applying engineering standards and engineering best practices is the role of engineering. The purpose of the process safety engineer is not to replace the role of engineering.

The process safety engineer will have an in-depth knowledge of engineering principles and a basic knowledge of the engineering standards that apply to his project or process. The process safety engineer will look at the engineering standards in a different way and he will, on a number of points, complete the standards with his insights in process safety. The process safety engineer will study the failure mechanisms of the subjects for which the engineering standards are applied.

For example, during engineering, the sizing of pressure safety valves is often done according to the rules in the API Recommended Practice 521 "Guide for Pressure-Relieving and Depressurizing Systems," (latest edition is fourth edition, 1997). One of the scenarios to be covered is an external fire. The RP 521 gives the equations to be used for the sizing of the pressure safety valve. Once the sizing is finished, each of the vessels of the unit can be equipped with its pressure safety valve. The different pressure safety valves are connected to the flare system which in many cases will be designed according the API Standard 537, Flare Details for General Refinery and Petrochemical Service, (latest edition of third edition 2020). The design flare capacity is the maximum design flow to the flare.

In-depth knowledge with the use of these standards is the job in engineering. There are number of topics which are important from a safety point of view but it is not clear in the standards how one should deal with these topics. A first example is the size of the fire that should be considered when calculating the design flare capacity. API RP 521 was used to perform the sizing of the pressure safety valve for each vessel but there is no rule to decide on the maximum size of the fire. This is important because the design flare capacity will be determined by the sum of the pressure safety valves that will open when there is a fire. Assuming that all pressure safety valves connected to the flare system will open can be feasible in some situation, but for a large refinery, this will result in an almost impossible flare design.

A pressure safety valve can fail to open. Hence, another question is whether a pressure safety valve should be redundant or triple redundant. It will be a case-by-case decision based on an analysis of the risk (potential consequence and likelihood) of the failure of a pressure safety valve.

Some other topics for which the process safety engineer will be consulted are:
- The outlet of a pressure safety valve to atmosphere has to go to what is indicated as "to safe location." What is meant by "safe location?"
- In case of a process upset, fail-safe devices (e.g., automatic closure of valves that feed heat to the reboiler of columns) will be activated. API RP 521 stipulates that "in the design of some components of a relieving system, such as the blowdown headern flare and flare tip, favorable instrument response of some percentage of instrument systems can be assumed. The percentage of favorable instrument responses is generally calculated based on the amount of redundancy, maintenance schedules, and other factors that affect instrument reliability." In practice, it means that a number of equipment will not need to be pressure relieved (because the instrumentation avoided the overpressure in the equipment). API RP 521 does

not specify the percentage. This is normal because it will be a case-by-case study. Determining this percentage is a job typical for the process safety engineer who will again verify the risk of taking a certain assumption.

- Pressure safety valves can fail according to different failure modes: (1) Fail to open, (2) opens above set pressure, (3) valve opens partially/fails to relieve required capacity, (4) valve stuck open, (5) spurious/premature opening and (6) leakage past valve. The process safety engineer will have good knowledge of these failure modes and he will possess reliability data for each failure mode. He will update his data as more information becomes available.

- There have been a number of major accidents in the past due to chattering of a pressure relief valve. Chattering is the rapid opening and closing of a pressure relief valve and can result in severe vibration and, if prolonged, mechanical failure of valve internals and associated piping. The cause of chattering is a very high pressure drop between the vessel and the pressure safety valve. Underlying causes are (1) an inlet pipe to a relief valve that is smaller than the valve inlet, (2) a very long pipe between the vessel and the relief valve, a piping with a lot of bends, many valves, fittings and other obstructions or (3) evidence of line plugging observed when removing a relief valve for maintenance. Farris Engineering, a leading manufacturer of pressure relief valves (PRVs), estimates that 25% of PRVs in refineries and chemical processing facilities experience high pressure drop at the inlet of the valve. Their estimate is based on calculations performed over 32,000 PRVs over 12 years (Farris Engineering Services, 2018)

The examples above concern only one particular topic. A similar discussion can be held for all type of technical standards (rotating machines, pressure vessels, piping, etc.). The regular engineer will use the technical standard to engineer the installation, while the process safety engineer will complete the engineering with specific knowledge about the safety aspects associated with the engineering and in particular with the risks associated with the failure of meeting the desired engineering purpose. The process safety engineer will need to study much more books about the subject than the engineer who will mainly apply the standards. For pressure relief systems, this would include standard books like Parry (1992) and Hellemans (2009).

5.5 Hazard identification

5.5.1 Hazard, danger and hazardous event

The meaning of the word hazard can be confusing because many people use the terms hazard, harm, danger and risk interchangeably. The word "hazard" has existed at least since the fourteenth century; even in the nineteenth century, it was more common to use the word dangerous.

The Factory Act of 1833 used the word danger and dangerous machines. Mather describes in 1853 *The Coal Mines: Their Dangers and Means of Safety* (Mather, 1853). Calder in 1899 writes about dangerous machines (Calder, 1899). Beyer (1916) and Vernon (1936) do not use the word hazard. From the 1930s onward, we increasingly see the use of the term "hazard" in the specialized literature.

Heinrich defines a hazard as "a condition with the potential of causing an injury or a damage" (Heinrich et al., 1959).

Bird defines hazard as "a condition or practice with the potential for accidental loss" (Bird et al., 1985).

Marshall discusses the need for an agreed terminology on the words "hazard" and "risk" (Marshall, 1987). He defines a hazard as "a physical situation with a potential for human injury, damage to property, damage to the environment or some combination of these."

ISO 17776 (2000) defines hazard as "potential source of harm" and adds a note: "In the context of this international standard, the potential harm may relate to human injury, damage to the environment, damage to the property, or a combination of these."

The definition of "Hazard" according CCPS is sometimes different for different publications:

- A hazard is a physical or chemical condition that has the potential for causing harm to people, property or the environment (CCPS, 2000 and CCPS, 2008).
- "Hazard is the way in which an object or a situation may cause harm: A hazard exists where an object or situation has the ability to cause harm. Such hazards include uneven pavement, unguarded machinery, an icy road, a fire, an explosion and a sudden escape of toxic gas. When a hazard triggers a series of escalating hazards, these escalating hazards are referred to as cascading hazards" (CCPS, 2010).

Table 5.1 gives some examples of hazards and the associated harm.

Table 5.1: Some examples of hazards and associated harm.

Hazard	Harm
Knife	Cut
Benzene	Leukemia
Electricity	Shock, electrocution
Wet floor	Slips, falls
Welding	Metal fume fever

The term "hazard" in other words refers to "everything that has the potential to cause harm to life or to cause damage to life, the environment or property."

A **hazard** is any source of **potential** damage, harm or adverse health effects on something or someone.

From the definition of "hazard," it can be concluded that "corrosion" is a hazard but that "a hole in a pipe due to corrosion" is also a hazard because both have the potential to cause harm. Some people make a difference between a hazard and a danger, while others consider both words as interchangeable. In the previous example, "corrosion" would be a hazard while a hole in a pipe would present a real danger.

More important for the purpose of risk management is the definition of a *hazardous event* which is "the manifestation of a hazard in a particular place during a particular period of time which results or could result in harm."

5.5.2 Hazard identification techniques

The purpose of the hazard identification is to establish a clear view on all potential hazardous events and to understand how these events can occur and when they occur how these events can evolve to a disaster. There are broadly spoken three categories of hazard identification techniques:
1. Knowledge and experience
2. Literature and documentation
3. Specific methods

Each of these methods has advantages and difficulties.

Knowledge and experience is the most powerful identification technique. An organization or an individual identified a hazard and accident or a near-miss situation and hence is very aware of that hazard. There is no expert advice needed to convince the organization to take the necessary actions to avoid repetition of the accident. The main difficulty is that organizations have no memory. It is therefore very important that a site has a comprehensive list of all accidents that happened and that the people at the site are regularly trained on these accidents. The importance of learning from previous accidents has been recognized for decades.

Figure 5.2 is an extract from Safety Newsletter No 5 from Trevor Kletz in 1968. Some large companies have in place an internal database with accidents. Information under the form of REX about (potential) accidents is shared via a companywide network. The purpose is to share lessons learned and to avoid recurrence of these accidents. This information is communicated to concerned staff (own as well as contractors) and, when needed, recommendations are implemented.

The American Institute of Chemical Engineers started in February 1967 with its first Loss Prevention Symposium. Miller and Doyle write the following in the introduction of the first symposium (Miller et al., 1967):

5/10 A QUOTATION

Incidents similar to those described in this Newsletter have happened before in the Division, some of them several times. The fact that they have happened again shows that we have still got a long way to go before we can be sure that the lessons of the past are not being forgotten.

"It should not be necessary for each generation to rediscover principles of process safety which the generation before discovered. We must learn from the experience of others rather than learn the hard way. We must pass on to the next generation a record of what we have learned" (From the Foreword to the series of 10 booklets on hazards in refining and related operations, published by The American Oil Co. If you have not seen these booklets they are strongly recommended. They are available from The American Oil Co., Whiting, Indiana. I will gladly give you a free sample.)

Many thanks to all those who replied to the message at the end of the last Newsletter. I have been encouraged .to learn that so many of my bottles landed on friendly shores.

4th November 1968

Figure 5.2: Extract from Safety Newsletter No 5 from Trevor Kletz in 1968.

Each year accidents in the process industries take a big toll in injury to people, damage to property, and reduction of profits. Experience is a dear teacher. In the field of Safety and Accident Prevention the price we must pay to learn each lesson the hard way is too great. We can all profit by sharing our experience. We can all benefit from the lessons we learn from our own operations.

Literature and documentation is nowadays an almost unlimited source of information. There are many articles, books and specialized magazines available on accidents (e.g., Loss Prevention Bulletin, reports from CBS). International databases (e.g., ARIA, FACTS) contain thousands of accidents that can be selected with keywords. The main difficulty is to study all these documents and to keep an oversight. Internet (Wikipedia, Google) and social media are very powerful resources as well. The only thing to look out for is the validity and cross-checking of the available information.

Specific techniques are needed when one wants to study more complex installations and in particular when it is necessary to identify specific hazards. Techniques that are often applied in the process industries are technical audits, checklists, hazard and operability study (HAZOP), failure mode and effect analysis (FMEA), fault tree analysis (FTA) and event tree analysis (ETA). The main advantage of these methods is that the hazard identification efforts are very specific. The main inconvenience is that the use of the techniques requires a lot of experience and that it is easy to be very concentrated on the method itself while being blind for the purpose which is "the identification of the hazards." A good starting point to read more about the different hazard identification methods is the booklet from the European Process Safety Center (Crawley et al., 2003), the US Department of Energy's Handbook on *Chemical Process Hazards* (DOE, 2004) or *the Guidelines for Hazard Evaluation Procedures of the Chemical Center for Process Safety* (CCPS, 2008).

5.5.3 Overview of some specific hazard identification techniques

5.5.3.1 Prior to starting to use a specific hazard identification technique

The first thing a company should do is to appoint a leader for the preparation of the use of the method. The goal is to identify hazards and potential hazardous events that are specific for the installation or subject under study. The method itself is secondary to the goal.

The responsible will coordinate the exercise. The responsible will typically be knowledgeable in the use of specific hazard identification techniques. He will at minimum have participated in about 10 important studies (HAZOP, FMEA, etc.) before taking the role as the leader.

By preference, the leader shall be a senior charismatic person (10+ years of experience) with outspoken competency in facilitating group discussions.

The leader will in a first phase:

- Collect technical information about the installation (product data, piping and instrument diagrams, process descriptions, etc.).
- Collect information about previous accidents or incidents with this installation.
- Perform a literature search on accidents with similar products and installations.
- Review the available people and resources for the study.
- Check availability of accommodation (meeting rooms, timing, etc.) at the site or at a remote location.

Once the leader has an overview of available documents and resources and had achieved a general knowledge about the process, he will:

- Establish the objectives of the study.
- Decide upon the overall scope.
- Select a technique that will meet the objectives.
- Organize the study team.
- Organize various aspects of the study.
- Conduct the study.
- Document the results of the study.
- Ensure that the recommendations made are followed and implemented. (When the leader is an external consultant, he can do a handover to an employee of the site of the company.)

5.5.3.2 Most recommended method

The leader will decide on the proposed method once he has a good idea of the available resources and information. A number of criteria that will be important to take into account are:

- Experience of the company with a certain method. When people are used, for instance, to perform their studies with a HAZOP method then there will be a tendency to continue to use this method.
- The complexity of the installation in combination with the experience with that type of installations. Some major oil companies have ten thousands of service stations all over the world that are built according to their standards. There is no reason to perform a detailed HAZOP study for a new service station.
- The available resources and the available documentation. Figure 5.3 shows an example of the documentation that becomes available during the different phase of a project. A HAZOP study requires a minimum of detailed information. During the conceptual design phase, it will be more recommended to perform a preliminary hazard analysis (PHA) to identify all "large-scale" potential hazards in the project very early in the design stage and to evaluate the severity of these hazards rather than a HAZOP study.

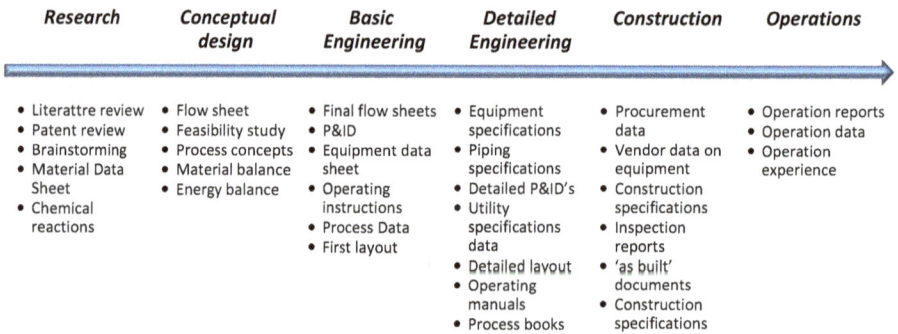

Research	*Conceptual design*	*Basic Engineering*	*Detailed Engineering*	*Construction*	*Operations*
• Literattre review • Patent review • Brainstorming • Material Data Sheet • Chemical reactions	• Flow sheet • Feasibility study • Process concepts • Material balance • Energy balance	• Final flow sheets • P&ID • Equipment data sheet • Operating instructions • Process Data • First layout	• Equipment specifications • Piping specifications • Detailed P&ID's • Utility specifications data • Detailed layout • Operating manuals • Process books	• Procurement data • Vendor data on equipment • Construction specifications • Inspection reports • 'as built' documents • Construction specifications	• Operation reports • Operation data • Operation experience

Figure 5.3: Availability of documentation during the time line of a project.

Table 5.2 gives a noncomprehensive overview of hazard identification techniques that can be used during different phases of the life of a project.

Table 5.2: List of hazard identification techniques that can be used in different phases of the life of a plant.

Phase of the project	Methods that can be applied
Research and development	Ranking methods such as Dow's fire and explosion index Literature study and accident casuistry Brainstorming Checklist

Table 5.2 (continued)

Phase of the project	Methods that can be applied
Conceptual design	Ranking methods such as Dow's fire and explosion index Literature study Brainstorming Structured what if method
Basic engineering	Ranking methods such as Dow's fire and explosion index Course HAZOP Maximum credible accident review Hold-up consequence analysis
Detailed engineering	HAZOP FMEA Fault tree analysis Event tree analysis Cause-consequence analysis or bow-tie Human reliability analysis
Construction	Safety Review Checklist
Operations	HAZOP FMEA Fault tree analysis Event tree analysis Cause-consequence analysis or bow-tie Human reliability analysis

5.5.3.3 How to be sure that all hazardous events have been identified?

Hazard identification is a significant investment in people and resources. The purpose is to identify all hazards and to put in place an action plan in order to avoid occurrence of an accident.

Given the significant effort, it is always frustrating to have an undesired event and to observe that the event was not identified during the hazard identification exercise.

So how can we ensure that all hazards will be identified? When an accident happens people will check whether the hazard was noticed during the hazard identification studies. Experience shows that most of the time the study will not include the undesired event or the accentual scenario.

This of course is very frustrating in particular for the management. They made the resources available, time and resources were allocated to perform an in-depth hazard identification and when an accident happens, it turns out that the scenario was even not identified. The answer to that is in reality quite simple:

– The quality of the hazard identification technique will depend on the knowledge of the team and in a number of cases not all hazards will be identified.

- When hazards are known, they will be in the study. When hazards are unknown but the team is aware of the lack of knowledge they will further investigate and the undesired event or hazard will end up in the study.
- When certain hazards are unknown and the team does not even realize that they do not know, then the scenario will not be in the study.
- Last but not least, when an accident happens after a hazard identification, then there is a real chance that the team did not identify the scenario because otherwise they would have put in place measures to avoid the scenario.

It is of course also possible that the scenario was identified and that the barriers were put in place but that these barriers were not properly managed.

5.5.3.4 Preliminary hazard analysis

The PHA is initiated during the conceptual design phase. The intent of the PHA is to identify hazardous conditions for the system and personnel at an early stage in design (Dussault, 1983). The following steps define a general procedure for performing a PHA:

1. Review hazards identified in similar systems and previous designs. Determine those hazards which may be present in the systems.
2. Identify the events that could potentially create a hazardous condition.
3. Evaluate the effects of the hazardous condition.
4. Identify available compensation and control for hazard or suggest corrective action.
5. Provide the results of analysis of corrective actions undertaken, and any additional remarks.

The format most often employed is the tabular worksheet.

5.5.3.5 HAZID

HAZID is a structured brainstorming exercise with the aid of checklist. HAZID is particularly useful in the early stages of a development, either as a stand-alone exercise or as part of a more general review. The "prompt" or "checklist" approach guides the less experienced and prompts the experienced. Success when using the technique depends upon a properly constructed team being well managed and having the opportunity to think beyond the checklist and identify the unusual.

The overall purpose of HAZID is to focus on the general hazards and the mutual impact between the surroundings and the facilities. A HAZID is guided by a typical checklist and taking benefit from the previous experience of the team.

The HAZID should be implemented as soon as preliminary information (plot plans, environmental conditions and process flow diagrams) is available.

The HAZID team should consist of a HAZID team leader with general experience of hazard identification. HAZID team members should be selected for their knowledge of the technical and operational aspects of installations under study.

Many companies developed elaborated checklists with the minimum subjects that have to be addressed by the team.

HAZID is a brainstorming exercise. Table 5.3 gives an overview of a number of brainstorming techniques. These techniques are explained in more detail in the Federal Emergency Management Agency's Continuity Risk Toolkit (FEMA, 2017)

Table 5.3: Some brainstorming techniques (FEMA, 2017).

Delphi method	The Delphi method (also known as Delphi technique) is a forecasting method that relies on obtaining a consensus from a collection of experts.
Devil's advocacy	Devil's advocacy involves challenging a single, strongly held view or consensus by building the best possible case for an alternative explanation.
Divergent–convergent thinking	Divergent–convergent thinking is a form of structured brainstorming that generates new analytic ideas, hypotheses and concepts or helps discover previously unimagined hazards, vulnerabilities and risky situations through an unconstrained creative group process.
Outside-in thinking	Outside-in thinking is used to identify the full range of basic forces, factors and trends that would indirectly shape an issue.
Round-Robin brainstorming	Round-Robin brainstorming relies on ideas being generated in the absence of discussion for completely free-form thoughts unhindered by group trends or consensus.
Reverse brainstorming	Reverse brainstorming is a structured brainstorming technique that asks how and why a hazard might not occur, and uses the converse of these reasons to suggest how it might actually occur.

5.5.3.6 Hazard and operability study

According to Kletz, the basics for the HAZOP methodology was developed by Imperial Chemical Industries (ICI) in 1963/1964 when a team of three people met for 3 days a week for 4 months to study the design of a new phenol plant. They started with a technique called critical examination which asked for alternatives, but changed this to look for deviations. The method was further refined within the company, under the name operability studies, and became the third stage of its hazard analysis procedure (the first two being done at the conceptual and specification stages) when the first detailed design was produced (Kletz, 1997).

In the 1970s, several articles were published in the open literature on hazard analysis (Kletz, 1971) and on HAZOPs (Lawley, 1974 and Knowlton, 1976). Internal to ICI a report was made available to the sites to explain how HAZOPs had to be conducted (Figure 5.4).

(c) Report No HO/SD/760003/A, available from Division Reports Centres, is an introduction to hazard and operability studies. Compared with earlier reports, for example No HO/SD/740009/4A, the papers and discussion at a seminar on operability studies, this new report concentrates on the application of operability studies to small scale and batch plants. It will be published soon by the Chemical Industries Association.

Figure 5.4: Extract from the internal ICI Safety Newsletter No 87 from May 1976.

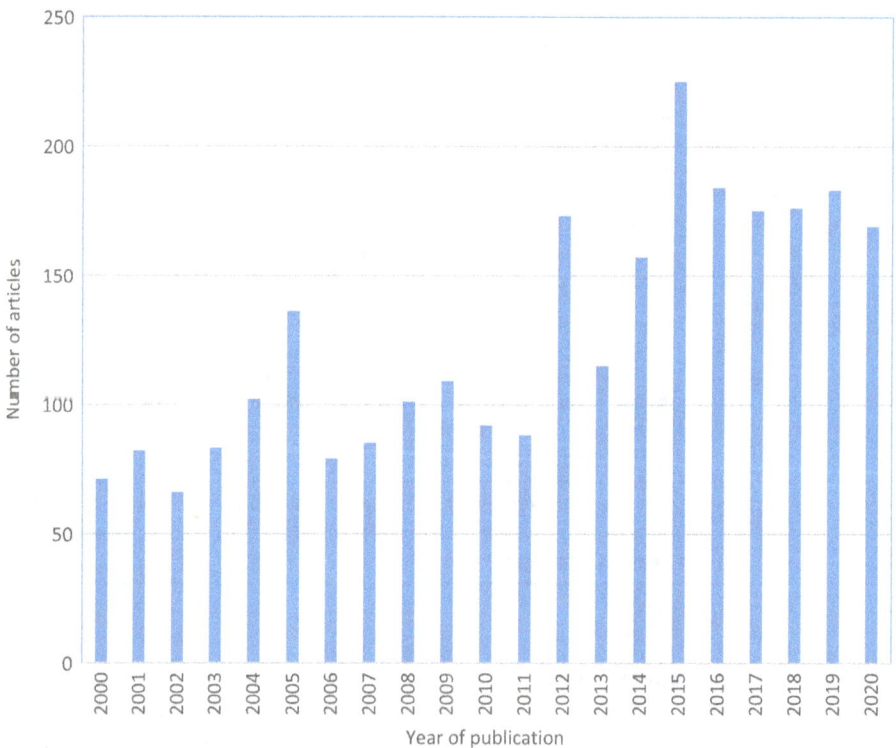

Figure 5.5: Number of articles in relation to HAZOP in Elsevier's ScienceDirect.

The HAZOP methodology is a very popular hazard identification technique. Elsevier ScienceDirect contains reference to 2,651 articles published in Journals over the last 20 years about HAZOP or at least referring to HAZOP.

The idea behind the technique was that many difficulties and hazards could be avoided if a multidisciplinary team would go in a systematical way through all the details of the process before the final design. During normal operations, they had discovered that many problems (operational problems and hazards) encountered could have been avoided if the right people would have discussed through it before.

The method consists of:

1. Definition of the purpose of the HAZOP;
2. Bringing together a multidisciplinary team;
3. Deciding on the important (process) parameters to be studied;
4. Developing a list of deviations by combining standard guide words with the (process) parameters;
5. The team will then, under the direction of a HAZOP leader, study in a systematic way the drawings by reviewing for each deviation:
   - The potential causes of the deviation;
   - The potential consequence of the deviation;
   - The risk of the deviation;
   - The required risk reduction measures.

HAZOPs are undertaken by the application of a formal, systematic and critical examination of the process and engineering intentions of a process design. The potential for hazardous situations is thus assessed and malfunction of individual items of equipment and the consequences for the whole system are identified.

HAZOP studies normally involve a team that has experience in the plant or design to be studied. These team members apply their experience of the design and their technical expertise in the HAZOP study sessions to achieve the aims of the HAZOP.

Each HAZOP has a set of objectives which are particular to that study and which are decided as near to the beginning of the study as possible. However, there are a set of four overall aims to which any HAZOP should be addressed:

- To identify all deviations from the way the design is expected to work, their causes and all the hazards and operability problems associated with these deviations.
- To decide whether action is required to control the hazard or the operability problem and, if so, to identify the ways in which the problem can be solved.
- To identify cases where a decision cannot be made immediately and to decide on what information or action is required.

To ensure that actions decided upon are followed through.

The study may not be able to resolve all the hazards that arise at the meeting, and so firm recommendations for change cannot always be guaranteed to result from deliberation at a HAZOP study meeting. The meeting may decide that it requires further information, or that a detailed study of a particular issue is necessary. Some of the questions may be answered by other personnel who did not attend the meetings, and some issues could require, for example, specific hazard analysis.

The standard guide words are given in Table 5.4.

The most common process parameters are given in Table 5.5 (CCPS, 2008).

Table 5.4: Standard guide words of a HAZOP study.

Guide words	Meaning
No or not	Negation of the design intent
Less or less of	Quantitative decrease
More or more of	Quantitative increase
Part of	Qualitative decrease
As well as or more than	Qualitative increase
Reverse	Logical opposite of the intent
Other than	Complete substitution
Early	Relative to clock time
Late	Relative to clock time
Before	Relating to order or sequence
After	Relating to order or sequence

Table 5.5: Most common process parameters for a HAZOP.

Flow	Time	Frequency	Mixing
Pressure	Composition	Viscosity	Addition
Temperature	pH	Voltage	Separation
Level	Speed	Information	Reaction

The standard guide words are combined with the process parameters to give HAZOP deviations. Table 5.6 gives some examples of HAZOP deviations and potential causes.

Table 5.6: Examples of HAZOP deviations and potential causes of these deviations.

No flow:	Wrong routing – blockage – incorrect slip plate -incorrectly fitted check valve – burst pipe – large leak – equipment failure (CV, isolation valve, pump, vessel, etc.) – incorrect pressure differential – isolation in error – etc.
Reverse flow:	Defective check valve – siphon effect – incorrect pressure differential – two-way flow – emergency venting – incorrect operation – in-line spare equipment – etc.
More flow:	Increased pumping capacity – increased suction pressure – reduced delivery head – greater fluid density – exchanger tube leaks – restriction orifice plates deleted – cross connection of systems – control faults – control valve trim changed – running two pumps – etc.
Less flow:	Line restrictions – filter blockage – defective pumps – fouling of vessels, valves, orifice plates -density or viscosity changes – etc.
More level:	Outlet isolated or blocked – inflow greater than outflow – control failure -faulty level measurement – gravity liquid balancing – etc.

Table 5.6 (continued)

Less level:	Inlet flow stops – leak – outflow greater than inflow – control failure – faulty level measurement – draining of vessel – etc.
More pressure:	Surge problems – connection to high pressure system – gas breakthrough (inadequate venting) – defective isolation procedures for relief valves – thermal overpressure – positive displacement pumps – failed open PCV's – design pressures, specifications of pipes, vessels, fittings, instruments, – etc.
Less pressure:	Generation of vacuum condition – condensation -gas dissolving in liquid – restricted pump/compressor suction line – undetected leakage – vessel drainage – blockage of blanket gas reducing valve – etc.
More temperature:	Ambient conditions – fouled or failed exchanger tubes – fire situation – cooling water failure – defective control – heater control failure – internal fires – reaction control failures – heating medium leak into process – etc.
Less temperature:	Ambient conditions – reducing pressure – fouled or failed exchanger tubes – loss of heating – depressurization of liquefied gas – Joule/Thompson effect – etc.
More viscosity:	Incorrect material or composition – incorrect temperature – high solids concentration – settling of slurries – etc.
Less viscosity:	Incorrect material or composition – incorrect temperature – solvent flushing – etc.
Composition change:	Leaking isolation valves – leaking exchanger tubes – phase change – incorrect feedstock/specification – inadequate quality control – process control upset – reaction intermediates/by-products – settling of slurries, etc.

The team will select a part of the installation to be studied (e.g., a piping between two equipment, a vessel, etc.) and the team will go through five steps:
1. The leader will announce the deviation to be studied (e.g., no flow in the pipe).
2. The team will examine possible causes.
3. The team will examine the ultimate consequences (consider hazards or operability problems).
4. The team will then list existing safeguards.
5. The team will decide upon action.

Figure 5.6 gives an example of a HAZOP on a truck unloading line with a typical recording sheet.

The team can of course add all type of other "Deviations" to be used during the HAZOP study. Table 5.7 gives some other deviations that can be used during the HAZOP study.

HAZOP is very common for process installations. However, the methodology is applicable to all types of topics and all types of installations.

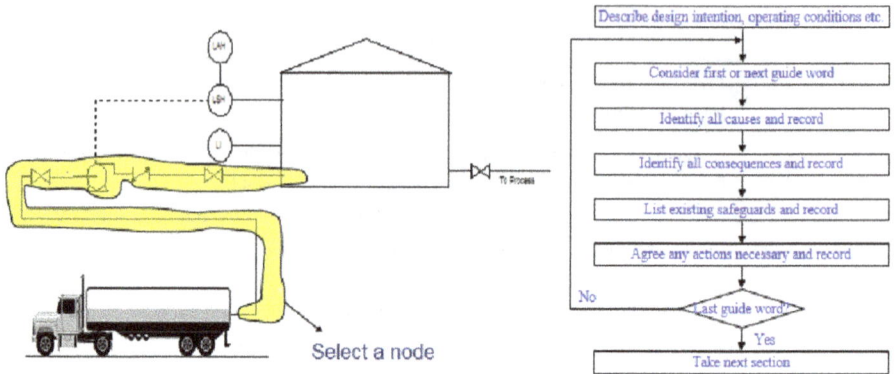

Select a node

Describe design intention, operating conditions etc.

Consider first or next guide word

Identify all causes and record

Identify all consequences and record

List existing safeguards and record

Agree any actions necessary and record

Last guide word?

No

Yes

Take next section

Project: Fuel Transfer System		Mode: 001		Page: 2	

Node Description:
The hose connection from the road tanker to the tank including the pump and manual valves on the pump inlet and outlet.

GUIDEWORD/ DEVIATION	CAUSES	CONSEQUENCES	SAFEGUARDS	REC#	RECOMMENDATIONS
No Flow	Manual valve on pump inlet closed	No transfer of fuel. Pump starved of liquid which could cause mechanical damage and possible leakage from the pump on to the road way.	Operators knowledge and training. Local stop which will be used by operator if pump leaks or is noisy.	R1	Provide low level/dry running cut out for the pump to prevent pump damage on loss of suction.
	Manual valve on pump outlet closed	No transfer of fuel. Pump runs against a closed discharge which could cause overheating, mechanical damage and leakage of left for too long.	As before. Also this valve is normally kept open.	R2	Develop a checklist for use by the operator when carrying out transfer transfers. Ensure checks of manual valve positions are included in this checklist for all stages of fuel delivery.
		As above			
	Line blocked		As above. Line blockage is unlikely since fuel delivered is clean.		

Figure 5.6: Example of HAZOP methodology on truck unloading line with typical recording sheet.

Table 5.7: Some additional HAZOP deviations that can be used during the HAZOP study.

Contamination:	Leaking exchanger tubes – leaking isolation valves – incorrect operation of system – interconnected systems (especially services, blanket systems) – effect of corrosion -wrong additives – ingress of air – shutdown and startup conditions – grade change – etc.
Relief:	Relief philosophy (process/fire etc.) – type of relief device and reliability – relief valve discharge location – pollution implications – two phase flow – effect of de-bottlenecking on relief capacity – inlet and outlet piping – etc.
Instrumentation:	Control philosophy – location of instruments -response time – set points of alarms and trips and ability to change – time available for operator intervention – alarm and trip testing -fire protection – trip/control amplifiers – panel arrangement and location – auto/manual facility and human error – fail safe philosophy – etc.

Table 5.7 (continued)

Sampling:	Sampling procedure and operator safety – time for analysis result – calibration of automatic samplers – reliability/accuracy of representative sample – diagnosis of result – loss of sample flow – etc.
Corrosion/ erosion:	Cathodic protection arrangements – internal/external corrosion protection – engineer-ing specifications – zinc embrittlement – stress corrosion cracking – fluid velocities – etc.
Service failure:	Failure of instrument air/steam/nitrogen/cooling water/hydraulic power/electric power/water or other – contamination of instrument air, nitrogen, etc. – telecommunications – heating and ventilating systems – computers – etc.
Special operation:	Purging – flushing – clearing blockages – steam out – startup – normal shutdown – emergency shutdown – emergency operations – inspection of operating machines – guarding of machinery – etc.
Maintenance:	Isolation philosophy – drainage – purging – cleaning – drying – slip plates – access – rescue plan – training – pressure testing – work permit system – condition monitoring – lifting and manual handling – etc.
Ignition:	Grounding arrangements – insulated vessels/equipment – low-conductance fluids – splash filling of vessels – insulated strainers and valve components – dust generation – powder handling equipment – electrical classification – flame arrestors – static electricity – hot surfaces – hoses – etc.
Spare equipment:	Installed/noninstalled spare equipment – availability of spares – modified specifica-tions – storage of spares – catalogue of spares – test running of spare equipment – etc.

The main differences are:
- The composition of the multidisciplinary team;
- The HAZOP deviations that will be used;
- The drawing on which the study will be done.

Table 5.8 gives an overview of some HAZOP deviations that could be used for the study of instrument loops.

Assume one would like to study an electrical distribution system. The team would of course consist of:
- Experts in electrical distribution systems (including experts in horizontal and vertical selectivity);
- Operators who can estimate the consequence of the failure of electrical power part of the units.

Similar to that of the control HAZOP but the following additional data is also required:
- Electrical distribution diagrams (i.e., one line diagrams);
- Documentation stating maximum electrical loads required for maximum process flow rates;
- General layout of equipment drawings, cable routes and area classification.

Table 5.8: Examples of HAZOP deviations to study instrument loops.

Current	No current	Failure of component, sensor, signal, power, connection, feedback signal, fuse or overload – control switched to manual or bypass for maintenance
	Reverse current	Incorrect loop connections – reverse intention received (high and low readings transposed) – initiator failure results in opposite reading
	More current	Short circuit, leak to earth – excess current protection does not work – excess protection policy – control switched to manual, interlocks not effective
	Less current	Defective limit switch (feedback) – incorrect calibration of sensor
Voltage	More voltage	Defective supply device – electrical overload protection philosophy
	Less voltage	Defective supply device – degradation of signal
Temperature	More temperature	External or internal fire – fire detection and protection, cable insulation and protection – cable routing – location of equipment resistance to ambient temperature and humidity variations
	Less Temperature	Effect of winter conditions, possible ice formation around electrical components – adjacent refrigeration plant – reliability of heating and ventilation system
Degradation	Contamination	Moisture, dust, flammable vapors, pollutants – interruption by external magnetic fields or signals
	Corrosion	Corrosion of electrical components by internal, external or galvanic means

Table 5.9 gives an overview of some HAZOP deviations that could be used for the study of electrical distribution systems.

Table 5.9: Examples of HAZOP deviations to study electrical distribution systems.

Current	No current	Distribution board component failure
	Reverse current	Wrong connections after maintenance
	More current	Plant being operated in excess of design parameters – overload settings too high
	Less current	Faulty load shedding arrangement – reduce generator capacity
Voltage	More voltage	Variable loading – electrical storm – defective transformer
	Less voltage	Voltage dip – defective contacts

A large pharmaceutical company wanted to construct a new research center. The building covered about 6 ha and was four stages high. The building would contain offices, meeting rooms restaurant and labs. Some of the labs were dedicated as biosafety level 4 laboratories, that is, labs used for diagnostic work and research on easily transmitted pathogens which can cause fatal disease. The various "objects/species" that enter the building at various times, move through the building and/or are brought outside the building are: people (staff – administrative people, technical people but also researchers, contractors and visitors), animals, food, office furniture, chemical products, different types of pathogens and so on. In addition to all standards and studies, the management decided to conduct a "HAZOP-study."

The different flows (of people, animals, goods, etc.) were identified and for each of the flows different HAZOP deviations guide words were developed (more people, less people, wrong people, contamination of people, etc.). The HAZOP study was then conducted within a multidisciplinary team. One of the hazards that was discovered via the study is that it was possible for a visitor to get lost and to achieve a level where biosafety level 2 labs were present. Biosafety level 2 is suitable for work involving agents of moderate potential hazard to personnel and the environment.

Once the study is done, a final report will be issued. It is very important that the reporting is as precise as possible. During the HAZOP, it can be decided to take short notes but the final report has to be clear and unambiguous. A HAZOP report is part of the assets in a similar way as the process book or the piping and instruments drawings. Table 5.10 gives an example of the correct recording of a HAZOP recommendation. This type of writing is however not limited to the recommendations but to the whole HAZOP report.

Table 5.10: Example of 3W ("Who, What, When") recording of a HAZOP study.

Acceptable as note but not as final recommendation	Final recommendation
Remove the manual valve	Remove the manual valve on the drain line from pH meter AT21015 to prevent the meter being from overpressurized if the valve is closed.
Add a check valve	Add a check valve on line 2" – 11015 – HD between E5041 and TV – 1015 to reduce the risk of backflow, which could lead to major damage to the compressor.

By the end of the 1990s, I was HAZOP leader in a large project. About 700 P&IDs were studied during a period of about 4 months. During the commission phase of the project, a check was done on the HAZOP recommendations. It was observed that a number of HAZOP recommendations were not correctly implemented because they were not fully understood. Sometimes the recommendation was correctly done but on the wrong location.

Another pitfall during HAZOP studies is that the consequences are not taken to the end. This means that the consequences are expressed taking into account some of the safeguards. The problem is that in that case the hazard is not identified.

Assume the following situation: a reactor has to be filled with two reactants. It is important that reactant 1 is added at least 30 min before reactant 2. The safeguards include: (1) the reactants are added via a programmable logic controller (PLC), (2) the level is measured once reactant 1 is added, (3) an independent operator verifies that reactant 1 is in the reactor and (4) a second independent operator checks the presence of reactant 1 and the 30 min before he authorizes to go to the next step which is adding reactant 2.

Table 5.11 shows an example of a wrong analysis. The recording under "consequences" takes into account the safeguards and hence it is still not clear what the ultimate consequence is by adding reactant 1 after reactant 2. It could be a quality issue or an explosion. The efficiency of the safeguards will depend on the ultimate consequence. For a quality issue, the safeguards might be considered as OK, while for an explosion, the same safeguards might not be sufficient.

Table 5.11: Example of a wrong HAZOP analysis.

Deviation	Description	Cause	Consequences	Safeguards
Wrong sequence	Reactant 2 is added before reactant 1	After an important overhaul, there has been a programming error in the PLC.	The process will be stopped via the level measurement or by an operator.	(1) The level is measured once reactant 1 is added (2) An independent operator verifies that reactant 1 is in the reactor (3) A second independent operator checks the presence of reactant 1 and the 30 min before he authorizes to go to the next step which is adding reactant 2.

There have been several attempts to automate HAZOP studies. Single et al. (2019) give a state of research in the automation of HAZOP studies.

More information on HAZOP can be found in the guide from IChemE (Crawley and Tyler, 2015).

5.5.3.7 Safety audit and checklist review

What an audit and a checklist review have in common is that a gap analysis is performed between the existing installation or equipment and a reference document. These reviews are broadly spoken of in two types:
- with open-ended points of attention or questions;
- with closed-ended points of attention or questions.

Closed-ended questions are those which can be answered by a simple "yes" or "no," while open-ended questions are those which require more thought and more than a simple one-word answer.

The idea of a safety audit or checklist review considers that the hazards are controlled except if something was forgotten or not properly looked at. An inspection with a checklist is also considered as a hazard identification technique. If substandard conditions or substandard acts are timely identified and corrected then accidents will be avoided.

The checklist has to be a formal controlled document of the organization. The topics to be checked have to be clear and unambiguous. The person who performs the gap analysis has to understand the topics on the list that he is checking and he needs to know what is OK and what is not OK.

This method is commonly used for project reviews. At different stages of a project, a checklist is used to ensure that nothing has been omitted.

5.5.3.8 Structured What If Checklist Technique

The SWIFT has been developed by Det Norske Veritas and General Electric Plastics in the 1990s as an efficient alternative to HAZOP for providing highly effective hazards identification when it can be demonstrated that circumstances do not warrant the rigor of a HAZOP. It is a combination between the checklist methodology and the HAZOP methodology.

The SWIFT is a thorough, systematic, multidisciplinary team-oriented analytical technique. HAZOP examines the plant line by line and vessel by vessel. The SWIFT, on the other hand, is a systems-oriented technique which examines complete systems or subsystems. To ensure comprehensive identification of hazards, the technique relies on a structured brainstorming effort by a team of experienced process experts with supplemental questions from a checklist. The questioning is done via 16 categories:
1. Material problems (MP)
2. External effects or influences (EE/I)
3. Operating errors and other human factors (OE&HF)
4. Analytical or sampling errors (A/SE)
5. Equipment/instrumentation malfunction (E/IM)
6. Process upsets of unspecified origin (PUUO)
7. Utility failures (UF)

8. Integrity failure or loss of containment (IF/LOC)
9. Emergency operations (EO)
10. Environmental release (ER)
11. Safety devices (SD)
12. Operability concerns (OC)
13. Quality factors (QF)
14. Process control (PC)
15. Reliability factors (RF)
16. Facility siting (FS)

A list of about 300 questions of the SWIFT methodology is given in appendix I.

5.5.3.9 Failure mode and effect analysis

From the 1940s, the US military developed a standard (MIL-P-1629, 1949) that applied to the acquisition of all designated Department of Defense (DoD) systems and equipment. The military standard was approved for use by all departments and agencies of the department of defense.

This standard "establishes requirements and procedures for performing a failure mode, effects, and criticality analysis (FMECA) to systematically evaluate and document, by item failure mode analysis, the potential impact of each functional or hardware failure on mission success, personnel and system safety, system performance, maintainability, and maintenance requirements. Each potential failure is ranked by the severity of its effect in order that appropriate corrective actions may be taken to eliminate or control the high risk items."

It primarily applies to the program activity phases of demonstration and validation and full-scale engineering development; for example, design, research and development, and test and evaluation. This standard also can be used during production and deployment to analyze the final hardware design or any major modifications.

The standard introduced a number of systematic analysis to be done as part of the demonstration the contractor had to deliver:
- Damage mode and effects analysis is analysis of a system or piece of equipment to determine the extent of damage sustained from a given level of hostile weapon damage. It examines the effect of that damage on the continued controlled operation and mission completion capabilities of the system or equipment (MIL-STD-1629A, 1980).
- FMEA is a procedure by which each a potential failure mode in a system is analyzed to determine the results or effects thereof on the system and to classify each potential failure mode according to its severity (MIL-STD-1629A, 1980).
- Failure mode, effect and criticality analysis (FMECA) consists of two steps: (1) FMEA and (2) a criticality analysis which is a procedure by which each potential

failure mode is ranked according to the combined influence of severity and probability of occurrence. Severity considers the worst potential consequence of a failure, determined by the degree of injury, property damage or system damage that could ultimately occur (MIL-STD-1629A, 1980).

The FMECA process was the first step and an integral part of a reliability process. The standard stipulates:

> The FMEA shall be initiated as an integral part of early design process of system functional assemblies and shall be updated to reflect design changes. Current FMEA analysis shall be a major consideration at each design review from preliminary through the final design. The analysis shall be used to assess high risk items and the activities underway to provide corrective actions. The FMEA shall also be used to define special test considerations, quality inspection points, preventive maintenance actions, operational constraints, useful life, and other pertinent information and activities necessary to minimize failure risk. All recommended actions which result from the FMEA shall be evaluated and formally dispositioned by appropriate implementation or documented rationale for no action.

In order to estimate the probability of failure of components, the DoD published the MIL-HDBK-217, the Military Handbook for "Reliability Prediction of Electronic Equipment" based on work done by the Reliability Analysis Center and Rome Laboratory at Griffiss AFB, NY. The MIL-HDBK-217 handbook contains failure rate models for the various part types used in electronic systems, such as integrated circuits, transistors, diodes, resistors, capacitors, relays, switches and connectors. These failure rate models are based on the best field data that could be obtained for a wide variety of parts and systems; these data are then analyzed and massaged, with many simplifying assumptions thrown in, to create usable models.

The following discrete steps shall be used in performing an FMEA (MIL-STD-1629A, 1980):

1. Define the system to be analyzed. Complete system definition includes identification of internal and interface functions, expected performance at all indenture levels, system restraints and failure definitions. Functional narratives of the system should include descriptions of each mission in terms of functions which identify tasks to be performed for each mission, mission phase and operational mode;
2. Construct block diagrams. Functional and reliability block diagrams which illustrate the operation, interrelationships and interdependence of functional entities should be obtained or constructed, for each item configuration involved in the system's use. All system interfaces shall be indicated;
3. Identify all potential item and interface failure modes and define their effect on the immediate function or item, on the system, and on the mission to be performed;
4. Evaluate each failure mode in terms of the worst potential consequences which may result and assign a severity classification category;
5. Identify failure detection methods and compensating provisions for each failure mode;

6. Identify corrective design or other actions required to eliminate the failure or control the risk;
7. Identify effects of corrective actions or other system attributes, such as requirements for logistics support;
8. Document the analysis and summarize the problems which could not be corrected by design and identify the special controls which are necessary to reduce failure risk.

The severity classes are four categories according the MIL-STD-882 as shown in Table 5.12.

Table 5.12: Severity categories according MIL-STD- 882 to be used in FMEA according MIL-STD-1629A.

Category		Description
I	Catastrophic	A failure which may cause death or weapon system loss (e.g., aircraft, tank, missile and ship).
II	Critical	A failure which may cause severe injury, major property damage or major system damage which will result in mission loss.
III	Marginal	A failure which may cause minor injury, minor property damage or minor system damage which will result in delay or loss of availability of mission degradation.
IV	Minor	A failure not serious enough to cause injury, property damage or system damage but which will not result in unscheduled maintenance or repair.

For the criticality analysis, MIL-STD-1629A considers five levels of criticality (Table 5.13).

Since the early 1960s, FMEA and FMECA became a standard method to improve the quality of a (manufacturing) process in different industries (civil aviation industry and car manufacturing industry).

Contractors used it for the US National Aeronautics and Space Administration (NASA) in several NASA programs such as Apollo, Viking, Voyager, Magellan, Galileo and Skylab.

The civil aviation industry was an early adopter of FMEA, with the Society of Automotive Engineers (SAE) publishing (Aerospace Recommended Practice (ARP)) ARP926 in 1967. After two revisions, ARP926 has been replaced by ARP4761 "Guidelines and Methods for Conducting the Safety Assessment Process on Civil Airborne Systems and Equipment." This ARP defines a process for using common modeling techniques to assess the safety of a system being put together. The first 30 pages of the document cover that process. The next 140 pages give an overview of the modeling techniques and how they should be applied. The last 160 pages give an

Table 5.13: Levels of criticality according MIL-STD-1629A.

Level	Description
A Frequent	A high probability of occurrence during the item operating time interval. High probability may be defined as a single failure node probability greater than 0.2 of the overall probability of failure during the item operating time.
B Probable	A moderate probability of occurrence during the item operating time interval. Probable may be defined as a single failure node probability which is more than 0.1 but less than 0.2 of the overall probability of failure during the item operating time.
C Occasional	An occasional probability of occurrence during the item operating time interval. Occasional probability may be defined as a single failure node probability which is more than 0.01 but less than 0.1 of the overall probability of failure during the item operating time.
D Remote	An unlikely probability of occurrence during the item operating time interval. Remote probability may be defined as a single failure node probability which is more than 0.001 but less than 0.01 of the overall probability of failure during the item operating time.
E Extremely unlikely	A failure whose probability of occurrence is essentially zero during the item operating time interval. Extremely unlikely may be defined as a single failure node probability which is less than 0.001 of the overall probability of failure during the item operating time.

example of the process in action. Some of the methods covered: Functional hazard assessment, preliminary system safety assessment, system safety assessment, FTA, FMEA, failure modes and effects summary and common cause analysis.

The Ford Motor Company introduced FMEA to the automotive industry after the Pinto affair. The Ford Pinto was manufactured by Ford Motor Company in North America and sold from the 1971 to the 1980 model years. Between 1974 and 1977, there were several articles about the fuel system design defects after reports from attorneys of three deaths and four serious injuries in rear-end collisions at moderate speeds. Approximately 117 lawsuits were brought against Ford in connection with rear-end accidents in the Pinto. The National Highway Traffic Safety Administration initiated an investigation and concluded in May 1978:

> 1971–1976 Ford Pintos have experienced moderate speed, rear-end collisions that have resulted in fuel tank damage, fuel leakage, and fire occurrences that have resulted in fatalities and non-fatal burn injuries . . .

Ford applied the same approach to processes (process FMEA (PFMEA)) to consider potential process induced failures prior to launching production. The Ford FMEA Handbook version 4.2 (2011) is publically available on internet. The handbook (286 pages) contains a chapter that provides general information about the FMEA process and three other chapters dedicated to respectively design FMEA, PFMEA and concept FMEA.

The Ford FMEA Handbook is consistent with the SAE Recommended Practice, SAE J1739 – "Potential Failure Mode and Effects Analysis in Design (Design FMEA) and Potential Failure Mode and Effects Analysis in Manufacturing and Assembly Processes (Process FMEA), and Potential Failure Mode and Effects Analysis for Machinery (Machinery FMEA)." SAE, founded in the USA in 1905, is a globally active professional association with principal emphasis on global transport industries.

DaimlerChrysler, Ford Motor Company and General Motors jointly developed the first release of this practice under the sponsorship of the United States Council for Automotive Research. SAE J1739 gives general guidance in the application of the technique. In 1993, the Automotive Industry Action Group first published an FMEA standard for the automotive industry. The SAE first published related standard J1739 in 1994. DaimlerChrysler, Ford Motor Company and General Motors representatives to the SAE have worked together to complete the latest revision of the SAE standards dated August 2002.

The purpose of FMEA according the Ford Handbook is given in Figure 5.7.

FMEA Purposes	General/overall purposes of an FMEA:
	• Improves the quality, reliability and safety of the evaluated products/processes.
	• Reduces product redevelopment timing and cost.
	• Documents and tracks actions taken to reduce risk.
	• Aids in the development of robust control plans.
	• Aids in the development of robust design verification plans.
	• Helps engineers prioritize and focus on eliminating/reducing product and process concerns and/or helps prevent problems from occurring.
	• Improves customer/consumer satisfaction.

Figure 5.7: Purpose of FMEA according Ford FMEA handbook version 4.2 (2011).

FMEA is commonly used as part of a reliability study for renewable energies such as hydropower (Lifar and Brom, 2019), photovoltaic systems (Colli, 2015) and windmills (Arabian-Hoseynabadi et al., 2010; Shafiee and Dinmohammadi, 2014).

The International Electrotechnical Commission (IEC) revised in 2006 its international standard IEC 60812 from 1985. The revised standard "Analysis techniques for system reliability – Procedure for failure mode and effects analysis (FMEA)" (IEC, 2006) contains the following main changes: (1) introduction of the failure modes effects and criticality concepts, (2) inclusion of the methods used widely in the automotive industry, (3) added references and relationships to other failure modes analysis methods, (4) added examples and (5) provided guidance of advantages and disadvantages of different FMEA methods.

Once the FMEA is done, the criticality analysis starts. Each failure mode gets a numeric score that quantifies (a) likelihood that the failure will occur, (b) likelihood that the failure will not be detected and (c) the amount of harm or damage the failure mode may cause to a person or to equipment.

The product of these three scores is the risk priority number (RPN) for that failure mode. The sum of the RPNs for the failure modes is the overall RPN for the process. As an organization works to improve a process, it can anticipate and compare the effects of proposed changes by calculating hypothetical RPNs of different scenarios. Just remember that the RPN is a measure for comparison within one process only; it is not a measure for comparing risk between processes or organizations.

While FMEA identifies important hazards in a system, for a process installation, its results may not be comprehensive as a HAZOP. In an FMECA, a "failure of the item" is considered in terms -of performance parameters and allowable limits for each specified output. A small leak on a pump is not necessarily a failure as long as the pump is able to continue to achieve its intended performance.

If used as a top-down tool, FMEA may only identify major failure modes in a system. FTA is better suited for "top-down" analysis. When used as a "bottom-up" tool FMEA can augment or complement FTA and identify many more causes and failure modes resulting in top-level symptoms.

Additionally, the multiplication of the severity, occurrence and detection rankings may result in rank reversals, where a less serious failure mode receives a higher RPN than a more serious failure mode. The reason for this is that the rankings are ordinal scale numbers, and multiplication is not defined for ordinal numbers. The ordinal rankings only say that one ranking is better or worse than another, but not by how much. For instance, a ranking of "2" may not be twice as severe as a ranking of "1," or an "8" may not be twice as severe as a "4," but multiplication treats them as though they are.

5.5.4 Selection of an appropriate specific hazard identification technique

There are many studies in the literature about the different hazard identification methods and comparison between them (e.g., Gould et al., 2000). All methods have advantages and inconveniences.

Table 5.14 gives some guidance for the choice of the method.

Table 5.14: Some advantages and inconveniences of different hazard identification techniques.

Method	Advantages	Inconveniences
PHA	– Easy to understand – Brainstorming – Can be used as a first screening – Can be used from concept phase	– Not a detailed method
Checklist	– Straightforward to use – Can be done by a small group or single person – The checklist incorporates experience and knowledge which does not has to be developed over and over again	– Only known hazards are checked – No cause and effect analysis – The identification will be as good as the checklist – May not apply to a particular situation
HAZOP	– Very effective for process installations; – Systematic and in depth analysis (line by line) – Provides hazard related to process deviations – Can be used for all type of situations	– The method seems simple but in reality is very difficult – Time and resource consuming
FMEA	– Systematic, element by element – Well established engineering practice	– Time consuming – Mainly about reliability, quality and failure to perform the mission

5.6 Developing scenarios

5.6.1 Undesired event versus accidental scenario

The identification techniques will be elementary for identifying hazards and potential undesired events. This information will have to be turned into an accidental scenario for further study.

The difference between a scenario (see Table 5.15) and an identified undesired event is that a scenario is much more elaborated and that it enables to estimate the associated risk (potential consequence and probability of occurrence). In generic QRA studies, calculations can be done on single undesired events (e.g., leak in a pipe) but the software (e.g., SAFETI from DNV) will the automatically generate generic scenarios by combining all potential outcomes (dispersion, rainout, pool formation, etc. – see further under the chapter generic scenario generation).

Table 5.15: Difference between an undesired event and a scenario.

Undesired event identified during a HAZOP	Scenario
Overfill of a tank with the potential for the formation of a flammable cloud and upon ignition there could be a fire or a vapor cloud explosion	During a filling operation of an atmospheric storage tank with winterized gasoline (contains 10% of butane), the high-level instrument fails and the tank is overfilled. The tank is equipped with one single overfill protection. The gasoline comes out at a rate of 10 m^3 per h during 30 min. It flows over the roof of the tank and fells down from 12 m height.

A number of textbook list generic scenarios for different type of equipment (e.g., CCPS, 1998).

The following techniques are very helpful in developing scenarios:

- FTA
- ETA
- Cause-consequence analysis
- Bow-tie analysis

5.6.2 Fault tree analysis

The fault tree technique was introduced in 1962 at Bell Telephone Laboratories, in connection with a safety evaluation of the launching system for the intercontinental Minuteman missile. The Boeing Company improved the technique and introduced computer programs for both qualitative and quantitative FTA.

The use of the technique spread rapidly through all type of industries. FTA was introduced in for instance Du Pont Company in 1966 and could immediately be applied to chemical process safety analysis because the mathematical evaluation methods for batch, nonrepairable systems had been developed. The appropriate equations for continuous processes were developed in the early 1970s (Prugh, 1981).

FTA is a deductive (top-down) method of analysis aimed at pinpointing the causes or combinations of causes that can lead to the defined top event. The analysis can be qualitative or quantitative depending on the scope of the analyzes.

> Fault tree analysis (FTA) is concerned with the identification and analysis of conditions and factors that cause or may potentially cause or contribute to the occurrence of a defined top event.
> (IEC Standard 61025, 2006 which cancels and replaces the first edition, published in 1990)

FTA requires the use of symbols, identifiers and labels in a consistent manner as shown in Figure 5.8.

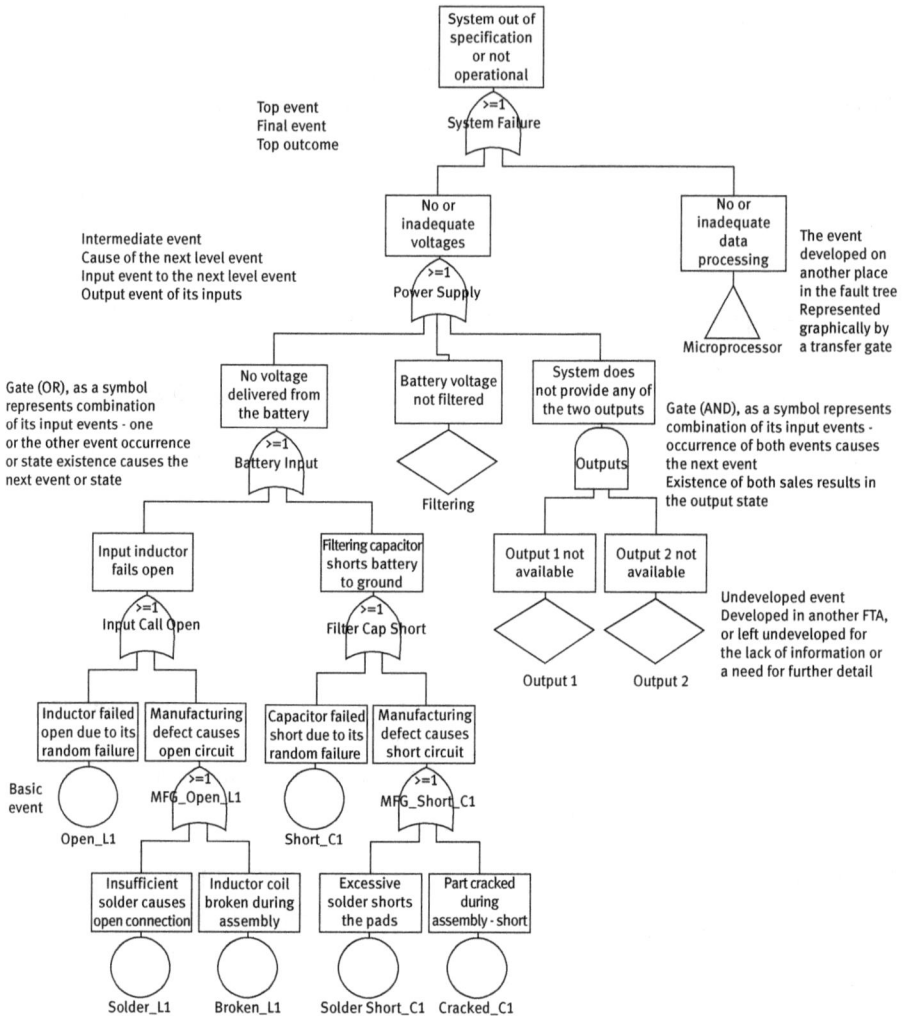

Figure 5.8: Use of symbols, identifiers and labels in fault tree analysis (IEC 61025, 2006).

In order to use the fault tree technique effectively as a method for system analysis, the procedure should consist of at least the following steps:
- Definition of the scope of the analysis;
- Familiarization with the design, functions and operation of the system;
- Definition of the top event;
- Construction of the fault tree;

– Analysis of the fault tree logic;
– Report on results of the analysis;
– Assessment of reliability improvements and trade-offs.

> A fault tree should be constructed in such way that it clearly represents the flow of events that cause occurrence of the top event. Once the top event is clearly defined as well as the system boundaries or analysis extent, then the fault tree construction flows from the top down. The top event has inputs and their combination is modelled and represented by the appropriate gate. Inputs into the top event are then developed systematically to be results of their own input events. Each of the inputs, going down the fault tree, is developed separately and this development is completed when the primary events are reached. (IEC 61025, 2006)

IEC 61025, 2006 (114 pages) contains more information:
– Details of the different steps in the construction of a fault tree
– Combinations with other reliability analysis techniques such as FMEA, reliability block diagram and Markov analysis
– The fault tree development and evaluation
– Use of quantitative FTA in system or product development for reliability improvement (series systems, parallel systems, active and passive redundancy)
– Logical analysis (Boolean reduction, identification of minimal cut sets, numerical analysis)
– Examples

FTA was extensively used in the 1970s and 1980s as a part of major risk studies:
– WASH-1400, "The Reactor Safety Study," was a report produced in 1975 for the Nuclear Regulatory Commission by a committee of specialists under Professor Norman Rasmussen.
– The US Coast Guard Vulnerability Model, a computerized simulation system for assessing damage that results from marine spills of hazardous materials (Eisenberg et al., 1975)
– *Risk Analysis of Six Potentially Hazardous Industrial Objects in the Rijnmond Area: a Pilot Study*, A Report to the Rijnmond Public Authority, 1981 (COVO study)
– *Canvey: An Investigation of Potential Hazards from Operations in the Canvey Island/Thurrock Area* (HSE, 1978 and HSE, 1981)

The conceptual adequacy of fault tree methodology has been challenged in the 1970s for the following reasons (see WASH 1400, Executive summary, page 149, 1975):
a. Fault trees cannot identify all potential causes of system failure and hence yield underestimates of system failure probability;
b. Fault trees are subjective because the analyst must decide which events are to be incorporated into the trees and which events are to be omitted;
c. The results of the quantification of fault trees cannot be relied on because of insufficient failure data.

However, FTA has increasingly been used in all types of industry. The US Nuclear Regulatory Commission published *Fault Tree Handbook* (NUREG-0492) in 1981 and NASA Office of Safety and Mission Assurance published *Fault Tree Handbook with Aerospace Applications*, Version 1.1 in 2002.

5.6.3 Event tree analysis

The event tree is complementary to the fault tree. The event tree starts with an undesired event and evaluates all potential sequences in a systematic way. Figure 5.9 shows an example of an event tree.

The event tree can be quantified in order to estimate the probability of each of the outcomes (see for instance the example on ETA in Wikipedia).

5.6.4 Cause-consequence diagram and bow-tie method

The fault tree can be combined with the event tree into a cause-consequence diagram. The first articles about the cause-consequence diagram date from the early 1970s (Nielsen, 1971).

The combination between the FTA and ETA is also discussed under Section 5.4.2 of the standard of IEC Standard 61025 (2006) which refers to it as cause-consequence analysis. A cause-consequence diagram gives an overview of all potential causes of an undesired event as well as all potential outcomes.

However, due to the complexity of modern facilities, it is difficult for the operators to envisage all possible interactions if something were to go wrong and hence there was a need for a simplified version. In the 1990s, Shell International Exploration and Production developed a number of tools as part of their HSE Manual. In EP 95-0300 (1995), Shell promotes the use of cause-consequence diagrams which were also called "bow-tie" analysis (Shell, 1995). Shell developed a software called THESIS to perform the bow-tie analysis (Trbojevic, 2008)

Bow-tie's assist the workforce to easily understand the main hazard issues and can recognize themselves as the "owners" of hazard barriers in their day-to-day tasks.

5.6.5 Generic scenario generation

In order to perform a risk assessment, it will be necessary to define a scenario. The hazard identification techniques allowed to identify the hazards (e.g., leak in a pipe, rupture of a vessel, overflow of a tank), but in order to perform a detailed consequence analysis and probability assessment, it will be necessary to develop a full scenario.

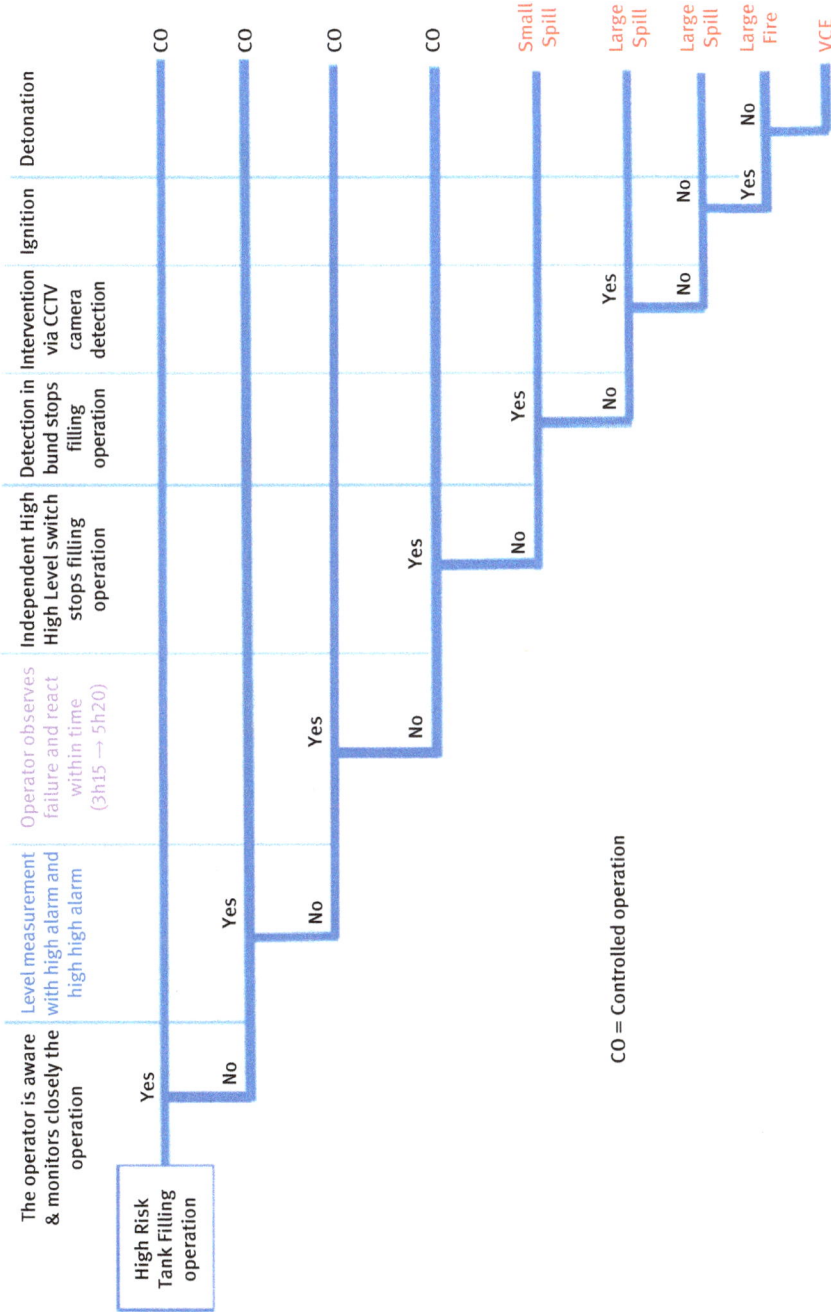

Figure 5.9: Example of an event tree.

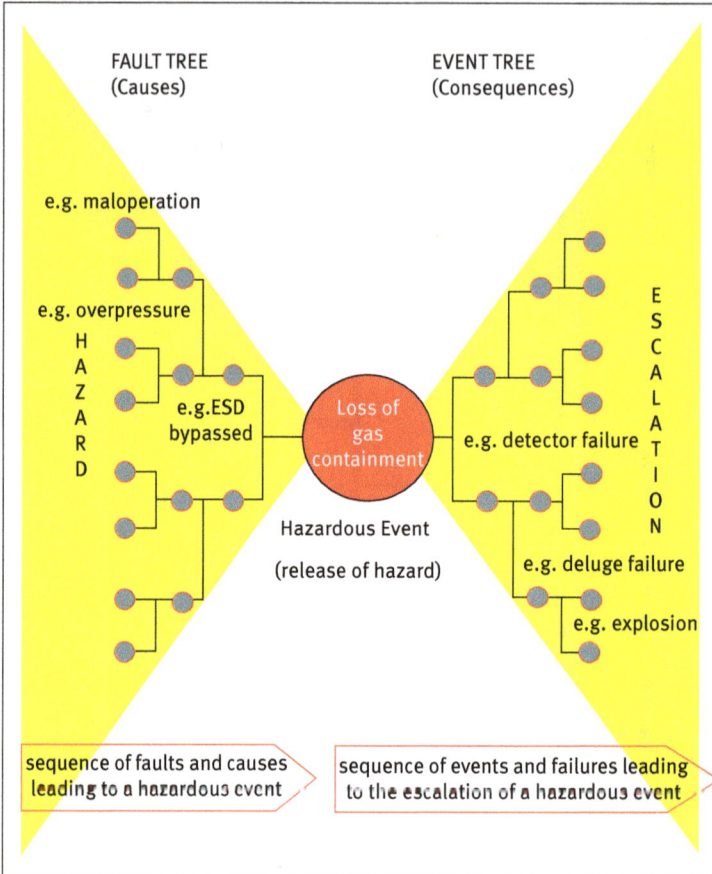

Figure 5.10: Cause-consequence diagram (bow-tie) (Shell International EP, 1995).

The characteristics of the leak will have to be specified (size, direction, product, temperature, pressure) and the possible outcomes will have to be defined.

Figure 5.11 shows an example of the potential outcomes of a leak of a flammable pressurized product such as propane.

Some software tools have in-built generic scenarios. The user can define the different probabilities but in the end the software will automatically calculate the generic scenarios.

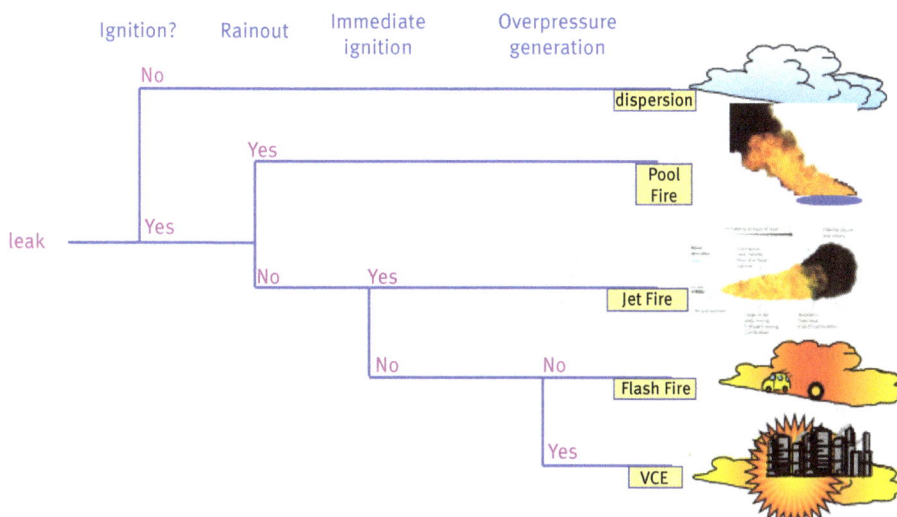

Figure 5.11: Possible outcomes of a leak of hazardous product.

5.7 Probability analysis

5.7.1 Understanding the use of probabilities in risk studies

If one wants to calculate the probability of finding the correct six numbers from 1 to 45 when no duplications are allowed then there is only one right answer which is one chance in 8,145,060. The result is obtained with the following formula $\left(\frac{45!}{6!(45-6)!}\right)$. There is only one mathematical correct answer.

In risk studies, the purpose is not the same. The purpose of risk studies is to "get a feeling for the level of uncertainty" before taking a decision. The probability analysis can be qualitative as well as quantitative. An example of a 5 level qualitative analysis of the probability of occurrence could be:
- Likely to happen;
- Unlikely to happen;
- Very unlikely to happen;
- Extremely unlikely to happen;
- Remote.

Such a qualitative approach can work very well but there might be lengthy discussions about the common understanding between, for instance, "unlikely" and "very unlikely."

Quantitative guidance was developed in order to help make a more objective selection of the level of uncertainty. An example of such guidance is given in Table 5.16.

Hence, the quantification of the probability of occurrence in risk analysis is to objectify the selection of the qualitative level of uncertainty.

Table 5.16: An example of quantitative guidance on qualitative level of uncertainty.

Qualitative level of uncertainty	Quantitative guidance (frequency per year)
Likely	More than 10^{-2}
Unlikely	More than 10^{-3} but less than 10^{-2}
Very unlikely	More than 10^{-4} but less than 10^{-3}
Extremely unlikely	More than 10^{-5} but less than 10^{-4}
Remote	Less than 10^{-5}

5.7.2 Types of probabilities used in quantitative risk analysis

There are three ways of classifying probabilities:
- Classical probability;
- Relative frequency of occurrence;
- Subjective probability.

The **classical probability** defines the probability that an event will occur as:

$$probability\ of\ an\ event = \frac{number\ of\ times\ the\ event\ occurs}{total\ number\ of\ possible\ outcomes}$$

A typical example is the probability of getting "head" when tossing with a coin.

The **relative frequency of occurrence** is a method where the relative frequencies of past occurrences are used as a probability that the event will happen again in the future. Supposing that we know from past actuarial data that of all males, 40 years old, about 60 out of every 100,000 died within a 1-year period, then using this knowledge, the probability of death for that age group for the coming year is calculated as:

$$\frac{60}{100,000} = 0.0006$$

The relative frequency of occurrence could also be used for tossing a coin a limited number of times (100 for instance). The relative frequency of occurrence will become more accurate as we increase the number of observations and it will become equal to the classical probability as the number of tosses increases.

Subjective probabilities are based on the beliefs of the person making the probability assessment. It can be defined as the probability assigned to an event based on whatever evidence is available.

Within risk analysis and risk assessment, the three types of probabilities are used interchangeable.

5.7.3 Sources of data

Statistical data on equipment has been collected for almost 80 years and is available in all types of formats.

According to Shewhart (Shewhart, 1939), it all started in 1870 with the introduction of the go, no-go tolerance limits. It became generally accepted practice to specify that each important quality characteristics of a given piece of product should lie within stated limits. Engineers started to perform quality control inspections.

In the late 1910s and 1920s, mass production was popularized by Henry Ford's Ford Motor Company, which introduced electric motors to the then-well-known technique of chain or sequential production. In mass production, it is not possible (or not desired) to inspect the important quality characteristics of every single piece. That is why, from 1920, the use of statistical process control was promoted and was taught at universities. Statistical techniques were used to decide on the number of inspections that had to be done. The course of Shewhart (Shewhart, 1939) was one of the most famous courses. Walter Andrew Shewhart is known as the father of statistical quality control.

The modern use of the word "reliability" was defined by the US military in the 1940s, characterizing a product that would operate when expected and for a specified period of time. In World War II, many reliability issues were due to the inherent unreliability of electronic equipment available at the time and due to fatigue issues.

The American Institute of Electrical Engineers, a US-based organization of electrical engineers founded in 1884 which became in 1963 the Institute of Electrical and Electronics Engineers (IEEE) formed the Reliability Society in 1948. In 1972, the IEEE published the first edition of Standard 352, which contains a basic methodology necessary to conduct a reliability analysis. At the time of publication, it was recognized that this methodology was incomplete without a supporting database. Work began to produce a database to support IEEE's Standard 352 and this culminated in 1977 with the publication of IEEE Standard 500.

In 1946, Aeronautical Radio Inc. (ARINC) was established by the airlines to collect and analyze defective tubes in commercial airline applications. One of the results of the 1950 ARINC study was the discovery that only 33% of navy electronic equipment on shipboard was operating satisfactorily at any one time (MIL-HDBK-338).

The focus on electron tube reliability continued in the early 1950s. In 1950, an Ad Hoc Group on electronic equipment was formed by the DoD. One of the by-products of this activity of importance to electronic part reliability was the formation of the

Advisory Group on Reliability of Electronic Equipment (AGREE) in August 1952. This group recommended three main ways of working:
- Improve component reliability.
- Establish quality and reliability requirements for suppliers.
- Collect field data and find root causes of failures.

In March 1954, the Panel on Electron Tubes was redesignated as the Advisory Group on Electron Tubes. In June 1954, the Advisory Group on Electronic Parts was formed.

One of the recommendations found in the AGREE report of June 1957 was that a permanent group be established at the DoD level to include representatives of industry and the three services to be charged with the task of developing military component specifications, testing component parts for design capability and developing inspection methods. This group would serve as the National Reliability Center. One year later, in July 1958, the Office of the Assitant Secretary of Defense (OASD), and the Ad Hoc Group on Parts Specification Management for Reliability was established to analyze the recommendations made by AGREE in order to advise the Assistant Secretaries of Defense Research and Engineering and Supply and Logistics.

In 1958, the Air Force issued the RADC Reliability Notebook which was revised several times until its final version in 1968. The RADC Reliability notebook also included human error as a contributor to the reliability: "The importance of the human element to achieve operational reliability varies with the proportion of the system functions that require human performance, and the sensitivity of the system to human error" (RADC Reliability Notebook pages 11–49, 1968). The method was published in 1965 and consisted of calculating the sum of the task rating (see Figure 5.12 and estimating the error rate as function of the sum of the task ratings (see Figure 5.13).

The requirements set by the US military prompted all supply companies to perform their own reliability studies. As from the mid-1950s, companies in different industries started to collect their own data. Most of this data remained in-house and was or is publically not available.

According to Akhmedjanov, it was in 1959 that the earliest broad-based widely distributed source of reliability data widely distributed source was published. It contained generic failure rates on a wide range of electrical, electronic, electromechanical and mechanical parts and assemblies. It was also the first known source to standardize the presentation of failure rates in terms of failures per 10^6 hours eliminating the necessity for conversions (Akhmedjanov, 2001).

Commercial nuclear power stations became popular in the 1950s. The first "commercial" nuclear power stations were taken in service in respectively 1956 in England (the 50 MW reactor at wind scale) and 1957 in the USA (the 60 MW Shippingport Reactor in Pennsylvania). The United States Atomic Energy Commission published in 1957 the WASH 740 report. The title of the report (see Figure 5.14) refers to "highly improbable" but the study does not contain a lot of data to support this statement. The estimate of probability was in the range of 1E-8 to 1E-5 per reactor year.

Task Element	Rating Mean	Rating S.D.	Reliability Estimate	Task Element	Rating Mean	Rating S.D.	Reliability Estimate
Read technical instructions	8.3	2.2	.9918	Fill sump with oil	4.3	1.6	.9981
Read time (Brush Recorder)	8.2	2.1	.9921	Disconnect flexible hose	4.2	2.0	.9982
Read electrical or flow meter	7.0	2.8	.9945	Lubricate torque wrench adapter	4.2	2.2	.9982
Inspect for loose bolts and clamps	6.4	1.9	.9955	Remove initiator simulator	4.1	1.9	.9983
Position multiple position electrical switch	6.3	2.4	.9957	Install protective cover (friction fit)	4.1	2.2	.9983
				Read time (watch)	4.1	2.1	.9983
Mark position of component	6.2	2.1	.9958	Verify switch position	4.1	1.9	.9983
Install lockwire	6.0	2.3	.9961	Inspect for lockwire	4.1	2.1	.9983
Inspect for bellows distortion	6.0	2.7	.9961	Close hand valves	4.0	2.6	.9983
Inspect Marman clamp	6.0	1.8	.9961	Install drain tube	4.0	2.1	.9983
Install gasket	6.0	2.1	.9962	Install torque wrench adapter	3.9	1.7	.9984
Inspect for rust and corrosion	5.9	2.1	.9963	Open hand valves	3.8	2.6	.9985
Install "O" ring	5.7	2.2	.9965	Position low position electrical	3.8	1.5	.9985
Record reading	5.7	2.3	.9966	switch			
Inspect for dents, cracks and scratches	5.6	2.4	.9967	Spray leak detector	3.7	2.0	.9986
				Verify component removed or	3.5	2.4	.9988
Read pressure gauge	5.4	2.2	.9969	installed			
Inspect for frayed shielding	5.4	2.3	.9969	Remove nuts, plugs and holts	3.5	1.7	.9988
Inspect for QC seals	5.3	2.6	.9970	Install pressure cap	3.4	1.6	.9988
Tighten nuts, bolts and plugs	5.3	2.6	.9970	Remove protective closure (friction	3.2	1.6	.9990
Apply gasket cement	5.3	2.3	.9971	fit)			
Connect electrical cable (threaded)	5.2	2.2	.9972	Remove torque wrench adapter	3.0	1.6	.9991
Inspect for air bubbles (leak check)	5.0	2.2	.9974	Remove reducing adapter	3.0	1.7	.9991
Install reducing adapter	4.9	1.6	.9975	Remove Marman clamp	3.0	1.7	.9991
Install initiator simulator	4.9	2.5	.9975	Remove pressure cap	2.8	1.8	.9993
Connect flexible hose	4.9	2.4	.9975	Loosen nuts, bolts and plugs	2.8	1.3	.9993
Position "zero in" knob	4.8	1.6	.9976	Remove union	2.7	1.4	.9993
Lubricate bolt or plug	4.7	2.7	.9977	Remove lockwire	2.7	1.5	.9993
Position hand valves	4.6	1.6	.9979	Remove drain tube	2.6	1.4	.9993
Install nuts, plugs and bolts	4.6	1.7	.9979	Verify light illuminated or	2.2	1.6	.9996
Install union	4.5	1.8	.9979	extinguished			
Lubricate "O" ring	4.5	2.5	.9979	Install funnel or hose in can	2.0	0.8	.9997
Rotate gearbox train	4.4	2.0	.9980	Remove funnel from oil can	1.9	1.4	.9997

Figure 5.12: Means and standard deviations of ratings and reliability estimates for the task elements (RADC Reliability Notebook, 1968).

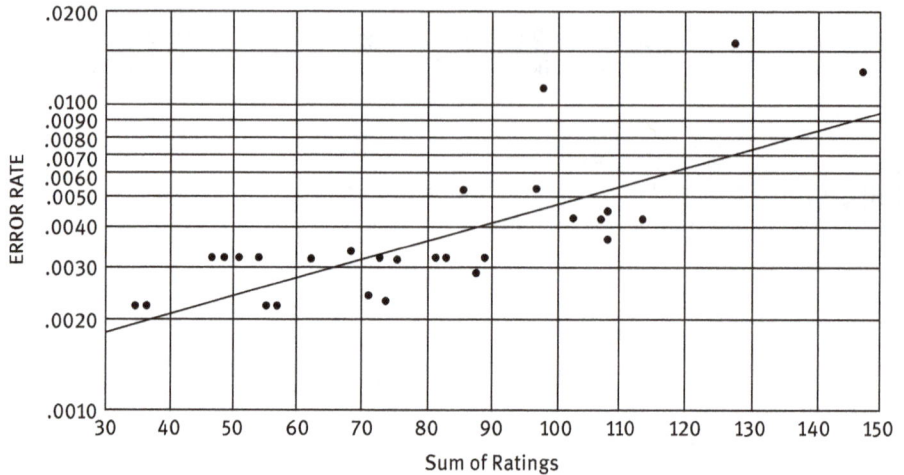

Figure 5.13: Error rate vs. error potential rating (RADC Reliability Notebook, 1968).

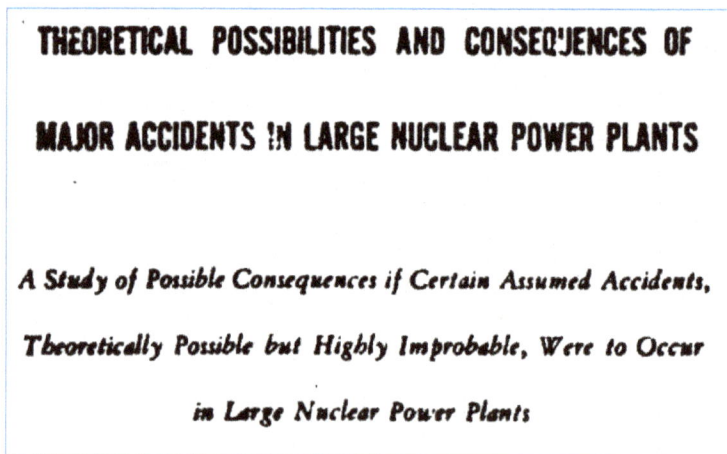

THEORETICAL POSSIBILITIES AND CONSEQUENCES OF

MAJOR ACCIDENTS IN LARGE NUCLEAR POWER PLANTS

A Study of Possible Consequences if Certain Assumed Accidents,

Theoretically Possible but Highly Improbable, Were to Occur

in Large Nuclear Power Plants

Figure 5.14: Title on front page of WASH 740 report (United States Atomic Energy Commission, 1957).

The probabilities used in WASH 740 report were estimates for which the origin of the date was not clear. In 1967 (10 years later), the discussion about the data needed to perform reactor safety assessment by probability methods was discussed by Crickmer at a Symposium on Fast Reactor Physics and related Safety Problems held by the International Atomic Energy Agency in Karlsruhe (Crickmer and Cave, 1968). Crickmer argued that the origin of the data and its validity was still a problem.

To perform Safety assessment of fast reactors by probability methods it will be necessary to assemble a considerable volume of information. Some of this data falls in the category of plant performance and reliability and some, of more interest to this Symposium, in the field of reactor physics. Most of this data is not available at the present time. The engineering data could probably be assembled from the records of manufacturers, electricity supply companies, etc., but the physics data has yet to be produced.

The WASH 740 report was revised by WASH 1400 report which was published in 1975. The WASH 1400 report makes extensive use of probabilities but here again, the probabilities are taken from all types of literature sources. The lack of hard data in developing the component failure rate estimates has been a major contributor to the large uncertainties and was highlighted by the well-publicized "Lewis Committee Report." Other studies, including WASH 1400, have clearly identified the large uncertainties inherent in using current data banks.

In the meantime, there is an abundance of articles in which reliability data is given (Philips and Warwick, 1969; Anyakora et al., 1971; Green, 1972; Smith and Warwick, 1974; Bush, 1975; Lees, 1976 and Arulanantham and Lees, 1981) but without any explanation on the origin of the data. Figure 5.15 gives an example of the date given in the article of Green (Green, 1972).

It is not clear where the data comes from, how it was collected and under which circumstances it can be used.

Item	Failure Rate (faults/year)
Bellows	0.05
Diaphragms, metal	0.05
Gaskets	0.005
Springs, lightly stressed	0.002
Unions and junctions	0.004
Nut and bolt	0.0002
Transistor, alloy germanium	0.01
Transistor, alloy silicon	0.005
Diode, silicon zener	0.001
Relays, each coil (general)	0.003
Relays, *each contact pair (general)*	0.002
Relays, P.O. type (general)	0.01
Milliameter, moving coil	0.15
Strip chart recorder	1.7
Electronic power supply unit	0.23
Electronic trip amplifier	0.95
Electronic servo trip unit	2.5
Pressure switch	0.14
Pneumatic valve, 5 part	0.97
Pneumatic valve shut-off	1.9
Overhead transmission line	5.4
Three phase, oil-filled transformer	0.09
Centrifugal pump, water	6.0
Diesel engine	8.8
Oil boiler, steam, raising	5.6

Figure 5.15: Example of equipment failure rate data (Green, 1972).

Since the publication of the Reactor Safety Study WASH 1400 in 1975, the interest in and need for the use of probabilistic techniques to assess the safety and reliability of nuclear power generating stations continued to grow and all published data was very welcome despite the uncertainty on its validity.

This data was needed and used by four categories of users:

- Risk analysts for analyzing and quantifying the level of risk of an installation or activity;
- Reliability analysts for analyzing the reliability and availability of complex systems;
- Maintenance engineers for measuring the maintenance performance;
- Component designers for analyzing the component performance.

All of these specialists need different types of data. The level of accuracy is less important for risk analysts than for the other users. Risk analysis has been done for decades by using these generic data from all type of sources.

Since the late 1970s, significant efforts were made in different industries to improve the accuracy of data and to make available industry specific data:

Nuclear industry
In the nuclear industry, a lot of effort has been done to make a database with specific data for nuclear power plants. Since 1978, the American National Standards Institute/Failure Incidents Reports Review committee has sponsored a voluntary program of visits to nuclear power stations for record collection at the plant site and the conversion of these records into a comprehensive database to be applied to risk and reliability analyzes (Drago et al., 1982).

The sources of the data are the in-plant maintenance work request records from a sample of nuclear power plants. This database is called the In-Plant Reliability Data system. Some data has been made publically available by the International Atomic Energy Agency under the form of NUREG or Technical documents

Offshore industry
The offshore industry started in the 1980s with the collection of offshore reliability data. The Norwegian Petroleum Directorate (now known as Petroleum Safety Authority) initiated the OREDA Project in 1981. The primary objective was to collect reliability data for safety equipment. It was agreed that OREDA was to be run by a group of eight oil companies in 1983. The objective of OREDA was subsequently expanded to collect experience data from the operation of offshore oil and gas production facilities to improve the basic data in safety reliability studies.

The first OREDA Reliability Data Handbook was 1984 edition which has been updated in 1992, 1997 (DNV, 1997), 2002, 2009 and 2015.

The database contains data from almost 300 installations, over 15,000 pieces of equipment, nearly 40,000 failure records and close to 75,000 maintenance records. Access to this data, and to the search and analysis functions of the OREDA software, is restricted to the OREDA member companies, though contractors working with member companies may be granted temporary access.

ISO 14224, 2016 Petroleum, petrochemical and natural gas industries ISO 14224 Petroleum, petrochemical and natural gas industries – This international standard provides a comprehensive basis for the collection of reliability and maintenance data in a standard format for equipment in all facilities and operations within the petroleum, natural gas and petrochemical industries during the operational life cycle of equipment. It describes data collection principles and associated terms and definitions that constitute a "reliability language" that can be useful for communicating operational experience. The failure modes defined in the normative part of this international standard can be used as a "reliability thesaurus" for various quantitative as well as qualitative applications. This international standard also describes data quality control and assurance practices to provide guidance for the user.

OREDA started to organize the data taking into account the guidelines from ISO 14224. The downstream industry (refineries, petrochemical plants, chemical plants) have no data collection program in place like the offshore. ISO 14224 could be a starting point.

The rise of renewable energy means that data files with reliability date and failure rates are also being created in that domain:

–   Data for components of photovoltaic systems has been published by many authors. However, the data still is very diverse. Sayed studies the reliability and availability for grid-connected solar photovoltaic systems (Sayed et.al., 2019). In his article, he gives the failure data that has been used in number of failures per year. Baschel studied the impact of component reliability on large-scale photovoltaic systems' performance. He lists the failure rate data that was used expressed in number of failures per hour (Baschel et al., 2018).
–   The UK offshore wind farm industry launched system performance, availability and reliability trend analysis (SPARTA) in 2014, which is a major collaborative project between ORE Catapult, The Crown Estate and offshore wind farm owner/operators. SPARTA is comparable for the offshore wind industry to what OREDA is for the offshore oil and gas industry.
–   The project is a database for sharing anonymized offshore wind farm performance and maintenance data.

- Sandia National Laboratories in the USA announced in 2008 the development of a national reliability database (Hill et al., 2008). The database would contain reliability data for all types of components needed for reliability studies on windmills (wind turbines, roller bearings, gears, lubrications pumps, couplings, gaskets, circuit breakers, AC motors, etc.).

Another type of statistical data that has been collected over the last 50 years is "frequency of occurrence of accidental events" in the major hazards industries. Failure rates are derived for all types of activities. Some examples:
- Accidents on cross-country pipelines are collected on a regular base by different organizations such as CONCAWE, UKOPA and EGIG. A recent summary has been made by Chaplin with recommendations of the failure rates to be used for land planning assessments in the UK (Chaplin and Howard, 2015).
- Data for hydrocarbon releases in UK offshore industry can be found in spreadsheet format on the website of https://www.hse.gov.uk/offshore/statistics/index.htm.
- Statistics on accidents with helicopters expressed as a frequency per 100,000 flight hours are available.
- Ship/platform collision incident frequencies (e.g., Robson, 2003).
- The International Association of Oil and Gas Producers (IOGP) published a report on storage incident frequencies (434-3, 2010).
- The World Offshore Accident Database contains information on over 6,700 incidents and technical information on 3,700 offshore units. The data source includes chains of events, causes and consequences. It is derived mainly from public domain sources such as Lloyds Casualty Reports, industry publications, newspapers and official publications. The majority of the data are from the North Sea (57%) and US Gulf of Mexico (26%).
- The SINTEF Offshore Blowout Database was released as an internet database in 2009. It includes good description, present operation, blowout causes and characteristics. The database also categorizes the blowout type and consist of worldwide data dating back to 1955.

A difficulty risk analysts encounter when using these studies is related to the sampling error. In statistics., sampling errors are incurred when the statistical characteristics of the sample are different from the characteristics of the entire population. Sampling is the selection of a subset of individuals or items from within a statistical population to estimate characteristics of the whole population. A main challenge is to define the population and finding a representative sample (or subset) of that population. The population from which the sample is drawn may sometimes not be the same as the population about which information is desired. Another challenge is to perform a probability sample which is a sample in which every unit in the population has a chance (greater than zero) of being selected in the sample, and this

probability can be accurately determined. The statistical studies that are available for the purpose of risk analysis are most of the time nonprobability sampling because the probability of selection cannot be accurately determined:
- Some elements of the population have no chance of selection;
- Some elements are selected based on assumptions regarding the population of interest, which forms the criteria for selection.

Nonrandom, nonprobability sampling does not allow the estimation of sampling errors. These conditions give rise to statistical feature of a statistical technique or of its results whereby the expected value of the results differs from the true underlying quantitative parameter being estimated.

Besides the fact that the origin of the collected data is not always obvious and the fact that the data was obtained by nonrandom and nonprobability and additional problem is that the data is not always consistent.

Idaho National Engineering and Environmental Laboratory made a report with a compilation of component failure rate and repair rate values that can be used for the International Thermonuclear Experimental Reactor safety assessment. Collection of component failure rate data is one of the tasks in the International Energy Agency cooperative agreement on Environmental, Safety and Economic Aspects of Fusion Energy (Cadwallader, 1998). The study observes a factor of 10 difference between leakage/rupture failure rates for piping while Eide and Calley (1993) suggest that the difference could be larger, perhaps a factor of 25 or 100.

Risk of an undesired event was defined as a combination of the potential consequence and its associated probability of occurrence. A difference of a factor 10 or more on the probability of occurrence has a proportional impact on the estimated level of risk.

5.7.4 Engineering approach for selecting failure data

Since the 1980s, the situation is as follows:
- The quantity of statistical data and reliability data is overwhelming and continues to grow (exponentially).
- Several organizations have worked on collections of failure rate data and improved largely the quality of the data.
- Despite the important improvements, it is not possible to verify for each equipment and each failure mode whether the available data is representative for the population of interest.
- For a same event (e.g., leak in a pipe), the probabilities that one can find in these databases can have a difference of a factor up to 100.

In order to ensure consistency among the different risk assessments companies, associations and authorities published guidance for the data to be used (e.g., CCPS, 1989).

Technica Ltd. (which was integrated in DNV in 1991) was founded in 1983 and provided risk analysis, and risks assessment services to industries all over the world was confronted like many other consultants with the challenge of delivering consistent studies for similar installations.

The *Risk Assessment Equipment Failure Data Handbook*, prepared by the Technica's Houston Office, was issued in May 1990 and became a standard reference for all Technica offices. When DNV acquired Technica Ltd. in 1991, the *Risk Assessment Equipment Failure Data Handbook* was reviewed in *Onshore Failure Frequency Handbook*, ARF Technical Note T14 Revision 0, prepared by DNV Technica's Columbus Office, issued in November 1993.

Hoorelbeke used this Technical Note T14 as a basis to compose the first Handboek Kanscijfers which was issued by the Flemish authorities in 1994 (Hoorelbeke, 1994). This book was revised in 2001 and 2004 and a recent version has been published in 2019. Risk assessments done for major hazard activities in the Flanders region have to follow the rules set out in this book. Interesting is to see that the recommended failure rates to be used have changed (see Table 5.17) between 1994 and 2019 while most of the installations in the Flanders region are still the same.

Table 5.17: Evolution of failure rate of leaks on a tank to be used in quantitative risk studies in Flanders between 1994 and 2019.

Leak size	Failure rate for different versions of the "Handboek Kanscijfers"		
	1994	2004	2019
Small	$8,1 \cdot 10^{-5}$	$1,3 \cdot 10^{-5}$	$2,4 \cdot 10^{-3}$
Medium	$5,2 \cdot 10^{-5}$	$4,4 \cdot 10^{-6}$	$2,8 \cdot 10^{-4}$
Large	$9,7 \cdot 10^{-5}$	$3,0 \cdot 10^{-6}$	$1,4 \cdot 10^{-4}$
Rupture	$6,5 \cdot 10^{-5}$	$3,0 \cdot 10^{-7}$	$3,0 \cdot 10^{-6}$

Risk analysists in the Netherlands in the 1980s used mainly the COVO study of 1981 as reference for failure data. The failure data in the COVO study was based on data from the late 1960s and the 1970s (e.g., Philips and Warwick, 1969; Smith and Warwick, 1974; Bush, 1975). In the 1990s, the authorities in the Netherlands asked different consultants to perform a study on a fictive plant called "Rietschap." The study demonstrated that different consultants used different failure data and hence assessed the risks at different levels. This work resulted in a first document with guidelines to be used in quantitative risk assessment (IPO, 1994). Five years later, the Committee for the Prevention of Disasters revised the data and published the so-called Purple Book with the recommended failure rate data to be used in quantitative risk studies (CPR, 1999). The Purple Book was revisited in 2004 and confirmed in 2005 (Beerens et al., 2006; Pasman, 2011).

The Bureau of Safety and Environmental Enforcement (BSEE) published a guide to assist in the development of probabilistic risk assessment of offshore drilling facilities, in order to support decision-making (Cross and Youngblood, 2018). The document does not specify which failure rates has to be used but lists in its appendix D a list of sources that could be used:

- Process equipment reliability database
- Failure rate and event data for use within risk assessments (HSE PCAG)
- Failure frequency guidance process equipment leak frequency data for use in QRA
- Lees' *Loss Prevention in the Process Industries* (third edition)
- OGP risk assessment data directory
- OIR/12
- Offshore reliability data
- Pipeline and riser loss of containment report
- WellMaster reliability management system
- Worldwide offshore accident database

Each of the data sources is discussed more in detail in the BSEE guide.

Large companies have compiled their own data based on a critical review of failure data in literature combined with own experience. For in-house studies, these companies will use their own database. When a risk assessment is part of a legal obligation, the same companies will use the failure rate data that is imposed by the authorities.

One can of course question how it is possible to use different probability data for the same risks. This is the reason why the use of failure data is an engineering discipline. The probability of occurrence is based on scientific studies but at the end, it is an engineering discipline.

In engineering, this approach is quite common. Engineering is about applying science rather than practicing science to verify a hypothesis. The engineering of, for instance, a pressure vessel can done according several engineering codes of standards. For examples, BS 1500 and BS 1515 in the UK, ASME in the USA, CSA Standard B51 in Canada, ANCC code in Italy, Swedish Pressure Vessel Code in Sweden and so on (Stewart et al., 2013); in addition, the company will have its own standards. All these standards have the same goal which is engineering a pressure vessel that can be used in a safe way. However, depending on the code that will be used, the pressure vessel will have different characteristics (wall thickness, material selected, size, etc.).

The decision by the authorities or by a company to use a certain standardized set of data is to ensure consistency between the different studies. The goal is **not** to make a precise prediction of the probability that an event will occur. The goal is to apply consistently an engineering standard to perform risk analysis and risk assessment. This way of working is a common practice in all engineering disciplines. In the communication, it can be confusing with people who are not familiar with this engineering practice in a way that quantitative risk analysis and risk assessment gives the false impression "to be a scientific and objective calculation of the real risk."

5.7.5 Frequency, fractional dead time and failure on demand

The data available can be presented in different ways: per unit time or per demand. When doing risk calculations, it is sometimes needed to transform a failure rate per unit of time in a fractional dead time (FDT) or in a probability of failure on demand (PFD).

There are some simple rules to perform these calculations. Let us consider the following example:

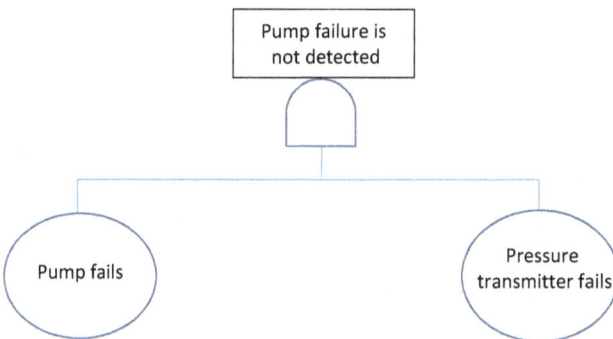

Figure 5.16: Example of simple FTA.

To calculate the probability that the pump failure is not detected, we need to have a failure and the pressure transmitter failure. An "AND" port means that we have to multiply the probabilities of failure.

In our database, we find the following data:
- Pump fails: 10^{-2} per year
- Pressure transmitter fails: 10^{-5} per h

Multiplying both figures does not make a lot of sense because we will obtain 10^{-7} per pump year-transmitter h. So what we would like to know is "what is the probability that the pressure transmitter fails when the pump is failing?" In other words, we would like to have an idea about the FDT which is the average fraction of time that the protective system is unavailable to do its assigned function.

There are three possible situations:
- The failure of the pressure transmitter is not monitored.
- The pressure transmitter is part of a regular inspection.
- The functioning of the pressure transmitter is continuously monitored by the DCS and when it fails the operators are alerted.

In case the pressure transmitter is not monitored, it can be assumed that the probability of failure at a certain moment in time behaves like:

$$p = 1 - e^{-\lambda t}$$

p Probability (-)
λ Failure rate (in the example 10^{-5} per h)
t The time in hours

The probability of failure will increase as time goes by. Hence, after 1 year (8,760 h) the probability of failure is 0.084 while after 2 years (17,520 h) it has increased to 0.16.

When the pressure transmitter is part of a regular inspection, it will be assumed that the probability of failure is:

$$p = \frac{1}{2}\lambda t^*$$

p Probability of failure (-)
λ Failure rate (in the example 10^{-5} per h)
t* Time in hours between two tests

The calculation is based on the following idea:
- The instrument could in the worst case fail just after the inspection (the FDT would then be the test period).
- The instrument could in the best case fail at the beginning of the next inspection (the FDT would be almost zero).
- By convention it is assumed that the instrument fails in the middle of the test period.

In case the functioning of the instrument is continuously monitored, the FDT is calculated as follows (see also Figure 5.17):

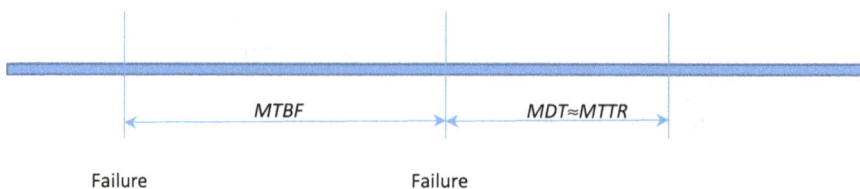

Figure 5.17: Visualization of fractional date time for a repairable item.

$$FDT = \frac{MDT}{MDT + MBF} \approx \frac{MTTR}{MTTR + MBF} = \frac{MTTR}{MTTR + \frac{1}{\lambda}}$$

FDT   The fractional dead time
λ Failure rate (in the example 10^{-5} per h)
MTTR Mean time to repair in hours
MTBF Mean time between two failures
MDT   Mean down time

The difference between MDT and MTTR is explained, for instance, by Smith (2001)
but for risk analysis the required accuracy is not the same as for reliability studies
and it is common to use MTTR instead of MDT.

5.8 Consequence analysis

5.8.1 Qualitative approach

The potential consequences of an accidental event can be estimated on a qualitative
basis. The consequences could be harm to life or to the environment but it could
also be damage to property, financial losses or media impact.

Table 5.18 gives an example of a qualitative five levels of ranking of potential
consequences.

In many cases, it will be necessary to estimate the consequences in much more
detail. In particular, when there is a need to identify additional protection meas-
ures. Detailed consequence calculation consists of three parts:
–   Determination of the source term;
–   Calculation of the physical effect (overpressure, heat intensity, etc.);
–   Calculation of the impact of the physical effect on life, the environment or property.

Table 5.18: Example of a qualitative classification of consequences.

Consequence level	Description
Negligible	– No measurable consequences
Minor	– First aid, medical treatment or restricted work case – Leak to the environment that can be cleaned up within 24 h without lasting damage to the environment – Repairable light damage to property
Major	– Lost time injury – Environmental damage for which intervention of a specialized company will be needed – Moderate damage to property
Significant	– Permanent disability or single fatality – Environmental damage will be visible for a longer time (< 6 months) and will need intervention of specialized company – Destruction of some property
Disastrous	– Multiple fatalities – Irreparable environmental damage or damage that will be visible and last for many years – Large destruction of property

Calculation of the physical effects and calculation of the associated impact is a highly specialized and very broad domain. The next chapters are limited to a brief description and some references to standard textbooks and methods.

5.8.2 The use of (mathematical) models

The phenomena involved in the physical release of products or the release of energy are chaotic in nature because small difference in initial conditions can yield widely diverging outcomes. The phenomena themselves are deterministic, meaning that their behavior follows strict physical laws but the uncertainty on the initial conditions make them unpredictable.

An approximation of the initial conditions does not deliver any information on the error that will be made on the final outcome. This is a well-known topic since Edward Lorenz and Ellen Fetter in 1963 developed a simplified mathematical model for atmospheric convection. The model is a system of three ordinary differential equations known as the Lorenz equations.

From a technical standpoint, the Lorenz system is nonlinear, nonperiodic, three-dimensional and deterministic, but small differences in the initial conditions deliver complete difference results. The Lorenz equations have been the subject of hundreds of research articles. The equations are as follows:

$$dx = \sigma \, (y - x) \, dt$$

$$dy = [x \, (\rho - z) - y] \, dt$$

$$dz = [xy - \beta z] \, dt$$

Lorenz used the values $\beta = 8/3$; $\sigma = 10$ and $\rho = 28$

Let us assume that two people use the same equations but there is a small difference on the location of the origin:

	X	Y	Z
Location of origin 1	0.1	0.1	0.1
Location of origin 2	0.1001	0.1	0.1

Solving the Lorenz equations delivers the following results for X, Y and Z after 20 s (dt was taken as 0.01):

	X	Y	Z
Starting in origin 1	−14.10	−10.83	37.73
Starting in origin 2	4.36	6.52	16.92

The Lorenz system is a demonstration of how a small difference on the initial conditions can give complete different results while using exactly the same set of equations.

With regard to quantitative consequences calculations, this means that it is **simply impossible** to predict **accurately** the effects for a particular scenario because of the inherent uncertainties of the initial conditions and the boundary conditions. In theory, however, it is possible to perform all possible calculations. The variability of the parameters and boundary conditions is huge. Let us assume that one wants to perform a calculation for the "dispersion of a leak in a pipe." The boundary conditions or parameters that need to be defined include:

1. The location of the leak
2. The direction of the leak (horizontal, vertical, at an angle of 10 degrees, etc.)
3. Size of the leak (1 mm, 5 mm, etc.)
4. Geometry of the leak (circular hole, square, etc.)
5. Thermodynamic conditions at the moment of the leak

6. Meteorological conditions (temperature, wind speed, humidity, wind direction, etc.)
7. Duration of the leak

If we assume that for each of the conditions there are eight possibilities then the number of possible combinations (or calculations) is 2,097,152 (8^7). Even with only four possibilities, the number of combinations is 16,385 (4^7).

This type of problem is a typical engineering problem. In all disciplines of engineering, there are a number of calculation methods available and there are a number of parameters with a certain variability. For simple quasistatic bending of beams, there is the Euler–Bernoulli bending theory, the extended Euler-Bernoulli bending theory and the Timoshenko bending theory. For each of these theories, some parameters have to be decided such as the Young's modulus, the shear modulus and the Poisson's ratios.

In engineering, this type of problems is solved by making conventions on how the calculation will be performed and which conditions will be used.

– The authorities in Flanders for instance issued the "Handboek Risicoberekeningen" (331 pages) which specifies the conditions that are mandatory for the risk calculations (Departement Omgeving, 2019).
– In France, for instance, the equations to use to calculate the thermal thresholds used for the definition of the lethal and irreversible effects in case of a fireball are defined by the decree of September 29, 2005.
– In the Netherlands, RIVM has published in April 2020 the guidelines for the quantitative risk analysis to be performed as part of the *Risk Calculations Manual for the Decree on External Safety Establishments* (VROM, 2020). The document (402 pages) defines the principles and models to be used and specifies that "Therefore, QRAs shall be based on the same principles, models and basic data executed."

The basis in the different guidelines is not always the same and sometimes there might be a significant difference. Table 5.19 compares the mass in the fireball to be considered according respectively the Flemish and Dutch rules. Assume that a

Table 5.19: Comparison on the mass in the fireball according the Flemish and Dutch rules.

Criteria	Flanders, 2020	Netherlands, 2020
Mass in the fireball in case of a BLEVE	The mass m of the fireball is equated to the released amount if the adiabatic flash exceeds 33%. Otherwise, the mass equated to 3x the adiabatic flash fraction, calculated at the failure pressure P of the container. (§ 19.5.2 of the reference)	The mass involved in the fireball is equal to 3 × the flash fraction at the failure pressure, with a maximum of the entire system content (§ 3.4.4 of the reference)

vessel contains 200 tons and that the calculated flash fraction is 35%. In Flanders, the consequence calculation will be done with 70 tons in the fireball while the same scenario will be modeled in the Netherlands with 200 tons in the fireball. Hence, the radius of the fireball will be 42% larger in the Netherlands. The safe distance in Flanders will be about 350 meters while in the Netherlands for the same scenario it will be in the order of 500 m.

The IOGP published a guide on consequence modeling for its members (OGP Report 434-7, 2010). The IOGP document specifies in relation to the fireball calculation that *one possible assumption is that the fireball mass is calculated assuming 3 × the adiabatic flash fraction at the burst pressure, constraining this to be ≤ 1.0* which is similar to the guidelines of VROM. The OGP document proposes the following equation for the diameter of the fireball $D = 6.48 \text{ m}^{0.325}$ which is different from the equation from the equation used in Flanders which is $D = 5.8 \text{ m}^{1/3}$

The French Industry association Union Française des Industries Pétrolières developed, for their members, a guidance document (525 pages) to perform consequence calculations (UFIP, 2002). The radius of the fireball is calculated by only considering the mass in the vapor space and the flash fraction.

The examples show that the purpose is not to determine scientifically the correct mass in the fireball nor to determine the exact radius of the fireball. There is uncertainty involved in assessing the risks of hazardous installations and Flanders as well as the Netherlands developed an engineering method to take decisions. These methods are prescribed in order to obtain consistency between different studies. The equations that are used have a scientific foundation but because of the uncertainties associated with the equations a choice has been made on the equations, assumptions and models to be used. This is called an engineering approach.

The engineering approach that implies that consequence calculations are done based on a number of conventions has been adopted by the industry and in-house guidelines exist in all large companies.

5.8.3 Effect analysis

5.8.3.1 Terminology
There is some confusion in literature about consequence, effect and impact. One can find the following sentence which seems to have the same meaning
- The radiation leak has had a disastrous effect on the people;
- The radiation leak has had a disastrous impact on the people;
- The radiation leak has had a disastrous effect on the people.

For the purpose of this book, the following distinction is made:

Consequence The consequence of an accidental event is a measure and description of the expected results of the impact of an effect on life, the environment or property.

Effect The effect is the physical result of an accidental event. This is a concentration at a certain distance during a certain time or a radiation flux at a certain distance during a certain time.

Impact The impact describes the influence of the effect on life, the environment or property. Examples are first degree burns and destruction of a building.

An example:

Undesired event Vapor cloud explosion
Effect 30 kPa on a building at 300 m
Impact Complete destruction of the building
100% fatality for people in the building
Consequence The consequence of the vapor cloud explosion is the destruction of a building at 300 m and the death of 10 people inside the building.

The purpose of the effect analysis is to estimate a physical quantity in the surroundings of the accidental event due to the accidental event. This will mainly be at different locations around the origin of the accidental event:
- A concentration (ppm or mg/m^3) during a certain time (seconds, minutes or hours);
- A heat flux (W/m^2) during a certain time (seconds, minutes or hours);
- An positive or negative pressure (Pascal) during a certain time (seconds);
- A fragment (kg and shape) arriving at a certain velocity (m/s).

The concentration at a certain distance will however not be constant because of the influence of the meteorological conditions. If we watch a chimney plume for a while, we see that the material is constantly evolving and changing shape. This is because the flow is turbulent. A calculation of the instantaneous concentration would be a fruitless exercise, for a single realization of the plume could never match the calculation. However, if we observed the plume with a time-lapse camera, then merged all the snapshots together, a smooth picture of the plume would result. In dispersion models, an averaging time of say 10 min or 1 h is used to provide a smoothing or averaging plume.

The effect calculations comprise:
- Source term modeling
- Dispersion modeling
- Heat radiation modeling
- Explosion overpressure modeling
- Modeling of missiles

5.8.3.2 Methods and models for the calculation of the physical effects

There is an overwhelming quantity of articles available with methods for the calculation of the physical effects due to releases of hazardous materials. Two standard references are:

– The "Yellow book": The first version of the yellow book was issued in 1979 and a first revision was published in 1988. The Committee for the Prevention of Disasters, Subcommittee Risk Evaluation started an in depth revision between June 1993 and March 1996 and published a complete revision of the Yellow Book in 1997. This project was carried out by TNO Institute of Environmental Sciences, Energy Research and Process Innovation, TNO Prins Maurits Laboratory and TNO Centre for Technology and Policy Studies. The current version of the Yellow Book dates from 2005 (VROM, 2005) and results from an extensive study and evaluation of recent literature on models for the calculation of physical effects of the release of dangerous materials.

– The second edition of the CCPS book *Guidelines for Chemical Process Quantitative Risk Analysis*, is published together with a set of excel programs to calculate source term characteristics (CCPS, 2000).

A number of software packages to perform consequence calculations have been commercialized. Examples are EFFECTS (TNO), PHAST (DNV GL), FRED (developed by Shell but commercialized by Gexcon) TRACE (SAFER), CANARY Hazard Models (Quest Consultants) CIRRUS (BP) and SafeSite3G (Baker Engineering and Risk Consultants, Inc.). Tausseef (Tausseef et al., 2017) makes a review of the strength and weaknesses of a number of software models.

The mathematical models used for the calculation of the physical effects can be classified in (Hoorelbeke, 1992):

– Semiempirical correlations (statistical approach);
– Mathematical equations which describe the physical phenomena.

Mathematical models vary from simple analytical equations over a set of differential equations up to the partial differential equations (PDEs) which describe the motion of viscous fluids. Table 5.20 shows some examples of equations that are used for the calculation of physical effects.

The perception may exist that the more complicated the model is the more it will accurately predict the effects. This perception is wrong. All models are by definition a mathematical attempt to model reality. The main difference between the models is their capacity to include all influencing parameters. The Wertenbach formula for instance can only take into account the mass release rate. The governing equations of transport phenomena of fluids can be described by a system of coupled nonlinear PDEs which includes the fundamental principles of conservation of mass, momentum, enthalpy and so on. In many cases, the nonlinearities in the governing

Table 5.20: Examples of mathematical equations that can be used for the calculation of physical effects.

Example	Description
$L = 18.5\,Q^{0.41}$	The formula of Wertenbach (Wertenbach, 1971) is a correlation between the length of a turbulent jetflame (L in meter) and the release rate Q (in kg.s)
$\dot{m} = AC_D\sqrt{2\rho g(P_1 - P_2)}$	This analytical equation calculates the discharge of nonflashing liquids through a sharp-edged orifice as described by the classical work of Bernoulli in which $\dot{m}$ is the discharge rate (kg/s), A is the area of the hole (m^2), C_D is the discharge coefficient (-), ρ is the density of the fluid (kg/m^3), g is the gravitational constant (9,81 m/s^2), P_1 is the pressure upstream of the hole (Pa) and P_2 is the pressure downstream of the hole (Pa)
$\frac{du}{dt} = -\left(u\,\frac{du}{dr} + g\,\frac{d(h)}{dr}\right)$	The equation is an expression of the spreading of a pool in case of axial symmetry and instantaneous release, ignoring the vaporization. The change in radial velocity du (m/s) per unit time dt (s) is function of the change of the radial velocity du as a function of the change in radius dr and the change of the liquid pool height dh as function of the change of the radius dr.
$\partial p/\partial x = \rho X - \rho[\partial u/\partial t + u\partial u/\partial x + v\partial u/\partial y + w\partial u/\partial z] - (2/3)\partial/\partial x[\mu(\partial u/\partial x + \partial v/\partial y + \partial w/\partial z)] + 2\partial/\partial x(\mu\partial u/\partial x) + \partial/\partial y[\mu(\partial u/\partial y + \partial v/\partial x)] + \partial/\partial z[\mu(\partial u/\partial z + \partial w/\partial x)]$	Navier–Stokes equation for the conservation of momentum in X direction

PDEs cannot be solved with analytical methods, which is the situation for most engineering flows, then numerical methods are needed to obtain solutions.

Computational **f**luid **d**ynamics (CFD) is concerned with obtaining numerical solution to fluid flow problems by using computers. CFD is the *art* of replacing the differential equation governing the fluid flow, with a set of algebraic equations (the process is called discretization), which in turn can be solved with the aid of a digital computer to get an *approximate* solution (Hoorelbeke, 2004).

In principle, CFD is capable to handle all influencing parameters (e.g., chemical reactions, obstacles). The results will however be influenced by the mathematical numerical method that will be used to solve the nonlinear PDEs. Different mathematical models have been proposed to solve the so-called Reynolds stresses. These turbulence models (i.e., the models used to represent the Reynolds stresses or effective viscosity to be used in the three flow equations) can be classified in a number of ways. In literature, they are often presented in function of the number of equations: zero-equation turbulence models, one-equation turbulence models and so on. In present day engineering calculations, two-equation models are often used. A very popular two-equation model is the k-ε model. This model includes an equation for transport of turbulent kinetic energy (k-equation) and equation for the dissipation of turbulent kinetic energy (ε equation) (Hoorelbeke, 2004).

Computer capability continues to increases exponentially and there is interest to make use of large eddy simulation or even direct numerical simulation, but these methods remain difficult to apply over a large domain due to the calculation time.

Another challenge with CFD is the discretization (grid definition) and the definition of the boundary conditions.

5.8.3.3 Current challenges with effect modeling

For a number of phenomena, the models available have been in use for a long time and it is recognized that the effect calculations can be done with a good degree of accuracy.

Table 5.21 gives an overview of the different phenomena and points out for which situations models are widely available and well known. The "challenges" mean that additional expertise will be needed to perform the calculations and that specific models or methods (e.g., CFD) are required.

Table 5.21: Overview of current situation for effect modeling.

Phenomenon	Accurate models available	Challenges
Source term	– Single product liquid and gas discharge – Two-phase outflow (gas/liquid) with or without slip between the two phases	– Reactive products (e.g., evacuation of product from a runaway of a polymerization reactor) – Three phase flow (liquid, gas, solid)
Pool spreading and pool evaporation	– Spreading of a single liquid product on a flat surface (concrete, sand, gravel, water)	– Multiphase components – Uneven surfaces or complicated geometries
Dispersion	– Dispersion of single component pollutants (lighter, heavier than air or with a molecular weight close to air) – Dispersion on water	– Reacting chemicals. Models have been developed for some reacting chemicals such as hydrogen fluoride (e.g., Raj, 1990) – Short distance dispersion – Dispersion of liquid in the soil or inside waters – Meandering of the cloud (e.g., due to wind effects)

Table 5.21 (continued)

Phenomenon	Accurate models available	Challenges
Radiation	– Stable jet fire – Stable pool fires – Flash fires – Fireball	– Impacting fires (i.e., the target is hit by the flame) – Fluctuating fire (e.g., radiation level at a certain distance varies due to wind effect)
Explosion	– Overpressure outside the cloud for physical explosions as well as chemical explosions – Vapor cloud explosion with a uniform regime (e.g., uniform deflagration or detonation of a vapor cloud) – Solid state explosions	– Overpressures inside the cloud – Variable regimes (e.g., deflagration to detonation regime) – Multipocket explosion (i.e., some parts of the cloud detonates while others undergo deflagration)
Missiles	– Calculation of the maximum flight distance of a nontumbling well defined (e.g., geometry and mass) missile	– Determination of the characteristics of missiles

5.8.4 Impact analysis

5.8.4.1 Methods for impact analysis

There is an overwhelming quantity of articles available with methods for the calculation of the impact.

The effect analysis will give information on the expected physical quantity (e.g., heat flux, overpressure, concentration) at different locations. The impact analysis will determine how this effect affects the target being life, the environment or property.

Accidental events can impact the terrestrial ecology and the biodiversity on short term and on longer term. These studies will however be very specific in particular because the fauna and flora has a great variety and can be very different from one location to another location. In some locations, models have been developed but these models are particular for a certain area and a certain type of accidents. In 2019 for instance, The Norwegian Oil and Gas Association, a professional body and employer's association for oil and supplier companies, introduced the ERAAcute methodology as an industry standard for environmental risk assessment (ERA) on the Norwegian Continental Shelf in 2019, and the ERAAcute software was developed to support this methodology. The ERAAcute method includes four environmental compartments: the sea surface, shoreline, water column and seafloor.

ERAAcute methodology uses input data from an oil spill trajectory model and biological resource data, and calculates the potential environmental risk (impact and recovery time) for biological resources in all compartments (Libre et al. 2018).

What follows is restricted to the impact studies on humans, buildings and process installations. There are mainly three approaches used to perform the impact analysis:
- Single value criteria;
- Statistical approach (probit equations);
- Mathematical analysis of impact on buildings and structures.

5.8.4.2 Single value criteria for impact analysis

Single value criteria consists of assuming certain damage when an effect is achieved. The use of single value criteria to estimate the potential damage for explosions goes back to the 1950s.

Societies had since very long history with sad experiences with the destructive power of heavy explosions:
- The explosion of 30 tons of gunpowder stored in barrels on 12 October 1654 in Delft ("the Delft Thunderclap") killed hundreds of people and damaged almost all buildings in the city center. At least 500 houses were damaged beyond repair.
- On 31 August 1794, about 30 to 150 tons of gunpowder exploded in the powder store of the Château de Grenelle, near Paris. The blast shook homes in the surrounding area, cracked bridges and broke the seals on the underground doors of the observatory. The catastrophe resulted in the death of more than 1,000 people.
- On 12 January 1807, a ship carrying hundreds of barrels of black powder explode in the town of Leiden in the Netherlands. The disaster killed 151 people and destroyed over 200 buildings in the town.
- On Saturday 10 July 1926, a lightning strike at the Naval Ammunition Depot, Dover (Lake Denmark). More than 600,000 tons of explosives stored inside the depot detonated, resulting in one of the most catastrophic man-made explosions in the USA. The blast completely destroyed nearly 200 buildings in a half-mile radius, resulting in 21 deaths and dozens more injuries. The explosion was so powerful that people reported finding debris nearly 22 miles away.

Based on the work of the special committee of the Association of Manufacturers of Powder and High Explosives, the American Table of Distances for inhabited buildings and public railways was established in December 1910. When it became apparent that the distance table should also contain minimum safe distances for the location of explosive storage and manufacturing buildings from public highways, the special committee, in conjunction with the Institute of Makers of Explosives (founded in 1913), conducted additional studies. The highway distances were approved and adopted by the Institute of Makers of Explosives in 1914. After the adoption of the American Table of Distances, the collection of data on explosions was continued. The table was reviewed in 1919 and again in 1939 to evaluate it and consider the data accumulated since the table was established. No significant revisions were made after either review (IME, 2011).

In 1945, Col. C. S. Robinson who was attached to the Army/Navy Explosive Safety Board published a report in which he questioned the accuracy of the inhabited building distance tables. His primary concern was for large quantities of explosives. He believed the distances specified were inadequate (Lyman, 1979).

The devastating power of the bombing of Hiroshima (6 August 1945) and Nagasaki (9 August 1945) had a major impact on congress and the American public. Postwar discussions on the control of the US atomic-energy program produced consensus that some special statutory control over atomic energy was necessary. Some people wanted continued tight control on all information related to nuclear weapons. Others were concerned that continued strict control of basic research in this area would hinder progress in the development of atomic energy, to the detriment of the nation. The Atomic Energy Act of 1946 (also known as the McMahon Act), which became law on August 1 1946, stringently controlled all atomic energy information. The US Weapons Effects Classification Board of the Atomic Energy Commission recommended giving the Los Alamos Scientific Laboratory the responsibility for preparing a handbook on the effects of atomic weapons. The recommendation was approved in 1948 and a first edition of the handbook was published in 1950 (Glasstone et al., 1950). About 100 scientists participated to the study. The book contains a chapter on "Air Blast Damage: Japanese Experience" and on "Thermal Radiation and Incendiary Effects."

On 1 July 1946, the USA had started their program Pacific Proving Grounds. It was the name given to testing of nuclear bombs on a number of sites in the Marshall Islands. The tests were organized in so-called operations (Operation Crossroad in 1946, Operation Sandstone in 1948, Operation Greenhouse in 1951, Operation Ivy in 1952, Operation Castle in 1954, Operation Redwing in 1956, etc.). During Operation Greenhouse, a number of target buildings, including bunkers, houses and factories were built on Enjebi and Mujinkarikku Island to test nuclear weapon effects.

On 11 January 1951, under President Truman, the US established the Nevada National Security Site, known as the Nevada Test Site, a 3,500-km^2 area of desert and mountainous terrain for the testing of nuclear devices. The effects of weapons were studied in a number of tests. Examples are the Sugar test on 19 November 1951 and the Uncle test on 29 November 1951 which were part of Operation Buster-Jangle. In a series of seven tests at the Nevada Test Site that followed Operation Greenhouse and the Encore tests on 8 May 1953 (Operation Upshot–Knothole), multiple objects were subjected to the blast, including trees. Many kinds of vehicles (ranging from cars to aircraft), nuclear-fallout and standard bomb-shelters, public-utility stations and other building structures and equipment were placed at measured distances away from "ground zero," the spot on the surface immediately under or over the center of the blast.

The handbook *Effects of Nuclear Weapons* was revised in 1957 to take into account the experiences of the Eniwetok Proving Grounds and the Nevada Test Site. Chapter IV of the book ("Structural Damage from Air Blast") describes the damage to industrial structures, to commercial and administrative structures, transportation and utilities and communications in 75 pages. In the Nevada tests in 1955, a complete 18,000 US gal bulk storage plant (≈ 68 m^3) containing propane with pump, compressor, piping, valves and fittings, and with a cylinder filling building, was located at the point where the overpressure was 5 psi (≈ 345 mbar = 34.5 kPa). The building was demolished and the filling line was broken at the point where it entered the building, but otherwise the installation received only superficial damage.

Chapter VII ("Thermal Radiation and Its Effects") describes in 55 pages the impact of thermal radiation on humans and damage to materials. The document uses single values to explain the damage.

Figure 5.18 shows an example from the 1957 report. For the unreinforced brick house subjected to 117 mbar, it adds that "The condition was such that it could be made available for habitation by shoring and some fairly inexpensive repairs." A revision of the handbook *Effects of Nuclear Weapons* was reprinted in 1964 and a third edition has been published in 1977 (Glasstone et al., 1977).

Unreinforced brick house after the nuclear explosion
(5 psi ≈ 345 mbar = 34.5 kPa)

Unreinforced brick house after the nuclear explosion
(1.7 psi ≈ 117 mbar = 11.7 kPa)

Figure 5.18: Examples of damage to buildings from nuclear explosion tests based on single value criteria (Glasstone, 1957).

Following the research into the effect of nuclear weapons, it became common practice in the high-risk industry from the 1960s onward to correlate the expected damage from explosions to the overpressure.

The US petroleum refining industry was in the 1960 regarded as critical to the continued national viability and in the event of damage by attack, petroleum refining is expected to be of primary interest, with emphasis on restoration of a petroleum refining level required for that viability. By the end of the 1960s, the US petroleum refining industry was made up of 267 refineries which processed more than 200 different types of crude oils. Over 100 individual refining processes and at least 50–100 types of equipment are used by these refineries to produce well over 1,000 different products. To plan the recovery of the petroleum refining industry, it is essential to be able to estimate (1) the extent of damage by blast overpressure levels, (2) the capability of individual refineries to produce products as they stand or with increments of repair effort and (3) the repair effort needed.

The Office of Civil Defense and the Office of the Secretary of the US Army requested Stanford Research Institute (SRI) to perform a study to estimate the production and repair efforts in (nuclear) blast damaged petroleum refineries. The study was delivered in July 1969 (Walker, 1969). The major conclusions of the study were:

– After 0.3–0.5 psi (2.1–3.5 kPa), a refinery can produce the same proportion of products but at about 70% of the initial capacity. This reflects the assumption that at this overpressure, refinery capacity is directly related to remaining cooling tower capacity.
– After 1.0 psi (6.9 kPa), a refinery temporarily shuts down, but with minor emergency repair to process controls, it can operate at about 50% of initial capacity.
– After 1.5 psi (10.3 kPa), a refinery is totally shut down, primarily because of process control damage by roof collapse in each of the numerous individual refining process control rooms.

Walker also produced a summary of the complete range of blast damage effects from 0.3 to 20 psi (2.1–138 kPa) for different refinery equipment. This summary which is shown in Figure 5.19 (in the original study the overpressure is given in psi) is based on earlier work of SRI in 1965 (SRI, 1965) and Van Horn in 1968 (Van Horn et al., 1968). The summary was copied and published by Stephens in an article in 1970 (Stephens, 1970).

Brasie and Simpson (Dow Chemical Company) examined three process explosion incidents:

– A release of 2,500 lbs (≈1,134 kg) of butadiene and styrene on 16 February 1950;
– A release of 10,000 lbs (≈4,536 kg) of vinyl chloride on 21 January 1964;
– A release of 1,400 lbs (≈635 kg) styrene on 13 October 1966.

Each of the incidents involved an observed and documented release of explosive vapors into an essentially confined but usually ventilated operating structure. In each case, a single polymerization reactor was involved in the initial vapor release. On the 2nd Loss Prevention symposium in 1968, they published data with observed damage levels versus estimated overpressures. The data from the original article from Brasie and Simpson was reworked and is presented in Table 5.22 (10 mbar = 1 kPa).

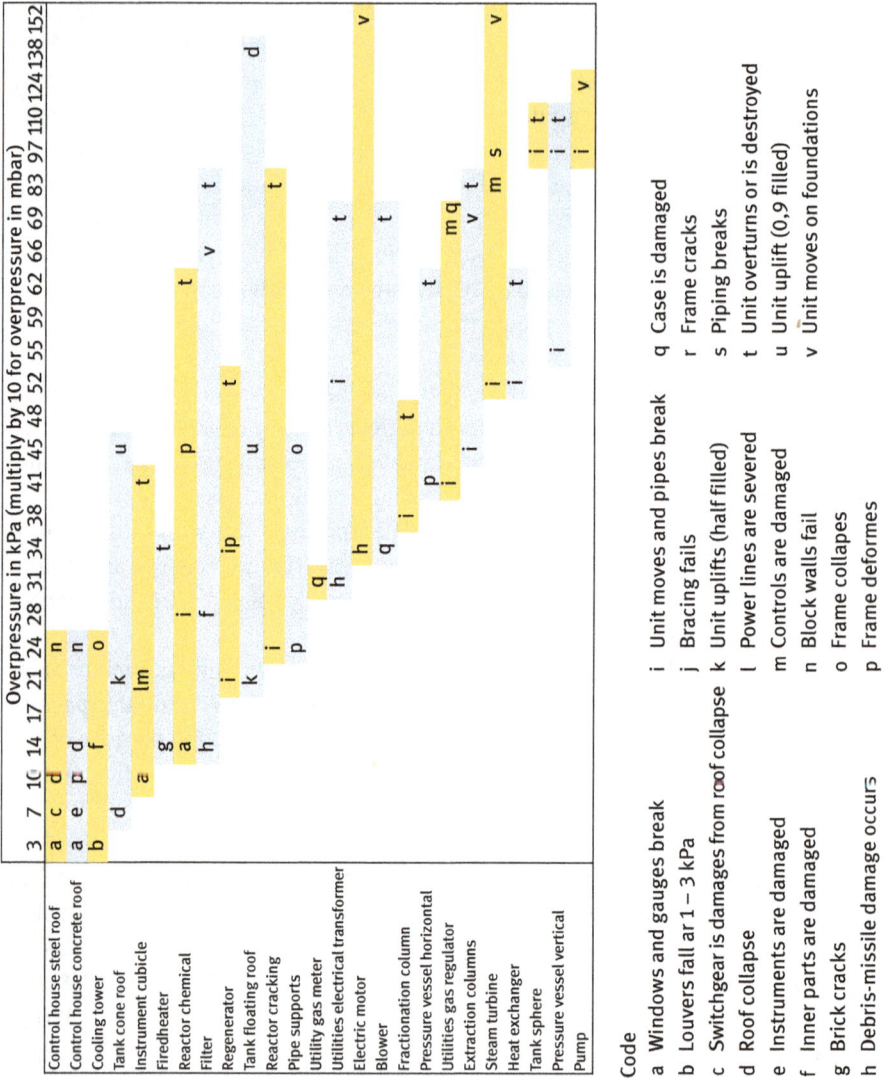

Figure 5.19: Correlation between overpressure and damage for refineries (Walker, 1969).

Table 5.22: Observed damage levels versus estimated overpressures according to Brasie (Brasie and Simpson, 1968).

Damage observed	Pressure (mbarg)	
	Min	Max
Glass broken	10	
Glass unbroken		34
Windows out, no structural damage	34	207
Windows blown in and out		276
Cracked partitions, windows in	138	276
Roof lifted and replaced, windows blown out	69	138
Parts of roof blown off from within		138
Roof caved in, block walls failed	207	345
Roof caved, serious wall damage, blocks out	207	276
Roof blown off, brick walls blown out, steel frame largely intact and undistorted	552	
Asbestos siding blown off	69	103
Panels loosened	69	138
Loosened brick		483
12-inch brick wall severely cracked		552
Brick missiles, maximum 50 ft		689
Steel and brick building destroyed	689	
Building demolished	414	689
Utility pole snapped	345	
Utility pole intact		345
Steel stack intact		345
Major blast destruction, minor steel frame distortion	345	689
Substation, severe damage	414	552
Box car overturned, sides bulged in	483	
50-ton tank car overturned	552	689
18-in × 33 ft beam bent about 4–6 in in center		689

More data reported in the literature for damage to process equipment due to pressure loads is given by Cozzani (Cozzani and Salzano, 2004).

In a similar way, as for explosion overpressures, single values were also determined for thermal radiation (see Table 5.23). The expected damage is a result of the thermal radiation and the time of exposure but it was a common practice to work with single value criteria.

Table 5.23: Single value criteria for damage due to heat flux.

Radiation flux (kW/m^2)	Description
0.7	Exposed skin reddens and burns on prolonged exposure
1.75	Pain threshold after 60 s
2.0	PVC-insulated cables damaged
2.3	Pain threshold after 40 s
5.0	Pain threshold reached after 15 s.
6.4	Pain threshold reached after 8 s. Second-degree burns after 20 s
9.5	Pain threshold reached after 6 s.
12.0	Plastic melts
12.5	Wood ignites on prolonged exposure
14	Intensity which normal buildings should be designed to withstand
15.0	Piloted ignition of wood
16.0	Severe burns after 5 s
20	Ignition of fuel oil in 40 s
25.0	Wood ignites on prolonged exposure
30.0	Spontaneous ignition of wood
37.5	Intensity at which damage is caused to process equipment

Single-value criteria are also used to analyze the risk of a potential exposure to a toxic chemical. The analysis of the potential impact of a toxic product in case of an accident is much more complicated than for explosions or heat radiation:
- Long-term health impact (on fauna, flora or humans) due to short-term exposure
- Long-term health impact (on fauna, flora or humans) due to long-term exposure
- Short-term health impact (on fauna, flora or humans) due to short-term exposure

The Seveso Directive (2012/18/EU) of the European Parliament and of the Council of 4 July 2012 on the on the control of major accident hazards involving dangerous substances, defined a "major accident" as "an occurrence such as a major emission, fire, or explosion resulting from uncontrolled developments in the course of the operation of any establishment covered by this Directive, and leading to serious danger to human health or the environment, immediate or delayed, inside or out-side the establishment, and involving one or more dangerous substances."

Taking into account this definition, it is obvious that long-term impact as well as short-term impact has to be calculated for humans, fauna and flora. In practice, most of the studies, if not all, are limited to short-term health impact on humans due to short-term exposure. Long-term impact of major accidents are monitored after the acci-dent but these impacts are not studied before the occurrence of the accidents. Examples are (1) the impact of the methylisocyanide leak at Bhopal accident on 3 December 1984 is still monitored and (2) the impact of the leak of 1,000,000 m^3 of liquid waste that affected about 40 km^2 due to the rupture of the dam of reservoir number 10 of the Ajkai Timföldgyar alumina plant in western Hungary is still monitored.

Hence, risk analysis of hazardous installations mainly look at acute toxicity. Table 5.24 gives some single-value criteria that are commonly used in safety studies. Other concentrations or doses (1% or 10%) are sometimes used.

Table 5.24: Single-value criteria for exposure to toxic products.

Toxic dose	Description
LC_{Lo}	The LC_{Lo} is the lowest concentration of a chemical, given over a period of time, which results in the fatality of an individual animal. LC_{Lo} is typically for an acute (<24 h) exposure. The LC_{Lo} is used for gases and aerosolized material.
LC_{50}	LC stands for "lethal concentration." According to the Organization for Economic Cooperation and Development Guidelines for the Testing of Chemicals, a traditional experiment involves groups of animals exposed to a concentration (or series of concentrations) for a set period of time (usually 4 h). The animals are clinically observed for up to 14 days. The concentration of the chemical in air that kills 50% of the test animals during the observation period is the LC_{50} value. Other durations of exposure (versus the traditional 4 h) may apply depending on specific laws.
LD_{Lo}	The lowest lethal dose (LDLo) is the least amount of drug that can produce death in a given animal species under controlled conditions. The dosage is given per unit of bodyweight (typically stated in milligrams per kilogram) of a substance known to have resulted in fatality in a particular species. When quoting an LDLo, the particular species and method of administration (e.g., ingested, inhaled, intravenous) are typically stated.
LD50	LD stands for "lethal dose." LD50 is the amount of a material, given all at once, which causes the death of 50% (one-half) of a group of test animals. The LD50 is one way to measure the short-term poisoning potential (acute toxicity) of a material.
ERPG-3	The maximum airborne concentration below which, it is believed, nearly all individuals can be exposed for up to 1 h without experiencing or developing life-threatening health effects. ERPG values are published by the American Industrial Hygiene Association (AIHA, 1989) to be used for emergency response purposes

Table 5.24 (continued)

Toxic dose	Description
ERPG-2	The maximum airborne concentration below which, it is believed, nearly all individuals can be exposed for up to one hour without experiencing or developing irreversible adverse health effects or symptoms which could impair an individual's ability to take protective action.
ERPG-1	The maximum airborne concentration to which nearly all individuals could be exposed for up to 1 h without experiencing other than mild transient health effects or perceiving a clearly defined objectionable odor.

5.8.4.3 Statistical approach (use of probits)

The single-value approach assumes that 100% of the injury or damage occurs at one single point. In reality, there will be a spread around that value. According to Table 5.23, the pain threshold for an exposure to a flux of 5 kW/m^2 is reached after 5 s. In reality, some people will feel pain after 3 s when exposed to 5 kW/m^2, while others might only feel pain after 8 s when exposed to 5 kW/m^2. There is in other words a distribution around the single value. Since there is actually a spread around the single-value criteria, this fosters the use of a statistical approach. The question we would like to answer is "What is the probability for a human to feel pain after 5 s when the human is exposed to 5 kW/m^2?"

If 100 people are exposed to 5 kW/m^2, we would like to calculate the probability that 10 will feel pain after 5 s, the probability that 20 feel pain after 5 s and so on. In the theory of probability, this is a very well-known and old problem.

The probability can be solved by using the binomial distribution:

$$P(x) = \frac{N!}{x!(N-x)!}\pi^x(1-\pi)^{N-x}$$

where

x The number of people that will feel pain
N The total number of people that are tested
π The probability of feeling pain (0.5 because the person will feel pain or will not feel pain)

This calculation will give the probability that exact x people will feel pain. But we would like to know the probability for 1, 2, 3 and so on up to 100. This will take a lot of calculations.

Abraham de Moivre (1667–1754), a French mathematician and statistician, was often called upon to make these lengthy computations. He discovered that the binomial distribution could be approximated by a continuous "normal" curve. Independently, the mathematicians Adrain in 1808 and Gauss in 1809 developed the formula for the normal distribution:

$$f(x) = \frac{1}{\sigma\sqrt{2\pi}}\; e^{-\frac{1}{2}\left(\frac{x-\mu}{\sigma}\right)^2}$$

where

μ The mean ("value with the highest probability of occurrence")
σ The standard deviation is a measure of the amount of variation or dispersion of a set of values

The importance of the normal curve stems primarily from the fact that the distributions of many natural phenomena are at least approximately normally distributed.

Some features of normal distributions are:

- Normal distributions are defined by two parameters, the mean (μ) and the standard deviation (σ).
- Normal distributions are symmetric around their mean.
- The mean, median and mode of a normal distribution are equal.
- The area under the normal curve is equal to 1.0 with 68% of the area of a normal distribution is within one standard deviation of the mean, Approximately 95% of the data lie within two standard deviations of the mean and approximately 99.7% of the data lie within three standard deviations of the mean.

There exist of course a lot of "normal curves" depending on the mean and the standard deviations (e.g., Figure 5.20).

A normal distribution with a mean of 0 and a standard deviation of 1 is called a standard normal distribution.

Pierre-Simon Laplace (1749–1827), a French mathematician, showed that even if a distribution is not normally distributed, the means of repeated samples from the distribution would be very nearly normally distributed, and that the larger the sample size, the closer the distribution of means would be to a normal distribution.

Lambert Adolphe Jacques Quetelet (1796–1874), a Belgian astronomer and mathematician who founded the Brussels Observatory introduced statistical methods in social sciences. Quetelet noted that human characteristics such as height, weight and strength were normally distributed.

Ernst Heinrich Weber (1795–1878), a German physician studied the human response to a physical stimulus in a quantitative fashion. Fechner, a student of Weber, conducted a number of experiments and stated that "the subjective sensation is proportional to the logarithm of the stimulus intensity." This so-called Weber–Fechner law has since then been applied in many fields of research (e.g., public finance, pharmacology) and remains very popular although other relationships have been proposed such as the Stevens's power law which describes a wider

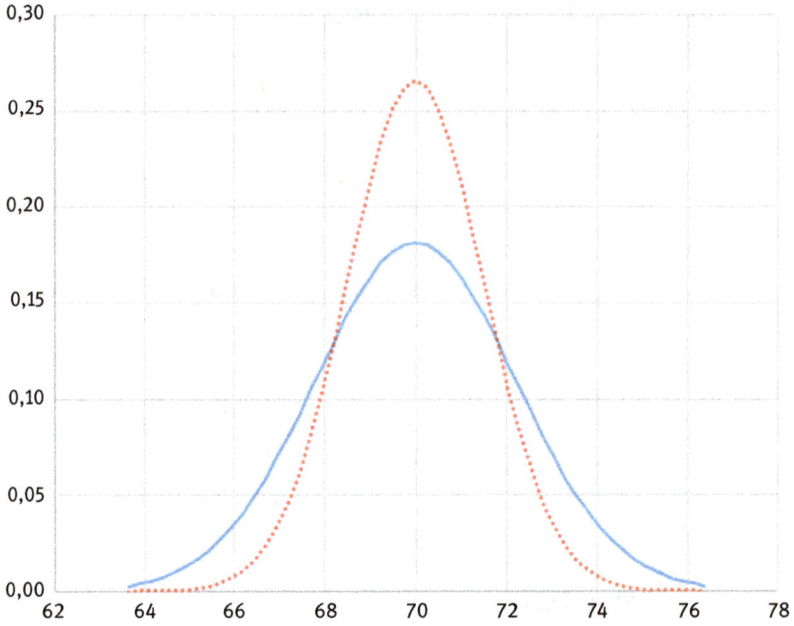

Figure 5.20: Examples of normal curves with mean at 70 and standard deviation at respectively 1.5 and 2.2.

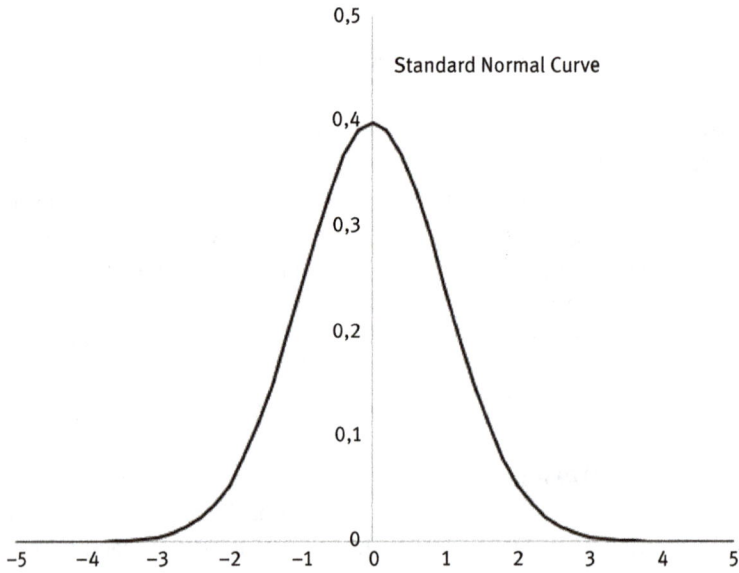

Figure 5.21: Standard normal curve (mean = 0 and standard deviation = 1).

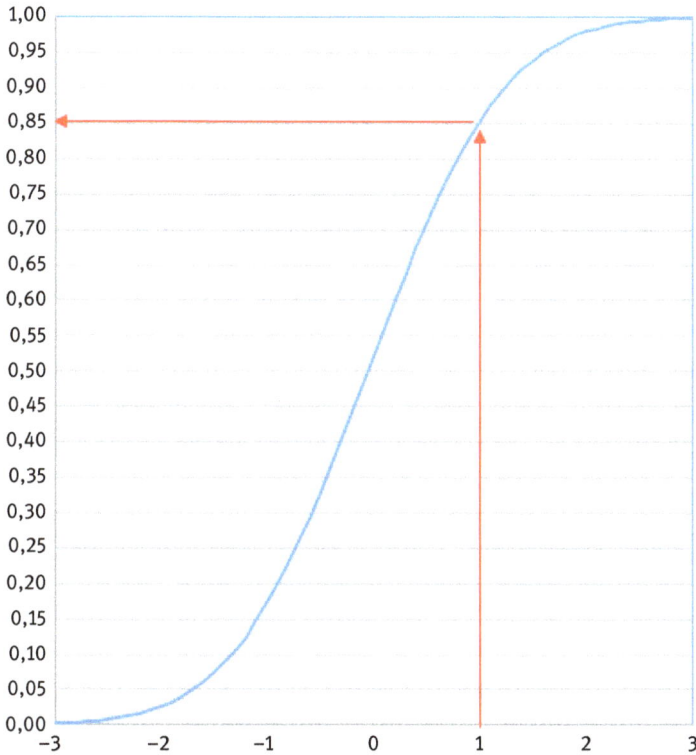

Figure 5.22: Cumulative distribution function of the standard normal distribution.

range of sensory comparisons. Thus, if the random variable X (i.e., the stimulus intensity) is log-normally distributed, then $Y = \ln(X)$ has a normal distribution.

The cumulative distribution function of a real valued random variable X evaluated at x is the probability that X will take a value less than or equal to x. The cumulative distribution function (CDF) of a normal distribution has an S-shape. On the CDF, we can read that for instance the area under the curve between ∞ and 1 (standard deviation) is about 85%.

In the nineteenth century, computers were not available and some authors proposed simpler approximations. Pierre Francois Verhulst (1804–1849), a Belgian mathematician and doctor in number theory, developed the logistic function in a series of three papers between 1838 and 1847, based on research on modeling population growth that he conducted in the mid-1830s, under the guidance of Adolphe Quetelet.

Figure 5.23 compares the cumulative distribution curve of the standard normal with de logistic function:

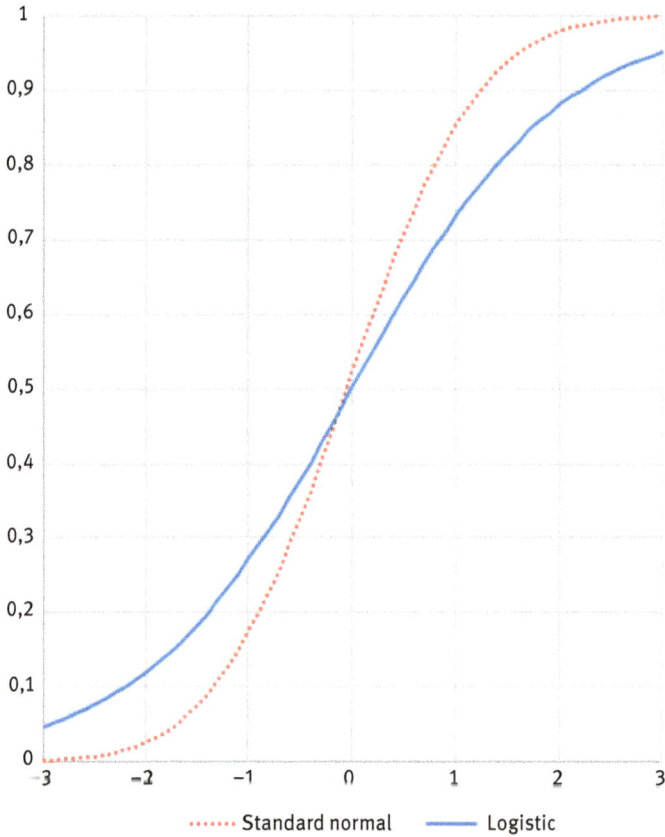

Figure 5.23: Comparison between cumulative distribution of the standard normal and a logistic function.

$$f(x) = \frac{1}{1 + e^{-x}}$$

The treatment of the logistic function is mathematically simpler but it deviates from the real standard normal curve.

In 1934, Charles Ittner Bliss (1899–1979) introduced an alternative to the logistic function. He proposed the probit, a specialized regression model of binomial response variables, which transforms the sigmoid dose–response curve to a straight line that can then be analyzed by regression either through least squares or maximum likelihood (Bliss, 1934).

Probit analysis is used to analyze many kinds of dose–response or binomial response experiments in a variety of fields. Probit analysis is commonly used in toxicology to determine the relative toxicity of chemicals to living organisms. This is done by testing the response of an organism under various concentrations of each of the chemicals in question and then comparing the concentrations at which one encounters a response.

As discussed above, the response is always binomial (e.g., injury or no injury, death or no death) and the relationship between the response and the various concentrations is always sigmoid. Probit analysis acts as a transformation from sigmoid to linear and then runs a regression on the relationship.

Once a regression is run, the researcher can use the output of the probit analysis to compare the amount of chemical required to create the same response in each of the various chemicals.

How does probit analysis work? A probit equation has the following form:

$$Probit = a + b\ln(x^n)$$

According to van Heemst, the probit for lethality due to an exposure to chlorine can be calculated by:

$$Probit = -10.1 + 1.11\ln\left(C^{1.65}\ xt\right)$$

In which, the concentration (C) is in ppm and the time (t) in min.

If a person is exposed to 1,000 ppm during 3 min, one can calculate the probit as

$$Probit = -10.1 + 1.11\ln\left(1,000^{1.65}\ x3\right) = 3.77$$

With the aid of the table in Figure 5.24 (Finney, 1947), one can read that a probit of 3.77 corresponds with an 11% lethality.

%	0	10	20	30	40	50	60	70	80	90
0	–	3,72	4.16	4.48	4.75	5.00	5.25	5.52	5.84	6.28
1	2.67	3.77	4.19	4.50	4.77	5.03	5.28	5.55	5.88	6.34
2	2.95	3.83	4.23	4.53	4.80	5.05	5.31	5.58	5.92	6.41
3	3.12	3.87	4.26	4.59	4.82	5.08	5.33	5.61	5.95	6.48
4	3.25	3.92	4.29	4.59	4.85	5.10	5.36	5.64	5.99	6.55
5	3.36	3.96	4.33	4.61	4.87	5.13	5.39	5.67	6.04	6.64
6	3.45	4.01	4.36	4.64	4.90	5.15	5.41	5.71	6.08	6.75
7	3.52	4.05	4.39	4.67	4.92	5.18	5.44	5.74	6.13	6.88
8	3.59	4.08	4.42	4.69	4.95	5.20	5.47	5.77	6.17	7.05
9	3.66	4.12	4.45	4.72	4.97	5.23	5.50	5.81	6.23	7.33

99.1	99.2	99.3	99.4	99.5	99.6	99.7	99.8	99.9
7.37	7.41	7.46	7.51	7.58	7.65	7.75	7.88	8.09

0.1	0.2	0.3	0.4	0.5	0.6	0.7	0.8	0.9
1.91	2.12	2.25	2.35	2.42	2.49	2.54	2.59	2.63

Figure 5.24: Table to transform a probit in a percentage.

Instead of using the transformation table, it is also possible to calculate the probability with:

$$Probability = \frac{1}{2} + \frac{1}{2}\mathrm{erf}\left(\frac{Probit - 5}{\sqrt{2}}\right)$$

In mathematics, the error function (also called the Gauss error function), often denoted by erf, is a complex function of a complex variable defined as:

$$\mathrm{erf}(z) = \frac{2}{\sqrt{\pi}} \int_0^z e^{-t^2} dt$$

The error function can be solved by a Taylor expansion

$$\mathrm{erf}(z) = \frac{2}{\sqrt{\pi}} \sum_{n=0}^{\infty} \frac{(-1)^n}{(2n+1)\ x\ n!} z^{2n+1}$$

$$\approx \frac{2}{\sqrt{\pi}}\left(z - \frac{z^3}{3} + \frac{z^5}{10} - \frac{z^7}{42} + \frac{z^9}{216} - \frac{z^{11}}{1320} + \frac{z^{13}}{9360} - \frac{z^{15}}{75600} + \cdots\right)$$

The Taylor expansion has to include about 10 terms (i.e., including $z^{21}/76{,}204{,}800$) to become stable and accurate.

A more detailed discussion on the origins of the logistics function and the probit function and the wider acceptance in statistics have been described by Cramer (2002). More details on the probit method can also be found in the books of Finney (1947 – the third revision of this book dates from 1971).

Cramer found that there were 629 articles in statistical journals between 1935 and 1994 containing the word probit (see Table 5.25).

Table 5.25: Number of articles in statistical journals containing the word "probit" (Cramer, 2002).

Period	Number of articles
1935–1939	6
1940–1944	3
1945–1949	22
1950–1954	50
1955–1959	53
1960–1964	41
1965–1969	43
1970–1974	48
1975–1979	45
1980–1984	93
1985–1989	98
1990–1994	127

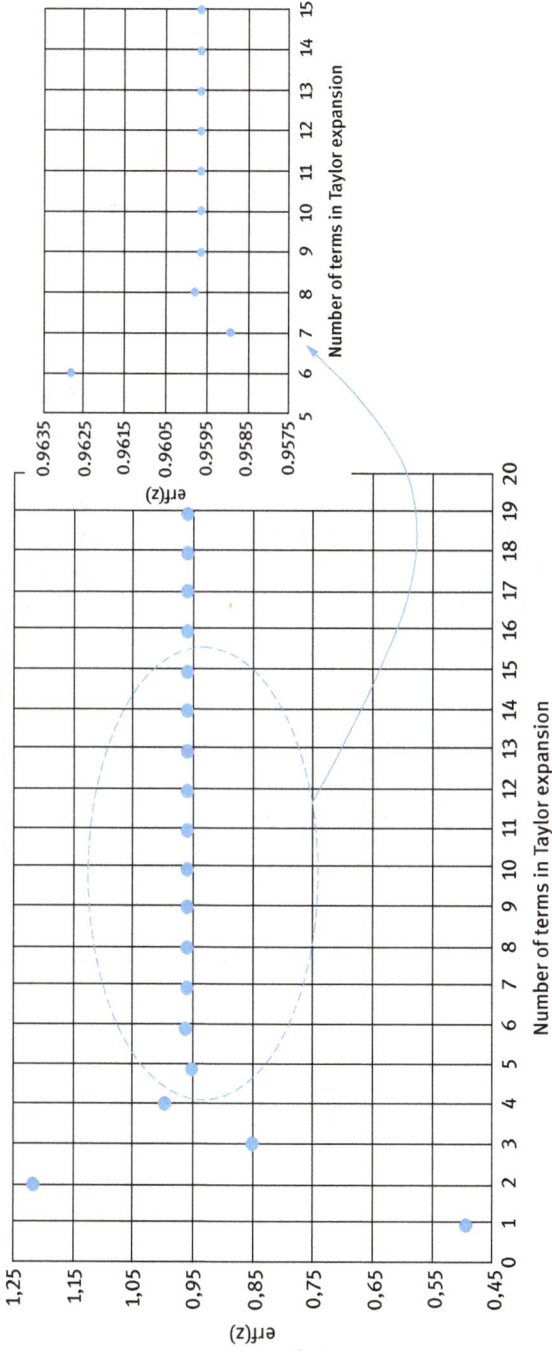

Figure 5.25: Influence of the number of terms in the Taylor expansion on the stability of the erf function.

Probit equations can be found in many literature sources. The Dutch Ministry VROM published in 2005 the second revision of the so-called Green Book (CPR, 2005) which contains probit equations for all type of impact on people and objects. Two examples taken from the Green Book are:

Probit	Description
$Pr = -17.56 + 5.30\ln(\frac{1}{2}\ m\ v_0^2)$	probit function for skull base fracture for a fragment mass of 0.1 < m < 4.5 kg with a velocity of v_0 in m/s
$Pr = -12.9 + 1.54\ln(P_s)$	Probit function for ear drum rupture

In a large number of cases, multiple probits will be available and a choice will have to be made as to which probit to use. As already mentioned many times, risk analysis is an engineering science. So it is not about determining exactly which probit the right one is. The point is to perform the analysis of the risks (uncertainties) in a consistent manner.

It may seem strange to the uninitiated that there is so much difference between various probit equations for one particular product (see Table 5.26 for chlorine for instance). The reason for this is that various experiments do not always give the same results. And as mentioned, analyzing risks is by definition "dealing with uncertainties." Back (Back et al., 1972) derived an LC50 of 293 ppm for a 1 h exposure with rats and 137 ppm for a 1 h exposure with mouse.

Table 5.26: Different probit equations for chlorine and lethal concentration calculations for respectively 1%, 50% and 90%.

Probit	Reference	Concentration (ppm) that results in a percentage fatality after an exposure of 5 min		
		1%	50%	90%
$Pr = -17.63 + 1.1\ln(C^{3.5}\ t)$	ten Berge et al. (1986)	123	225	384
$Pr = -4.4 + 0.52\ln(C^{2.75}\ t)$	HSC (1992)	78	399	1,671
$Pr = -8.29 + 0.92\ln(C^2\ t)$	Withers and Lees (1985)	173	613	1,867
$Pr = -11.85 + \ln(C^{2.3}\ t)$	de Weger et al. (1991)	274	755	1,840
$Pr = -5.036 + 0.5\ln(C^{2.75}\ t)$	ten Berge and van Heemst (1983)	151	824	3,657
$Pr = -10.1 + 1.11\ln(C^{1.65}\ t)$	Van Heemst (1990)	402	1,435	4,394
$Pr = -23.76 + 2.78\ln(C^{1.04}\ t)$	Zwart and Wouterson (1988)	1,986	4,445	9,036

The US Department of Health and Human Services published in 2010 a comprehensive document (269 pages) on the toxicological profile of chlorine (U.S. Department of Health and Human Services, 2010). Figure 5.26 shows some excerpts from this comprehensive document. The purpose of these excerpts is only to demonstrate the challenge toxicologists are facing when they have to derive a probit equation.

"Work conducted at the U.S. Army's Medical Research Laboratory of the Chemical Warfare Service cited by DOA (published in 1933) indicates that acute exposures to concentrations >870 ppm were usually lethal to dogs, whereas concentrations below 656 ppm were rarely fatal.

Weedon et al. conducted lethality studies in unspecified strains of rats and mice. In a group of eight rats exposed to 1,000 ppm chlorine, the first death occurred in 20 minutes and all were dead in 1.7 hours. The exposure level of 1,000 ppm was the LC50 in 53 minutes, and 250 ppm was the LC 50 in 440 minutes.

In mice, the first death occurred in 21 minutes, and all eight were dead in 50 minutes. The 1,000 ppm exposure level was a 28-minute LC50, whereas 250 ppm was a 440-minute LC50. Mice that died immediately after exposure had slight brain congestion; lungs partly collapsed and hemorrhagic; moderately distended heart; liver congestion; distended and hemorrhagic stomach; slightly distended intestines, and congested kidneys.

In lethality studies conducted by Zwart and Woutersen, 5,484 ppm was a5-minute LC50 in Wistar rats, whereas 447 ppm was a 60-minute LC50; in Swiss-Webster mice, 1,032 ppm was a 10-minute LC50 and 516 ppm was a 30-minute LC50."

Figure 5.26: Excerpts for toxicological profile of chlorine (US Department of Health and Human Services, 2010).

5.8.4.4 Mathematical analysis of impact on buildings and structures

The single-value criteria as well as the statistical approach give an indication on the generic potential damage.

In a number of cases, however one can be interested in the damage for a particular structure taking into account the specific characteristics of the building or structure. In that case, a more detailed analysis will be needed. Structural mechanics is the computation of deformations, deflections and internal forces and stresses within structures. It is a subset of structural analysis.

To perform an accurate analysis, a structural engineer must determine information such as structural loads, geometry, support conditions and material properties.

The difficulty with response of buildings and structures to explosions is twofold:
- The load is dynamic;
- The explosion curve has a very jagged shape.

In addition, the exact structure of older buildings is not always precisely known.

Figure 5.27 gives a characteristic shape of a shock wave and a pressure wave. The load on a building or structure is characterized by:
- A peak overpressure (P_s);
- A positive phase duration (t_p);
- An impulse which is defined as the area under the time-pressure diagram:

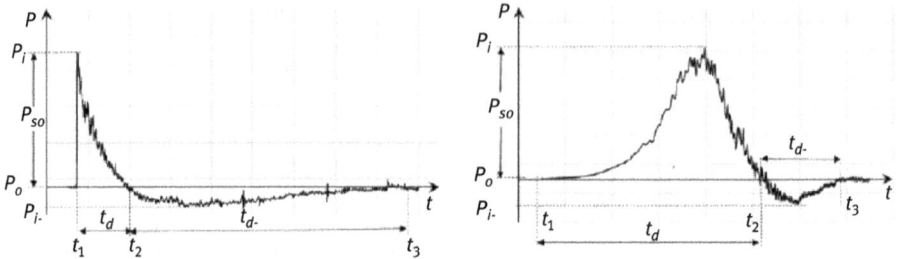

Figure 5.27: Characteristic shape of a shock wave (left) and a pressure wave (FABIG, 2018).

$$i_s = \int_{t_p} P_s(t)dt$$

- Reflected overpressure (P_r). When the blast wave meets an obstacle, it will initially reflect against the obstacle and a reflected wave begins to move in a direction opposite to the incident wave. The surface on which the incident wave is reflected becomes loaded by a reflected overpressure which is higher than the incident peak pressure.
- Dynamic pressure (Q_D). The blast is accompanied by an air displacement in the direction of the blast wave. This pressure of this air displacement multiplied by the so-called drag coefficient will create a dynamic pressure (Q_D) on the obstacle.
- The negative phase of the explosion wave with a sometimes important impulse.

The dynamic load of an explosion will result from the peak overpressure, the impulse, the reflected overpressure and the dynamic pressure. For small piping, the damage will mainly be the result of the drag while for large obstacles the reflected pressure will be an important parameter.

In order to perform calculations, the shape of the explosion will be approximated by an idealized forms as shown in Figure 5.28.

The building or the structure have an natural period (T) which is the time it takes for the building or structure to complete one cycle to naturally vibrate back and forth. Natural periods are expressed in seconds or fractions of a second.

- For buildings, it will depend on the structural system, materials and geometric proportions but some generic figures are: about 0.05 s for a piece of equipment such as a filing cabinet, about 0.1 s for a one-story building, 0.5 s for a four-story building, 1 to 2 s for a 10- to 20-story building and up to 7 s for a 60-story building;
- For steel structures, some simple formulas are given in Table 5.27 (Nassani, 2014).

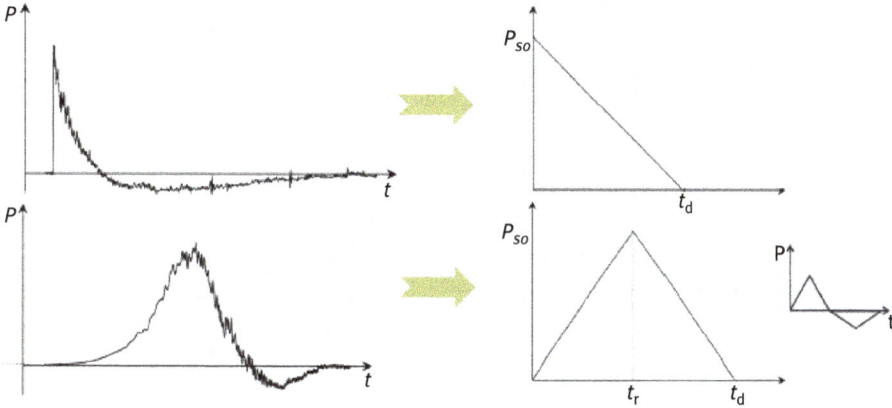

Figure 5.28: Transformation of a pressure or shock wave in an idealized form (Burgan, 2020).

Table 5.27: Natural period of steel structures (T) as function of the structural height (h) in meters.

$T = 0.0724\ h^{0.8}$	Steel moment-resisting frames
$T = 0.0466\ h^{0.9}$	Concrete moment-resiting frames
$T = 0.0731\ h^{0.75}$	Eccentrically braced steel frame
$T = 0.075\ h^{0.75}$	Framed structures

The ratio of the positive phase duration (t_p) to the natural period of the structure (T) will determine whether the load is impulsive, dynamic or quasistatic as shown on the sketch in figure (Shi et al., 2008).

Some studies consider the boundaries as:
- $t_p/T < 0.3$ impulsive
- $0.3 \le t_p/T \le 3$ dynamic
- $3 < t_p/T$ quasistatic

NORSOK (NORSOK, 2000) considers the boundaries as:
- $t_p/T < 0.3$ impulsive
- $0.3 \le t_p/T \le 3$ dynamic
- $3 < t_p/T \le 6$ quasistatic
- $t_p/T > 6$ static

Bowerman (Bowerman et al., 1992) proposes:
- $t_p/T < 0.4$ impulsive
- $0.4 \le t_p/T \approx 2$ dynamic
- $t_p/T > 2$ quasistatic

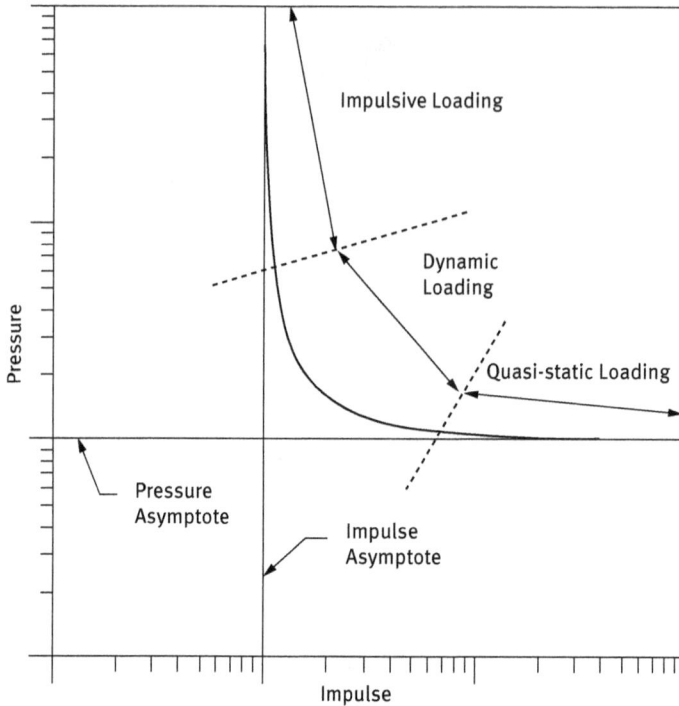

Figure 5.29: Sketch of a pressure–impulse diagram (Shi et al., 2008).

Czujko (Czujko, 2001) proposes:
- $t_p/T \ll 1.0$ impulsive
- $t_p/T \approx 1.0$ dynamic
- $t_p/T \geq 2$ quasistatic
- $t_p/T > 5$ static

Pressure–impulse diagrams have been in use since the 1940s. Chernin (Chernin et al., 2019) published a fundamental review of 70 years of history on the use of pressure–impulse diagrams.

Pressure–impulse diagrams can be established by experiments or by performing a load-response calculation. The most common methods to calculate the response to a dynamic load are:
- Single-degree-of-freedom model (SDOF). SDOF can be applied as a "simple" hand method to calculate elastoplastic bending and plastic bending as well as shear and buckling.
- Multi-degree-of-freedom model (MDOF). MDOF is a full analysis of the structure. The most common computer programs are conventional linear finite element analysis and nonlinear finite element analysis.

Examples of calculations with SDOF are given by FABIG in the technical notes 7 ("Simplified method for analysis of dynamic response") (FABIG, 2002) and 10 ("Advanced SDOF model") (FABIG, 2002). The use of SDOF for a building is detailed in FABIG Technical Note 14 (FABIG, 2018). More information can also be found in the following references:
- UFC 3-340-02, *Structures to Resist the Effects of Accidental Explosions*, DoD, USA, December 2008 (1943 pages)
- ASCE, *Design of Blast Resistant Buildings in Petrochemical Facilities*, American Society of Civil Engineers, 1997

Nourzadeh (Nourzadeh, 2017) shows in his PhD that the number of degrees of freedom in the analytical model of the columns and second-order deflections (P-δ) both have important effect on the calculated response to blast load. On the other hand, shear deformations have only a minor effect on the response. Also, it is shown that the analytical response of columns to blast loads, when they are modeled individually can be different from when they are modeled as part of the lateral load resisting system of the building. He observed that the simplified methodologies used for the analysis of roof and side beams spanning in a direction perpendicular to the shock front lead to large overestimation of the response. For a more reliable estimate of the response, analysis should be carried out using MDOF models of such elements and the blast loading imposed by a travelling wave.

A dynamic nonlinear finite element study was performed in 2004 on a polypropylene installation. The pressure-impulse signature was calculated with a CFD model (FLACS) and the dynamic, non-linear, finite element analyzes were carried out using the LS-DYNA model. LS-DYNA was originally developed as a 3D finite element analysis program DYNA3D, at Lawrence Livermore National Laboratory in 1976. It was acquired by the company ANSYS who released it version 18 in 2017. Its core-competency lies in highly nonlinear transient dynamic finite element analysis which is used for automobile industry (e.g., crash simulations), aerospace, construction and civil engineering and manufacturing.

The study was presented at the FABIG Technical Meeting number 42 (Hoorelbeke et al. 2005).

Finite elements is a method which to some extent is comparable with CFD. Many engineering problems can be described by PDEs. These PDEs in most cases cannot be solved analytically. Mathematicians have developed numerical approaches that make it possible to perform "accurate approximations." Examples of these mathematical approaches are:
- FEM: finite element method
- DEM: discrete element method
- BEM: boundary element method
- FVM: finite volume method
- FDM: finite difference method

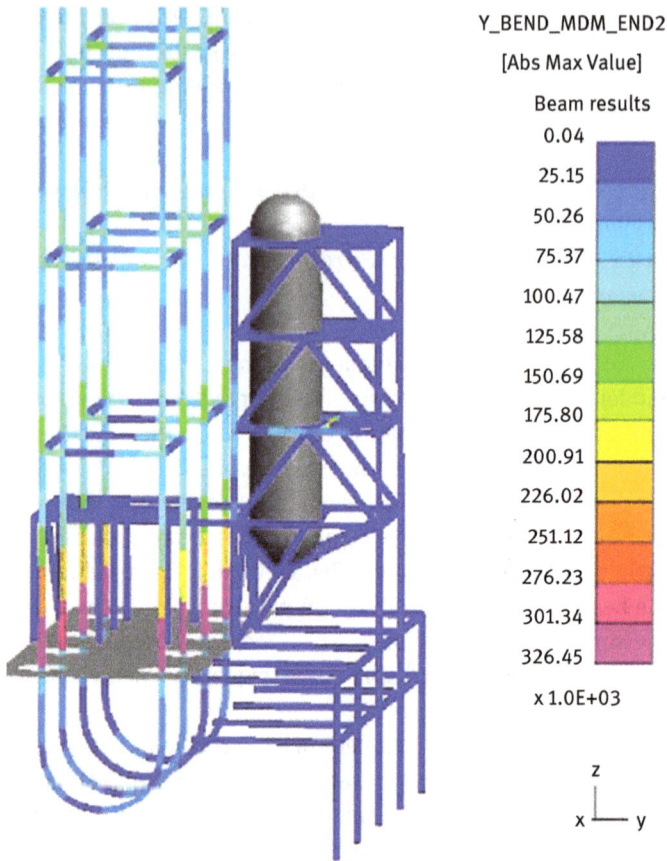

Figure 5.30: Dynamic response of a part of a polypropylene installation (Hoorelbeke et al., 2005).

CFD (see for instance under Section 5.8.3.2) most often uses the FVM and FDM to solve fluid-flow problems while in structural analysis it will be more logic to use FEM. A list with more than 40 commercial finite element programs can be found on Wikipedia.

5.9 Risk analysis and risk assessment for individual scenarios

5.9.1 Risk analysis versus risk assessment

The purpose of the risk analysis is to get insights in the risks. As explained earlier, the engineering approach considers that risk is an objective property of a potential undesired event. This property is defined by the combination of the potential consequence and the likelihood that the event and its potential consequence will manifest.

Once the risk is calculated, it has to be decided whether the risk should be further reduced or whether the risk as calculated is considered as acceptable.

The evaluation of the acceptability of the risk is called the risk assessment. Risk assessment can be done for a single scenario but should also be done for the aggregate risk profile because the sum of all risks could be unacceptable while each individual risk is acceptable.

5.9.2 Criteria for risk assessment

There are many possibilities to express the risk:
- Individual risk
  - Localized individual risk per annum is the probability that an average unprotected person, permanently present at a specified location, is killed during 1 year due to a hazardous event at an installation
  - Personal individual risk or individual risk per annum is the probability that a specific or hypothetical individual will be killed due to exposure to the hazards or activities during 1 year
  - Maximum individual risk
  - Average individual risk
- Societal risk
  - The potential loss of life (PLL) is the expected number of fatalities within a specific population per year.
  - The fatal accident rate (FAR) is defined as the expected number of fatalities per 100 million exposed hours. This exposure is sometimes illustrated by 1,000 persons that work 2,000 h per year for 50 years – or 1,000-life-years exposure (with current working hours, it will be closer to 1,300 life years.). FAR requires explicit definition of exposure and does not distinguish between accidents with single fatalities and accidents with multiple fatalities. It is therefore suited for comparison across activities.
  - fN = cumulative fatality frequency curve
- Risk matrix
- ICAF = investment cost to avert a fatality
- Unicohort
- PLM = potential loss of money

The reason why different expressions are needed is that the different measures give a different insight. Let us take two examples:
- Risk related to circulation on the routes in France
- Risk related to a nuclear power plant

Table 5.28 gives an overview of the number of accidents and the number of victims due to accidents on the road in France. The number of people exposed to the risk of an accident on the road in France is difficult to determine. The population on France increased from 63.96 million in 2008 to 66.89 million in 2018. However, some French nationals do not participate (e.g., small children, people in hospitals, people in retirement house) while there are people not living in France who are on the road in France (e.g., tourists, cross-border workers, international traffic that drives through France). In order to keep it simple, it is assumed that 70 million people are exposed to the risk of an accident on the roads in France.

Table 5.28: Number of accidents on the road in France in the period 2008–2017.

Year	Number of accidents	Number of people	
		Killed	Injured
2008	76,767	4,443	96,905
2009	74,409	4,172	93,993
2010	69,379	4,111	87,173
2011	66,974	3,842	83,872
2012	62,250	3,495	78,209
2013	58,397	3,557	72,645
2014	59,854	3,557	75,142
2015	58,654	3,616	73,384
2016	59,919	3,738	75,819
2017	61,224	3,684	76,840

With the data from Table 5.28, we can then calculate the probability for an individual to have an accident and the probability for an individual to be killed. The personal individual risk for having an accident decreased from 1.1×10^{-3} in 2008 to 8.7×10^{-4} in 2017 while the personal individual risk for being killed in an accident on the road in France decreased from 6.3×10^{-5} in 2008 to 5.3×10^{-5} in 2017. The PLL can for 2018 can be estimated based on the evolution between 2008 and 2017. Based on the trend between 2008 and 2017, the PLL for 2018 is estimated at 3,384 fatalities. The correct figure for 2018 was 3,488 deaths. Hence, when studying the risks of accidents on the roads in France, we will need to consider the PLL. The individual risk could be considered as acceptable compared to other types of risk an individual is exposed to but the total number of victims per year is difficult to accept.

Let us now have a look at the risks of commercial nuclear power plants. There are in the world 442 reactors in operation. The oldest nuclear power plant still in operation is the Beznau nuclear power plant in Northern Switzerland. Construction began in 1965 and Beznau 1 started producing on 1 September 1969. At the end of 2019, the worldwide cumulative reactor operating experience amounted to over 18,329 reactor years of experience (IAEA 2019 Data). We consider that only two accidents caused multiple fatalities due to the emission of radiative material outside the nuclear power plant: Chernobyl in 1986 and Fukushima in 2010. There is a lot of discussion about the number of fatalities caused by these accidents:

- **Chernobyl.** Thirty-one people died as a direct result of the Chernobyl accident; two died from blast effects and a further 29 firemen died as a result of acute radiation exposure (where acute refers to infrequent exposure over a short period of time) in the days which followed (NEA, 2002). The number of people who were impacted over long-term radiation exposure is more difficult to discern and remains highly contested. Part of this difficulty lies in the methodology used to estimate long-term deaths from low-level radiation exposure. In its 2005/06 assessment, *Chernobyl's Legacy: Health, Environmental and Socio-Economic Impacts* the World Health Organization (WHO) estimated that the total number of long-term deaths will be around 4,000 (source: IAEA, WHO (2005/06). *Chernobyl's Legacy: Health, Environmental and Socio-Economic Impacts*). However, this figure is related only to the proximate populations of Ukraine, Russia and Belarus which were exposed to high radiation levels; if extended to estimates of those exposed to low-level radiation across the region, this number rises to 9,000. Radiation scientists Fairlie and Sumner provide some of highest estimates, predicting 30,000–60,000 deaths (Fairlie and Sumner, 2006).
- **Fukushima.** In the case of Fukushima, although 40 to 50 people experienced physical injury or radiation burns at the nuclear facility, the number of direct deaths from the incident are quoted to be zero. In 2018, the Japanese government reported that one worker has since died from lung cancer as a result of exposure from the event. However, mortality from radiation exposure was not the only threat to human health: the official death toll was 573 people – who died as a result of evacuation procedures and stress-induced factors.

The 99% confidence interval for the probability of a nuclear accident like Chernobyl or Fukushima, based on historical experience is 8.1×10^{-6}–5.1×10^{-4} per reactor. year (average = two accidents in 18,329 reactor years which gives 1.1×10^{-4} per reactor.year). The 99% confidence interval for the probability of having 10,000 fatalities is 6.0×10^{-7}–4.0×10^{-4} per reactor.year (average = one accidents in 18,329 reactor years which gives 5.5×10^{-5} per reactor.year). The **personal individual risk** for being killed due to a nuclear accident for people living close to a nuclear power plant is difficult to estimate but certainly much lower than 10^{-5} per reactor.year. The **PLL** (i.e., the number of people expected to be killed every year) is beneath 1 per year. But the number of people than can be killed in one accident is several thousands and could be as high as 60,000. When evaluating the risk of nuclear power plants, it is not the personal individual risk nor the PLL which is a good measure.

5.9.3 Risk matrix

Many companies make extensive use of risk matrices. Because the method is widely distributed, this section discusses the uses and limits of the method in more detail.

The use of risk matrix, among others, is described in the international ISO Standard IEC/FDIS 31010, Risk Management Techniques, 2009. Risk matrix is used in all types of industries. Figure 5.31 is an example of a risk matrix.

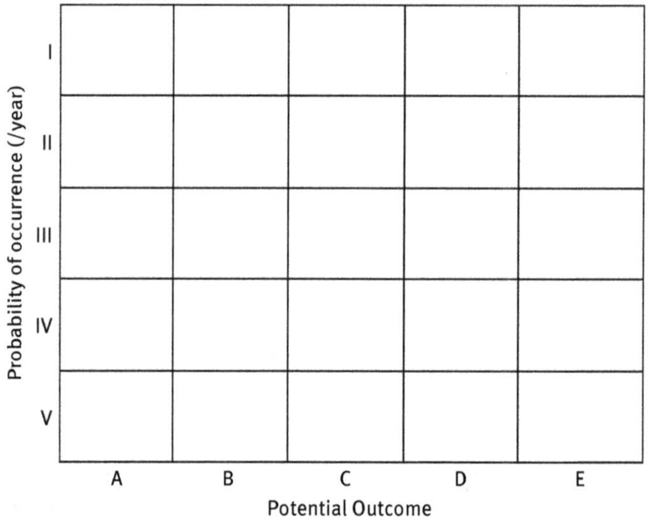

Figure 5.31: Example of a risk matrix.

The basic assumption is that the risk associated with an undesired event can be represented by the potential outcome (e.g., loss time accident) and the probability of occurrence of this outcome (e.g., one chance in 100 in a period of 1 year). Once this assumption is accepted, it becomes possible to localize the "risk" of an undesired event in the matrix. In the example above, risks for all type of undesired events can be localized in one of the 25 boxes (e.g., C & III).

Although it is not mandatory, most matrices have an increasing outcome from left to right and a decreasing probability of occurrence from up to down. In the above matrix, the lowest risk category is in box (A,V) and the highest risk category is in box (E,I)

The matrix in the example is called a 5 × 5 matrix (five outcome categories and five classes for the probability of occurrence). The number of boxes differs from one company to another and sometimes it will differ within the same company or it will be function of the type of outcomes (health, safety, finance, etc.).

The number of boxes is **a deliberate choice** based on the consideration that keeping:

- the number of cases as low as possible helps in striving for a consistent approach among all users of the matrix (the ultimate choice would be a 1 × 1 matrix because everybody would always estimate the same level of risk – but the matrix would be useless);
- the number of cases relatively high (e.g., 10 × 10 matrix) which enables a very fine classification of the risks but which makes it more difficult to achieve a consistent approach among the users.

Most common is the use of four to six classes for the outcome and four to six classes for the probability of occurrence of the outcome. The matrix can be uneven (e.g., 4 × 6 matrix). Many companies try to have one common matrix for the evaluation of all types of risks. This is however not recommended in international standards. ISO Standard IEC/FDIS 31010 states "a matrix should be designed to be appropriate for the circumstances so it may be difficult to have a common system applying across a range of circumstances relevant to an organization."

Once the size of the matrix is decided, then the classes of the "probability" and of the "potential outcome" have to be defined. An example could be:

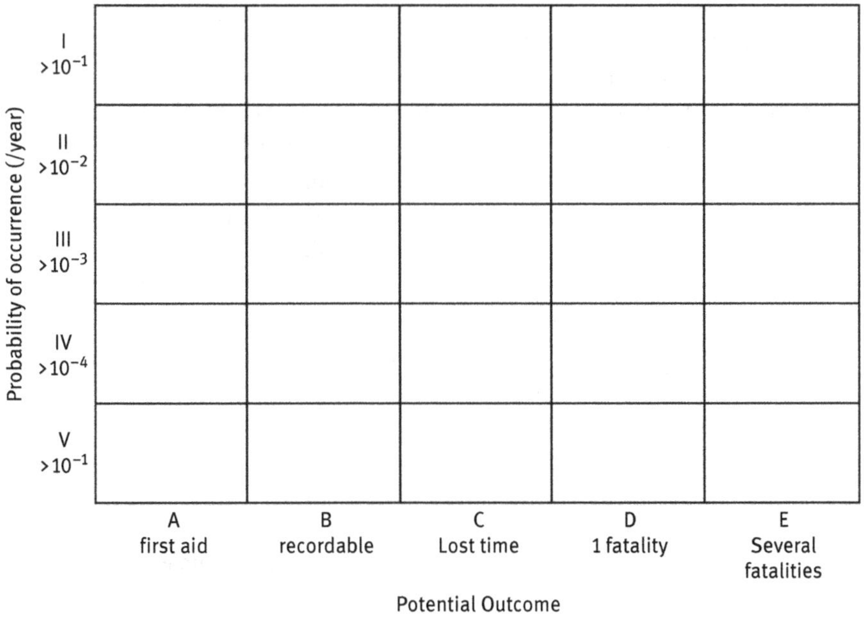

Figure 5.32: Example of a classification of a risk matrix.

Now that the classes have to be defined, a decision will be made on the different zones of the matrix.

Figure 5.33 gives two different ways of defining the areas in a similar risk matrix (red zone is considered as high risk while green zone is considered as low risk). The following can be concluded about the risk matrix:

– The size of the risk matrix is defined by the user (e.g., a company);
– The classification of the probabilities and the consequences are defined by the user;
– The zones with "high risk," "low risk" or "intermediate risk" is defined by the user.

Because everything is defined by the user, a risk matrix expresses the classification of risk as defined by the user. The result is that the risk matrix can be used as an engineering tool to prioritize risks according to the user's wishes. However, it cannot be used to identify risk in a universal way. What is referred to as "high" or "low" risk is a pure convention within a community that has been agreed to use these definitions.

It is extremely important that this community is aware of this. Within the community that uses a risk matrix, sometimes it is incorrectly assumed that a risk in the green zone is an "acceptable risk" when in fact it is "acceptable" because that was defined that way by the user. The same goes for the red zone. A risk in the red zone is considered unacceptable because it is defined as such by the user community. A risk matrix is therefore much more an agreement within a user community than a risk measurement. It is an engineering tool for prioritizing actions.

An additional problem is that it is not always clear which scenarios should be placed in the matrix. Suppose you want to place the scenario "leak at a pump" on the risk matrix and the company has 10,000 pumps worldwide. A point on the matrix can be set for each pump or one can set one point that corresponds with the scenario "a leak at a pump somewhere in the company". Both scenario's are a risk for the company. However in the latter case the probability is 10,000 times higher.

A risk matrix can be used within a community or company to prioritize the action plans. But the following conditions will be necessary:
1. Clear definition of the type of scenarios than can be assessed with the matrix;
2. The method on how to use of the matrix should be clearly established;
3. The matrix should only be used by people who received a training in how the matrix can and should be used.

If these conditions are not fulfilled then the matrix will give a false illusion for presenting the risks and there is a real chance that useless investments will be made.

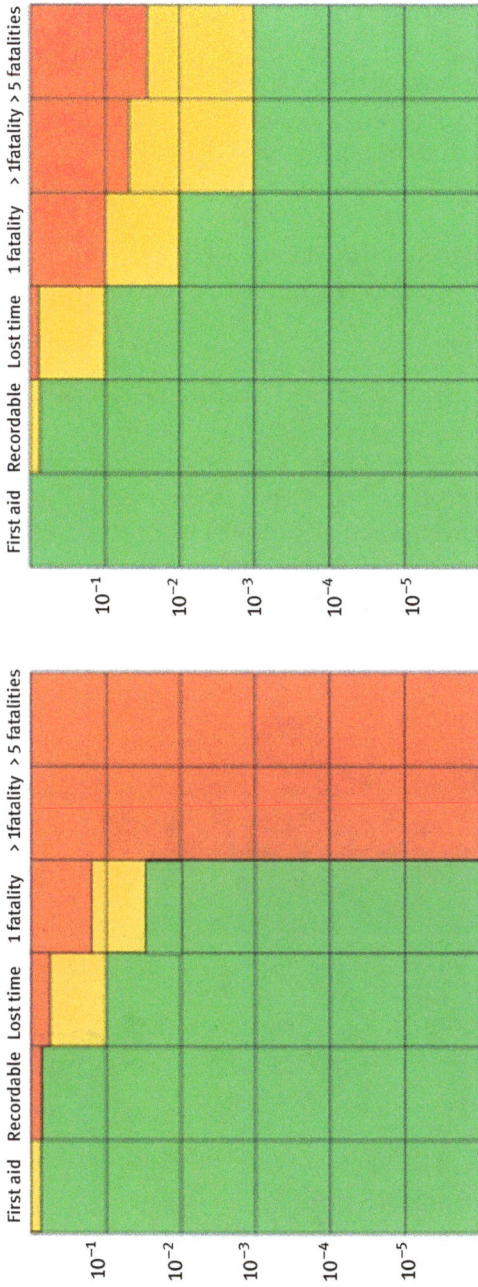

Figure 5.33: Two different ways of defining areas in a similar risk matrix.

5.9.4 Aggregate risk analysis and assessment

A risk matrix does not allow to assess aggregated risks. Let us consider the example of a fatal accident on the road in France:

- There are every day accidents on the road in France with fatalities;
- If we put one specific single accident on the matrix of Figure 5.33, then it will always be in the green zone;
- If we consider the aggregate probability of a fatal accident on the road in France (irrespective of who will have the accident), then it will always be in the red zone.

In order to get insight in the aggregate risk, we can use "Localized Individual Risk Contours," and "f-N curves."

Figure 5.34 is an example of Localized Individual Risk Contours on a petrochemical complex. For each equipment potential leak scenarios are listed. For each of the scenarios, the probability is determined and all potential outcomes (dispersion, ignition, all types of fires, etc.) are modeled taking into account wind directions and different meteorological conditions. In every point of a raster, the effect (radiation, overpressure, etc.) is calculated and the probability of being killed at

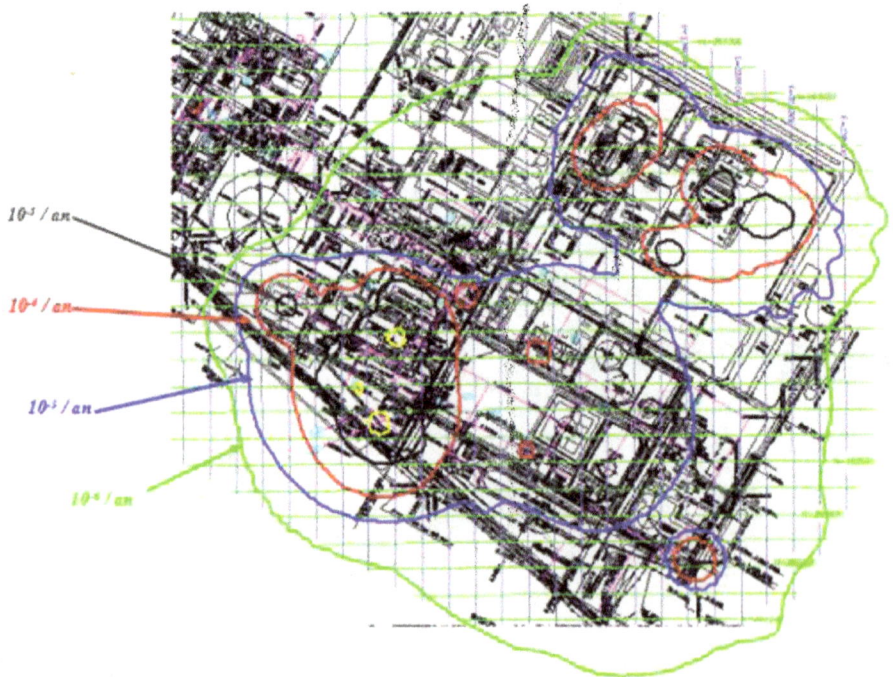

10^{-3} / an

10^{-4} / an

10^{-5} / an

10^{-6} / an

Figure 5.34: Example of localized individual risk contours on a petrochemical complex.

that location is determined (assuming that a person who would be present in that point 100% of the time). All the points with an equal probability are then connected by so-called iso-risk contours.

The final result is a plot with, for each location inside and outside the refinery, the probability of death due to the sum of all scenarios. These contours can be used for plot planning and to determine in which zones additional measures should be considered.

The individual risk contours do not give a view on the number of victims that could occur in one accident. To get insight in the maximum number of victims, an f-N curve can be drawn. The principle of an f-N curve is as follows:

Assume three outcomes (from one scenario or from different scenarios):
- Outcome 1 10 people killed probability of occurrence = $10-3$ per year
- Outcome 2 50 people killed probability of occurrence = $10-4$ per year
- Outcome 3 80 people killed probability of occurrence = $10-5$ per year

Hence, it can be stated that the probability of at least 10 people killed is $1.11 \ 10^{-3}$ per year and the probability of at least 50 people killed is $1.1 \ 10^{-4}$ per year. The points (10; $1.11 \ 10^{-3}$) and (10; $1.1 \ 10^{-4}$) are plotted on a cumulative curve called f-N curve.

Hoorelbeke and Roosendans studied 1,657 man-made disasters (minimum 10 fatalities) in the period 1945–2014. The f-N curve is given in Figure 5.35. The authors identified four categories of man-made disasters:
- **Transport of people.** People travel sometimes in large groups (airplanes, ferries, trains, etc.).
- **Meeting locations**. People meet in large numbers for different reasons. Examples are: churches, sporting events, theater and concerts. People within a society can choose to participate in these activities but are not obliged to do so.
- **Crowded locations**. This category comprises schools, hospitals and so on. The authors concluded that these crowded locations are different from meeting locations because they are a logical consequence of the organization of our society. All individuals are affected because that is the way a society is organized and hence people of a society are obliged to accept that risk.
- **Industrial activities**. It is a well-established paradigm that improving the well-being of people and society can be achieved by deploying industrial (economic) activities. The industrial activities increase the prosperity of society, thereby improving the welfare of the population. These industrial activities can be at the heart of a man-made disaster.

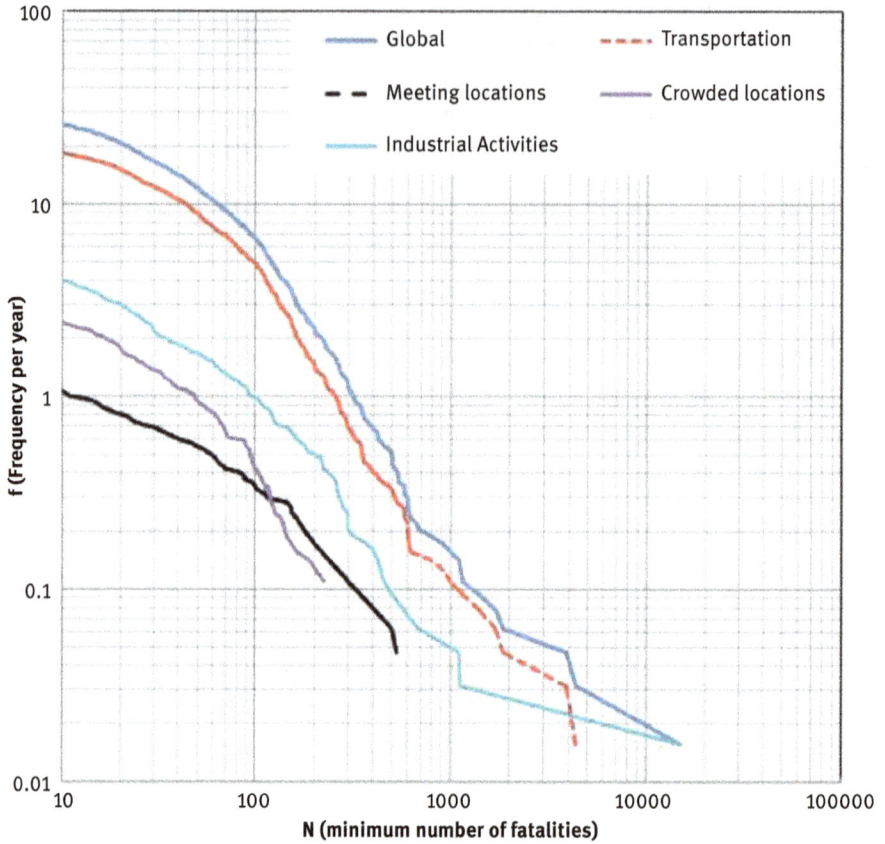

Figure 5.35: Historical f-N curve for man-made disasters in 70 years post–WWII period (Hoorelbeke et al., 2015).

5.10 Pitfalls in risk assessments

UK HSE requested health and safety laboratory to perform a study to look at the good practices and pitfalls in risk assessments (Gadd et al., 2003)

The objectives of this stage of the project were to (1) carry out a review of published critiques of both general and specific risk assessment methodologies, (2) identify examples of inadequate industry risk assessments that illustrate common pitfalls in the application of risk assessment and (3) carry out a brief review of HSE guidance for Inspectors on assessing the adequacy of risk assessments.

The identified pitfalls were as follows:

1. Carrying out a risk assessment to attempt to justify a decision that has already been made;
2. Using a generic assessment when a site-specific assessment is needed;

3. Carrying out a detailed quantified risk assessment without first considering whether any relevant good practice was applicable, or when relevant good practice exists;
4. Carrying out a risk assessment using inappropriate good practice;
5. Making decisions on the basis of individual risk estimates when societal risk is the appropriate measure;
6. Only considering the risk from one activity;
7. Dividing the time spent on the hazardous activity between several individuals – the "salami slicing" approach to risk estimation;
8. Not involving a team of people in the assessment or not including employees with practical knowledge of the process/activity being assessed;
9. Ineffective use of consultants;
10. Failure to identify all hazards associated with a particular activity;
11. Failure to fully consider all possible outcomes;
12. Inappropriate use of data;
13. Inappropriate definition of a representative sample of events;
14. Inappropriate use of risk criteria; Ÿ No consideration of ALARP or further measures that could be taken;
15. Inappropriate use of cost benefit analysis;
16. Using "Reverse ALARP" arguments (i.e., using cost benefit analysis to attempt to argue that it is acceptable to reduce existing safety standards);
17. Not doing anything with the results of the assessment;
18. Not linking hazards with risk controls.

5.11 Risk reduction measures

5.11.1 The basic assumption of risk reduction measures

It was assumed that risk is an objective physically property of a hazardous event and that it can be calculated as a combination of the potential outcome (consequence or impact) of an undesired event and its associated frequency of occurrence.

Risks can thus be reduced by decreasing the potential outcome, by decreasing the frequency of occurrence, or a combination of both. Figure 5.36 illustrates this principle on a risk matrix.

Roosendans (2018) performed research on the risk reduction of a system called cloud explosion mitigation by inhibition system – a TOTAL patented invention (VEMIS) on the maximum overpressures that can be generated in case of a vapor cloud explosion in a congested polypropylene unit. Calculations were done with FLACS, a CFDs model for simulation of flame accelerations. Figure 5.37 illustrates the risk reduction obtained with VEMIS: the area affected by high overpressures (red zone) for a given frequency of occurrence decreases significantly.

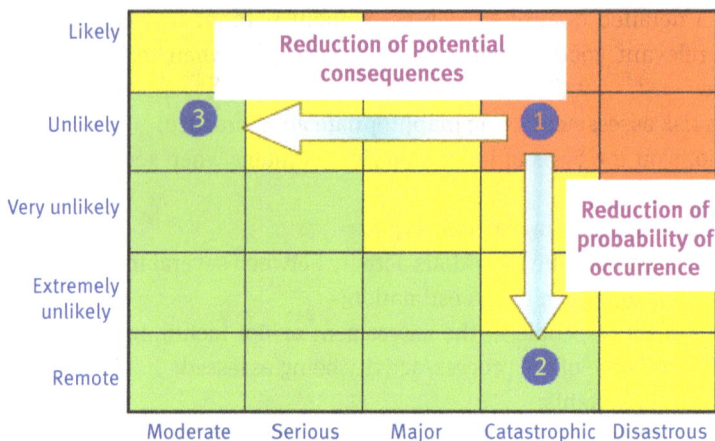

Figure 5.36: The principle of risk reduction on a matrix.

5.11.2 Defense-in-depth principle: multiple layer approach

Defense in depth is an old military strategy in which a defender deploy their resources, such as fortifications, field works and military units at and well behind the front line. Once attackers breach the defended front line, as they advance, they continue to meet resistance at new lines of defense.

In 1957, the US Atomic Energy Commission (USAEC, 1957) published a report called *Theoretical Possibilities and Consequences of Major Accidents in Large Nuclear Power Plants* (WASH-740 report), which estimated maximum possible damage from a meltdown with no containment building at a large nuclear reactor. The acting Chairman of the United States Atomic Energy Commission, Harold Sines Vance (1889–1959), writes in his foreword:

> Looking to the future, the principle on which we have based our criteria for licencing nuclear power reactors is that we will require **multiple lines of defense** against accidents which might release fission products from the facility.

In the chapter "The probability of Catastrophic Reactor Accidents" of the WASH-740 report, the examiners write:

> Should some unfortunate sequence of failures lead to the destruction of the reactor core with attendant release of fission product inventory within the reactor vessel, however expensive this would be to the owners, no hazard to the safety of the public would occur unless **two additional lines of defense** were also breached: (1) the integrity of the reactor vessel and (2) the integrity of the reactor container or vapor shell.

Since then, the notion of "defense in depth" and the concept "multiple layer approach" have progressively be used in the nuclear industry. A historical review of

Maximum blast overpressure (barg) at ground level with VEMIS.
(Inhibition of VCE with K_2CO_3)

Maximum blast overpressure (barg) at ground level without VEMIS.

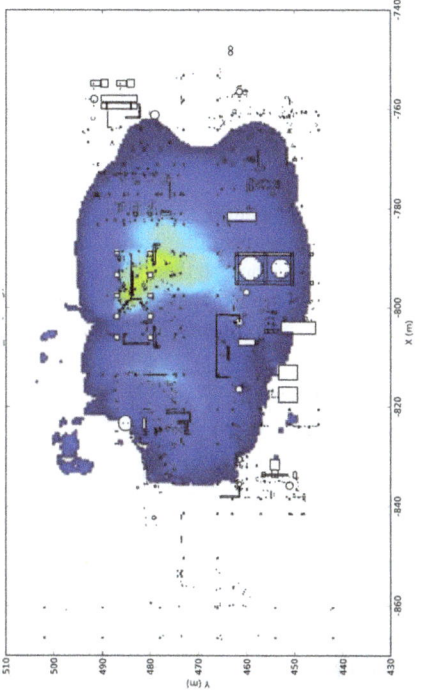

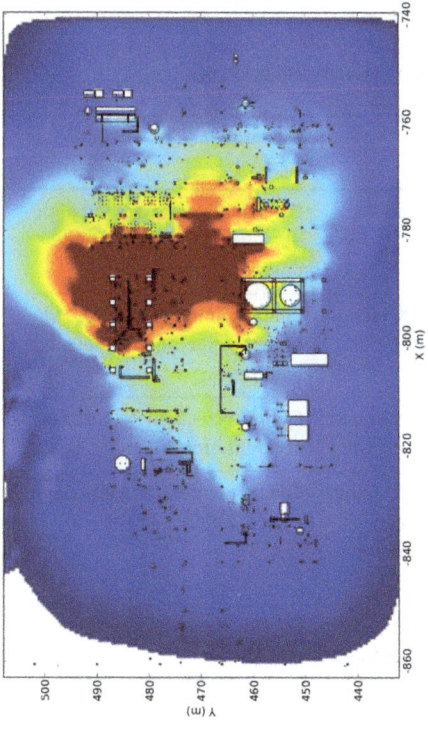

Figure 5.37: Overpressure risk contours in a PP unit with CFD modeling (Roosendans, 2018).

the concept in the nuclear industry has been published by the United States Nuclear Regulatory Commission (USNRC, 2016).

The International Atomic Energy Agency considers, since 1988, the "strategy of defense in depth" as one of the fundamental principles of basic safety principles for nuclear power plants (IAEA, 1988):

> "Defense in depth" is singled out amongst the fundamental principles since it underlies the safety technology of nuclear power. All safety activities, whether organizational, behavioural or equipment related, are subject to layers of overlapping provisions, so that if a failure occur it would be compensated for or corrected without causing harm to individuals or the public at large. This idea of multiple levels of protection is the central feature of defence in depth, . . .

The objectives of defense in depth were defined as follows:
- To compensate for potential human and component failures;
- To maintain the effectiveness of the barriers by adverting damage to the plant and to the barriers themselves;
- To protect the public and the environment from harm in the even that these barriers are not fully effective.

The International Nuclear Safety Advisory Group has published in 1996 an INSAG-10 report (IAEA, 1996) in which the approach to the defense in depth in five levels is explained.

Table 5.29: Five levels of defense in depth in the nuclear industry (IAEA, 1996).

Levels of defense in depth	Objective	Essential means
Level 1	Prevention of abnormal operation and failures	Conservative design and high quality in construction and operation
Level 2	Control of abnormal operation and detection of failures	Control, limiting and protection systems and other surveillance features
Level 3	Control of accidents within the design basis	Engineered safety features and accident procedures
Level 4	Control of severe plant conditions including prevention of accident progression and mitigation of the consequences of severe accidents	Complementary measures and accident management
Level 5	Mitigation of radiological consequences of significant releases of radioactive materials	Off-site emergency response

Defense in depth is a safety philosophy involving the use of successive compensatory measures (often called **barriers** or layers of protection or lines of defence) to prevent accidents or reduce the damage if a malfunction or accident occurs on a hazardous facility. Barriers should, as far as possible, be **independent**, meaning that the failure of one barrier does not affect the effectiveness of other barriers.

From the late eighties the concept of multiple layer defense was introduced in the chemical industry. CCPS begun in 1989 a project on Guidelines for Engineering Design for Process Safety, in which a group of volunteer professionals representing major chemical, pharmaceutical and hydrocarbon processing companies, worked with engineers of the Stone & Webster Engineering Corporation. The book was published in 1993 (CCPS, 1993) and deals in chapter 2 with inherent safe design (see next chapter). It states that:

> Process safety relies on multiple safety layers, or defense in depth, to provide protection from a hazardous incident. These layers of protection start with the basic process design and include control systems, alarms and interlocks, safety shutdown systems, protective systems and response plans.

The multiple layer principle is illustrated in Figure 5.38 (based on CCPS, 1993).

Risk reduction can be applied in each of the layers or by using a combination of the layers. However, the basic idea should be that an acceptable residual risk is achieved preferable in the first layers.

ANSI/ISA-S84.01-1996 is an American standard that deals with the fourth layer ("Automatic action SIS of Safety Instrumented Systems"). The standard specifies:

The desire is to provide appropriate number of non-SIS protection layers, such that SIS protection layer(s) are not required. Therefore, consideration should be given to changing the process and/or its equipment utilizing various non-SIS protection techniques, before considering adding SIS protection layer(s) (ANSI/ISA-S84.01-1996, 1996 – paragraph 4.2.4).

Each of the layers will be dealt with in the next chapters but the most important one is the first layer "Inherent Safe Design."

Risk reduction can be achieved via technical measures and/or procedural measures. The logic order when possible and reasonable practical will be (CCPS, 1998):
- Inherently safe;
- Passive protection, that is, the protection is a technical measure that is always present and does not have to be activated (e.g., the wall of a bund);
- Active protection, that is, the protection measure is technical but has to be activated;
- Procedural – using operating procedures, administrative checks and other management approaches to prevent the incident or to minimize the potential consequences of an incident.

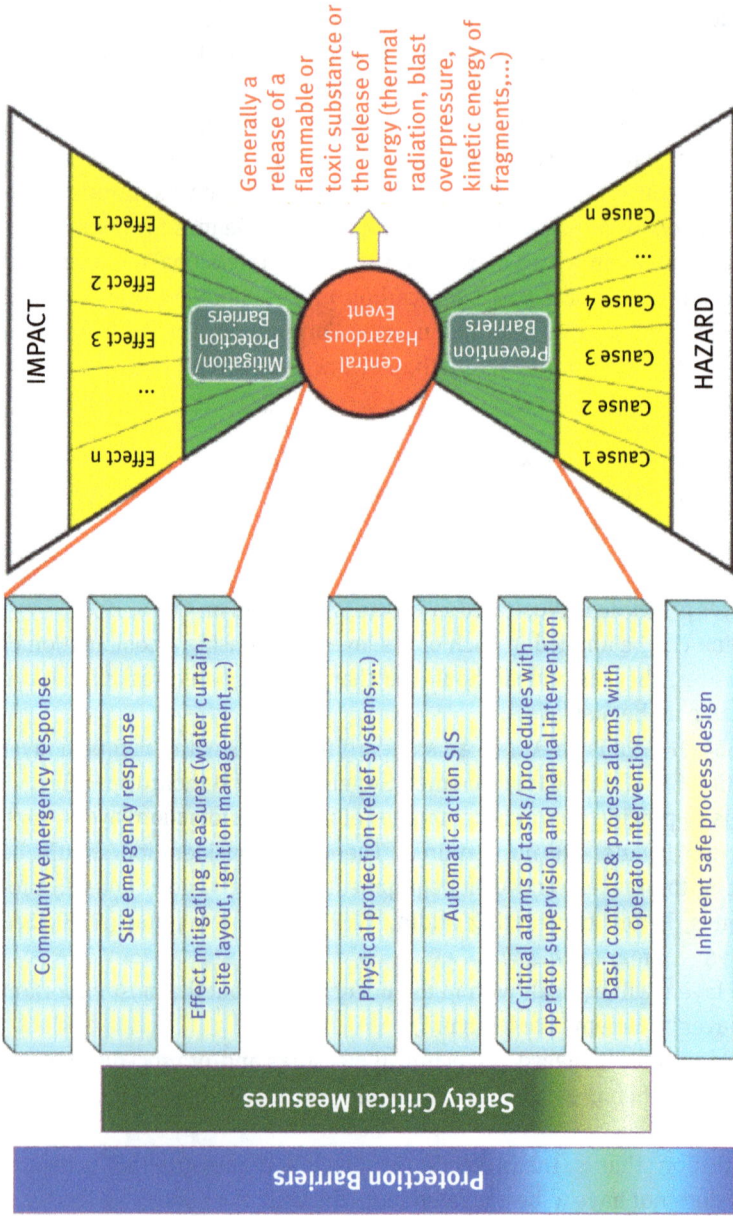

Figure 5.38: Illustration of the multiple layer principle (Roosendans, 2012 in EPSC publication).

The author prefers however a model with four layers. Each layer avoids undesired events and accidents. When one layer did not avoid a scenario from developing toward an accident the consecutive layer will enter in action. Each layer has the same purpose:
1. Avoid occurrence of undesired events;
2. When an undesired event occurs it is stopped, neutralized or the consequences are mitigated.

The four layers are:
1. Design and engineering;
2. Operations;
3. Automatic systems and passive protections;
4. Emergency response.

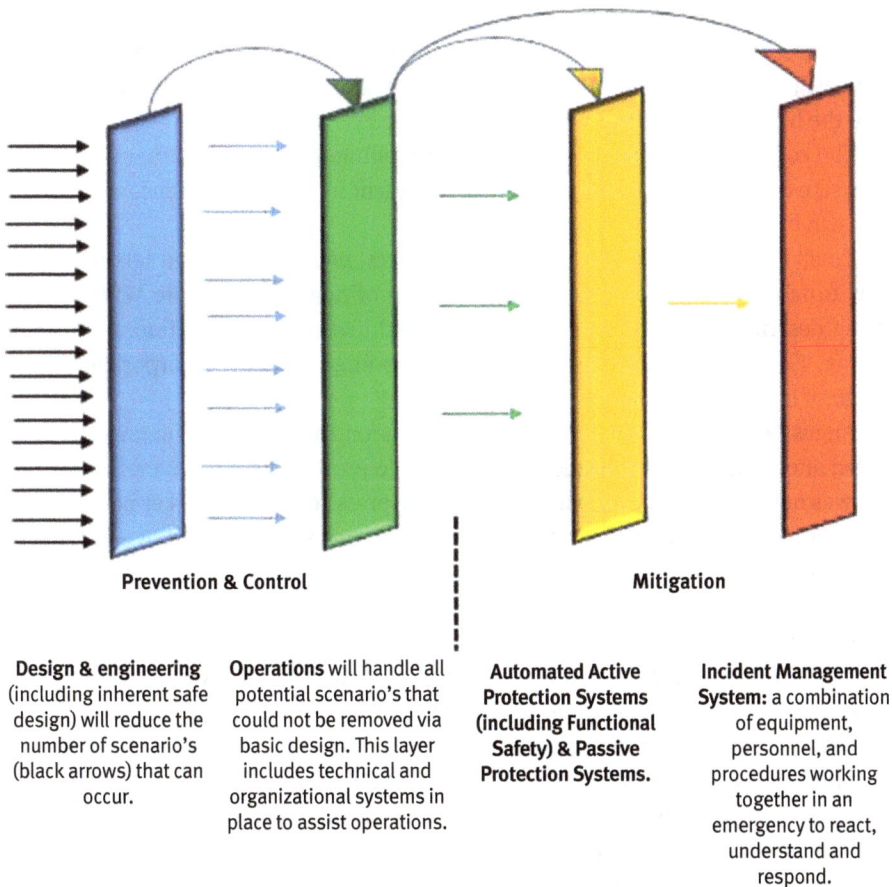

Prevention & Control		Mitigation	
Design & engineering (including inherent safe design) will reduce the number of scenario's (black arrows) that can occur.	**Operations** will handle all potential scenario's that could not be removed via basic design. This layer includes technical and organizational systems in place to assist operations.	**Automated Active Protection Systems (including Functional Safety) & Passive Protection Systems.**	**Incident Management System:** a combination of equipment, personnel, and procedures working together in an emergency to react, understand and respond.

Figure 5.39: The four-layer protection scheme.

The arrows on the top refer to the influence of one layer on the other layer. The design will influence the potential accidental scenarios in operations while operations will influence the status of the automated actions and the passive protection as well as the emergency response.

The layers "Design & engineering" and "Operations" will in essence prevent accidents from happening while the layers "automated actions & passive protection" and "Emergency response" will mitigate (and limit) the consequences of an undesired event.

Each of these layers will be discussed more in detail in the following chapters.

5.11.3 The first layer of protection: design and engineering, inherent safe design

The importance of the use of standards and engineering best practices was discussed in 5.4. During the design, engineering should also check whether inherent safe design principles can be applied. Inherent safe design principles strive to reduce the hazards rather than controlling them.

The concept of "reducing rather than controlling hazards" is not named "inherently safe design" but the concept is clearly present in the industry since the 1910 as is shown by many articles.

Companies (US Steel, Dupont, Lever Brothers, etc.) incorporated recommendations from safety engineers in the basic design of new plants before WWI. Robert Young describes that safety was being designed into new construction at US Steel (Young, 1912). Everett King describes how safety engineering is incorporated in the design of textile plants (King, 1921).

August Kaems, a safety engineer for the Wisconsin Industrial Commission, concluded after much experimentation that the way to prevent presses from removing fingers was not to install guards but to redesign the press so that the worker never had to put a hand under the ram. The redesign largely eliminated injuries and in addition the output rose from 6,000 pieces a day to 13,000 pieces a day. Kaems's findings were given wide publicity at the National Safety Council Meeting of 1919 (Kaems, 1919).

By the 1930s, design engineers were treating safety as a professional responsibility and "the need for a guard was a sign of poor design" (Aldrich, 1997). The first safety engineer at BASF is employed in 1929. His job was to study technical safety devices and educating employees about accident risks.

Although the topic of inherent safe design has not been left out, it was more explicitly brought back to the attention by a British chemical engineer Trevor Kletz. Trevor Kletz was appointed in 1968 as the first Technical Safety Advisor of the Petrochemical Division of Imperial Chemical Industries Ltd. Every month Kletz issued an internal newsletter.

Throughout the Safety Newsletters of Kletz, it is obvious that accident prevention through safer design was a theme that grew over the years within ICI. In his

> Locate diesels when possible in safe areas
>
> Make the engines safer, i.e. less likely to cause ignition
>
> 1. Be aware of the danger.
>
>   Men who have to authorise the entry of diesels into hazardous areas must be aware of the danger. They must realise that when they sign a permit to admit a diesel this is not a formality but similar to signing a permit for welding. Once this is accepted the rest of this note follows logically.
>
> 2. Avoid the use of diesels when possible.
>
>   Electric fork lift trucks of approved design can be used instead of diesel driven ones. (Unfortunately a trial on one of our Works is showing that the models at present available have disadvantages for certain uses – they are slow and are unstable on rough ground).
>
>   Compressed air pumps or eductors can be used instead of diesel driven pumps for pumping out drains and excavations.
>
>   Diesel engines should not be used at all during start-ups and at other times when leaks are more likely than usual.

Figure 5.40: Extract from Safety Newsletter No 23 from Trevor Kletz in 1970.

Safety Newletter No 23 of October 1970, Kletz writes about using electric forklift trucks instead of diesels.

In newletter number 67 of August 1974, Kletz question about "*safer process design.*"

> # IMPERIAL CHEMICAL INDUSTRIES LIMITED
> # PETROCHEMICALS DIVISION
>
> ## SAFETY NEWSLETTER No 67
>
> **67/1 SHOULD WE DESIGN SAFE PROCESSES OR SHOULD WE TRY TO MAKE PROCESSES SAFE?**
>
> When we are designing a new plant, we try to find out all the hazards and then we try to control them. For example, if the temperature can get dangerously high, we may use a high temperature trip to isolate the feed. Sometimes we have to add a lot of safety hardware on to the plant.
>
> Only very occasionally do we go back to the beginning and ask ourselves if there is a safer process.

Figure 5.41: Extract from Safety Newsletter No 67 from Trevor Kletz in 1974.

An famous quote from Trevor Kletz is "What you don't have can't leak" (Kletz, 1978).

In 1985 Kletz, won with his paper "Inherently Safer Plants" the Bill Doyle Award for the best paper presented at the 19th Annual Loss Prevention Symposium, sponsored by the Safety and Health Division of the American Institute of Chemical Engineers.

The concept gained more and more support throughout the eighties and numerous publications have appeared related to "Inherent Safe Design" (CCPS, 1993; Mansfield et al., 1996; CCPS, 1996, Kletz, 1998; Heikkilä, 1999).

In order to structure and organize, the approach teams will work around a number of "guide words." Heikkila (Heikkilä, 1999) refers to 12 basic principles but most authors go back to the four main words for achieving inherently safer design:
– Substitute;
– Minimize;
– Moderate;
– Simplify.

Substitute refers to replacing hazardous products or processes with less hazardous products or processes. The table below gives some examples. In practice, many aspects must of course be taken into account and the choice is not always trivial.

Alkylation units	An alkylation unit (alky) is one of the conversion processes used in petroleum refineries. It is used to convert isobutane and low-molecular-weight alkenes (primarily a mixture of propene and butene) into alkylate, a high octane gasoline component. The process occurs in the presence of an acid such as sulfuric acid (H_2SO_4) or hydrofluoric acid (HF) as catalyst. The HF alkylation process has been in commercial use since WWII, when it was developed by Phillips Petroleum Co to help supply the demand for high-octane aviation fuel. The first HFA unit entered in service in 1942. Literature about safety in HFA units concerned mainly corrosion problems but experts generally considered that HFA units could be operated safely without additional safety measures compared to other process units (e.g., Hill and Knot, 1960). Till 2009, there was a real competition between both technologies. In 2009, the worldwide capacity had an equal share for sulfuric acid based technologies and HF-based technologies. Since 2009 over 90% of the additional installed capacity was based on sulfuric acid technology because of major risk considerations.
Ethylene dichloride	Ethylene dichloride (EDC) or 1,2-dichloroethane is mainly used to produce vinyl chloride monomer (VCM) the major precursor for PVC production. EDC is made by the direct chlorination or oxychlorination of ethylene. Direct chlorination is performed in the liquid phase where liquid chlorine and pure ethylene are reacted in the presence of ferric chloride. The reaction can be carried out at either low (20 °C–70 °C) or high (100 °C–150 °C) temperatures. The low temperature process has the advantage of low by-product formation but requires more energy to recover the EDC. The high temperature process utilizes the heat of reaction in the distillation of the EDC, leading to considerable energy savings. However, the direct chlorination at high temperatures presents a risk of a large vapor cloud and a vapor cloud explosion in case of a leak. Hence, from a major risk point of view a direct chlorination at low temperature is preferred.

Carbaryl   Since 1954, Union Carbide works closely together with the Boyce Thompson Institute to develop a new pesticide. Several experiments with a newly discovered broad-spectrum insecticide 1-napthyl-*N*-methylcarbamate were very promising. Union Carbide considered this as a unique opportunity to compete DDT, under pressure because of its environmental impact, out of the market. Union Carbide applied for several patents (e.g., Patent 2,903,478, "*α-Napthol bicyclic aryl esters of N-substituted carbamic acids*" by Joseph A Lambrech patented on 8 September 1959; Patent 2,904,464, "*Insecticidal Compositions containing 1-napthyl-N-methylcarbamate and piperonyl butoxide as a synergist therefore*" by Herbert H. Moorefield patented on 15 September 1959 and Patent 3,009,855 "*Method and composition of destroying insects employing 1-napthyl N-methyl carbamate*" by Joseph A. Lambrech patented on 21 November 1961).

According to these patents the 1-naphtyl methylcarbamate (called "Sevin" for convenience by Moorefield) the production method is via the chloroformate route: "The compounds of this invention may be prepared, generally, by reacting 1-napthol with phosgene to form the corresponding chloroformate and reacting the chloroformate with a primary amine to form the corresponding carbamate and HCl."

In 1963, the French company Progil, applies for a patent for the preparation of alpha-naphtyl alkyl carbamate via another chemical route. The patent argues (patented on 4 February 1969 by United States Patent Office under number 3,426,064):

a further object of the invention is the provision of a process for manufacturing various alkyl carbamates of alpha-napthol in a simple and economical manner from complex mixtures." And further in the patent: "In a general way, the process of the invention consists in treating a mixture of alpha-napthol with an **alkyl isocynate** having 1 to 6 carbon atoms in the alkyl group, the mixture containing other naphatlene derivates.

Hence, at the end of the 1960s, when Union Cabide signs a contract with the Ministry of Agriculture to import Sevin and to construct a manufacturing plant for Sevin in India there exist two production routes for the manufacturing of 1-naphtyl methylcarbamate:

A 1-naphthol can be treated with excess phosgene to produce 1-naphthyl chloroformate, which is then converted by reaction with methyl amine to carbaryl.

B Methylamine and phosgene result in **methyl isocyanate** (MIC) which reacts with 1-naphtol to produce the carbaryl.

Method B seems to be an improvement over method A (it solves a number of technical problems, it delivers a more pure product and it is a simple and economical process).

Union Carbide built a plant in 1969 near Bhopal (in Madhya Pradesh, India) to produce carbaryl. Union Carbide decided to use the production process using MIC as an intermediate. An MIC production plant was added to the UCIL site in 1979. According to Lapierre (Lapierre, 2001) some companies alerted Union Carbide engineers on the risk they took by having the highly dangerous product MIC in their plant. Other manufacturers produced carbaryl without MIC, though at a greater manufacturing cost.

A leak in a MIC tank on the night of 2–3 December 1984 resulted in one of the worst man-made disasters ever. The number of fatalities is estimated between 16,000 and 30,000 and the number of injuries exceeds 500,000.

Minimize refers to minimizing the quantity of material or the quantity of energy:
- **Process equipment.** The need for large hold-up drums or intermediate storage are present (e.g., bottom of columns) is challenged.
- **Process piping.** Pipe length, pipe routing, piping diameter are challenged. Assume that the design requires that 15 m^3 of liquid chlorine is pumped from a storage vessel to a reactor through piping of about 100 m. A typical value for liquid chlorine through a steel pipe is about 1.5 m/s. The engineer can chose between a 2-in pipe (velocity is 2 m/s) or a 3-in pipe (velocity is 0.9 m/s). The holdup in a 2-in pipe is 200 l (about 294 kg) while in a 3-in line it is 450 l (660 kg). The decision for a 2-in line or a 3-in line can then be done by comparing both options in a risk assessment. While the word minimize invites to decrease the piping diameters, it will also be important to avoid too small diameters ("small bore risks").
- **Storage.** The inventory of raw materials impacts the autonomy of the plant. Hence the number of days of storage of raw material will be an evaluation between inherent safe design (storage as low as possible) and an operational interest (the more storage capacity the more flexibility).

Moderate also called attenuation, means using materials under less hazardous conditions. Moderation of conditions can be accomplished by strategies that are either physical (i.e., lower temperatures, dilution) or chemical (i.e., development of a reaction chemistry which operates at less severe conditions). Moderate can also refer to attenuation of potential consequences by spacing. Some examples:
- **Triethylaluminum** [Al$_2$(C$_2$H$_5$)$_6$], an organoaluminum compound, is used as a cocatalyst in the industrial production of polyethylene, polypropylene and for the production of medium chain alcohols. This colorless liquid has a boiling point of about 130 °C but is highly pyrophoric, igniting immediately upon exposure to air. It can be stored as a pure liquid but to avoid ignition in case of a leak

it is often stored in stainless steel containers either as a solution in hydrocarbon solvents such as hexane, heptane or toluene.

- **Alternatives to chlorine for disinfection of drinking water**. Chlorination is the process of adding chlorine to drinking water to disinfect it and kill germs in order to achieve good quality for drinking water. Chlorine was first used in the USA as a major disinfectant in 1908 in Jersey City, New Jersey. Chlorine use became more and more common in the following decades, and by 1995 about 64% of all community water systems in the USA used chlorine to disinfect their water (EPA, 2000). However, chlorine is a toxic gas and exposure to high levels of chlorine gas from a release can cause severe health effects, including death. Several less hazardous alternatives have been used for chlorine. Shah and Qureshi (2008) report the result of a study of the city of Lakeville, Minnesota, to evaluate alternative disinfection technologies. The city of Lakeville provides water to more than 50,000 residents. The treatment plant was built in 1998. The problem is avoided by the use of sodium hypochlorite (NaOCl), which eliminates the danger of large-scale gas leaks from high-pressure chlorine cylinders and containers. The main advantage of selecting NaOCl over chlorine gas is that the disinfectant is produced and stored in liquid form, eliminating the danger of large-scale gas leaks from high-pressure chlorine cylinders. With the plant's conversion to NaOCl, the city will require fewer safety regulations and will not be required to develop and maintain a risk management plan

- **Decrease congestion in process plants**. Process plants that handle large quantities of flammable products can be exposed to an accidental leak, followed by the formation of a flammable cloud which upon ignition could result in a vapor cloud explosion. The power of the explosion is proportional to the turbulent flame speed. A low speed flame can accelerate due to confinement and congestion of the equipment in the plant. The potential flame acceleration and hence destruction power of a vapor cloud explosion can be moderated by decreasing confinement and congestion in a plant.

- **Full containment for Refrigerated liquefied natural gas storage**. Liquefied natural gas (LNG) is stored at minus 160 degrees C for ease and safety of non-pressurized storage or transport. The East Ohio Gas Company that built a full-scale commercial LNG plant in Cleveland, Ohio, in 1940 experience an major disaster on 20 October 1944 when the 283,200 m^3 cylindrical tank ruptured spilling LNG over the plant and nearby neighborhood. The accident caused 130 fatalities and destroyed 2.6 km^2 area of the east side of Cleveland, Ohio. Refrigerated LNG storage tanks are nowadays built according to EN 1473 as a full double-containment tank. When the primary container fails, the secondary container is capable of holding the liquid inside (Figure 5.42).

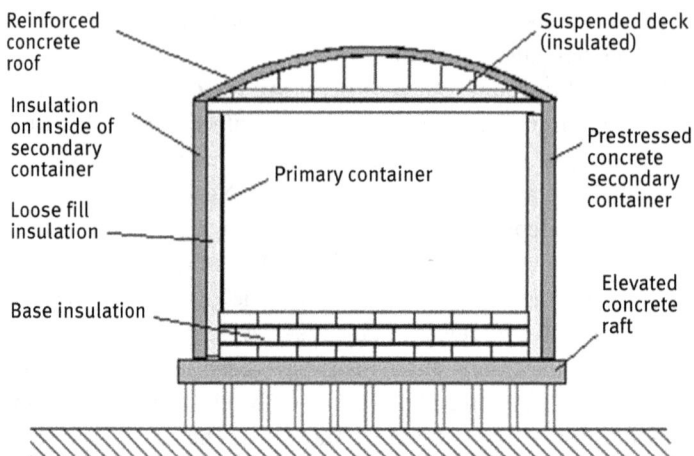

Figure 5.42: Full double containment LNG tank design (EN 1473).

Simplify means designing the process to eliminate unnecessary complexity, thereby reducing the opportunities for error and misoperation. In complex systems, unwanted outcomes do not occur solely due to individual component failure, but most often, they emerge from the unpredictable interactions *between* the components.

5.11.4 The second layer of protection: operations

The second layer is the most important one and the most difficult one. This layer is dynamic and evolves over time. It can evolve in both directions: increasing the risks or decreasing the risks.

The responsibilities and duties of chemical plant operators are:
- Excel in operations
  - Perform chemical operations as per standard operating procedures (SOPs) in a safe and effective manner;
  - Monitor and control operating parameters (temperature, pressure, etc.) of the process and the process equipment;
  - Handle, transport and process chemicals safely (including taking samples);
  - Keep the plant documentation up to date;
  - Clean equipment and report substandard conditions and request repairs to maintenance personnel;
  - Relate with technical inspection department: provide input on operating conditions and single point specificities. Alert on substandard conditions such as vibrating, visual corrosion and change in operating conditions.

- Report substandard conditions and substandard actions and propose corrective actions
    - Analyze and troubleshoot problems;
    - Identify and propose improvements to enhance safety, quality and productivity.
- Manage interventions of contractors
    - Elaborate planning of works;
    - Prepare each of the works;
    - Supervise interventions of contractors.
- Intervene in case of an undesired event
    - Activate safety devices;
    - Bring the installation to a safe shutdown when needed.

The quality of operations is at the heart of the management of risks of a chemical plant and their means to reduce the risks to a minimum are:
- Technical systems
    - High-quality documentation (engineering specifications, SOP, P&ID's, hazard studies, etc.);
    - Basic process control system (BPCS);
    - Safety hardware equipment: detectors, isolation valves, emergency buttons, leak intervention means and firefighting intervention means.
- Organizational measures
    - Knowledge and experience of operating staff;
    - Safety management system activities;
    - Supporting services: inspection, maintenance, training, process safety expertise, audits;
    - Appropriate resources: staffing and budget.

The occurrence of initiating events (corrosion) and the potential evolution from an initiating event to a hazardous scenario will to a large extent depend on the quality of the "operational layer."

In many risk studies, scenarios are studied in a reductionistic way while the operational layer should be taken into account in a holistic manner. A holistic approach is characterized by the belief that the parts of something are intimately interconnected and explicable only by reference to the whole.

What does this means in practice? In risk studies, extensive use is made of generic probabilities and scenarios are considered in a static way and this does not match with reality. An example will make this more clear.

End 2019 corrosion caused a hole with an equivalent diameter of 5 mm in a 80 mm gasoline product line in a crude unit of a refinery. Gasoline was spilled on the ground, formed a pool and ignited. There were no casualties but the fire lasted for about 24 h and caused important damage to the unit. The investigation revealed that the cause of the leak was corrosion on the line under a pipe support (trunion).

In a risk analysis, the scenario would be represented via a bow-tie as follows:

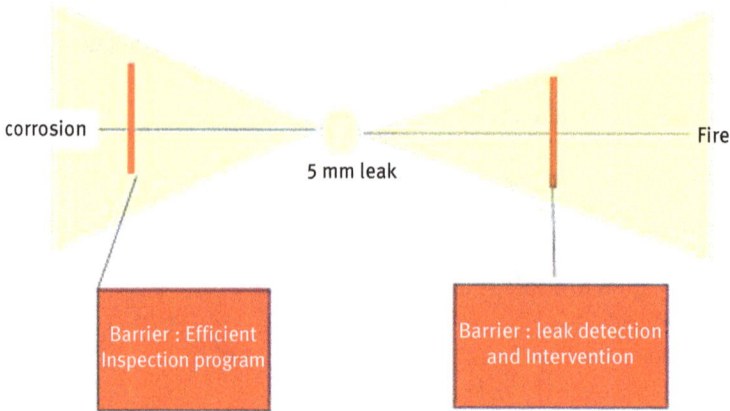

Figure 5.43: Simple bow-tie for gasoline leak.

The frequency of occurrence of the 5 mm leak would be estimated from a generic database and the risk would be the combination potential consequences ("major fire") and its generic frequency of occurrence. The analysis would underline the importance of an efficient inspection program (see Figure 5.43).

The underlying cause of the undetected corrosion is however much more complicated as was revealed by the investigation:

- The whole line was inspected in 2016 and corrosion of the support was identified.
- The inspection recommended "to remove the support and to bring it up to the new standard." There is no doubt that this work (removal of the trunion) would have made the corrosion on the product line visible.
- Based on ultrasonic thickness measurements, it was decided to perform the work during the next turnaround in 2019.
- The job got the number 1699 and was included in the list of 900 works that had to be done during the turnaround of the unit.
- In 2019, the 900 works to be done were attributed to a selected number of contractors.
- The contractor in charge of work 1699 (among many other jobs) confused the exact location of the corroded support and he prepared under number 1699 work on a support nearby.
- Nobody realized the confusion and once the work (on the wrong support) was done, it passed without difficulty through the quality control of maintenance and operations. The contractor unconsciously showed each time the work done (at the wrong location) and none of the people that performed the quality check

realized that the number 1699 corresponded to another support close by. Hence, the work was noticed as "done."
- During the turnaround, inspectors from the inspection department perform daily rounds to make a last verification of the list of 900 works. When a job is considered as done, the inspectors mark it on the (electronic) punch list. Several inspectors are the whole day in the field and each inspector verifies about 20 jobs per day. About two weeks before the startup an inspector verified the support of work 1699 and he mentioned on paper that the "job was not done." This means that work 1699 is still in the list of the works that have to be checked before startup.
- One week later, another inspector had job 1699 on his list, but this inspector was disoriented and he considered job 1699 as done. Hence, it went on the punchlist and was marked as finished.

This example demonstrates that the genesis of a scenario is not static. The generic frequency of occurrence of the event "5 mm hole in 80 mm pipe due to corrosion" might be correct but it does not help a lot to avoid the accident. The barrier "efficient inspection program" is correct but again it is not very helpful. The inspection department identified the problem in 2016 and requested removal of the support.

The example demonstrates the importance of the interactions between different people from different departments or companies at different times. The different interactions that could have avoided the accident are:
- In 2016, an inspector from a contractor company identified the corrosion on the support. He as well as the inspector from the Inspection Department could have indicated the work to be done more clearly (so that no confusion is possible).
- The contractor who prepared the work in 2019 could have avoided the confusion and the manager who is overseeing the contractor could have observed the confusion.
- Once the work was done, it went through a quality check by people from maintenance and people from Operations. All the people that performed quality check were mislead by the fact that nobody verified the original request. They all performed the quality check of the work that was presented (at the wrong location) by the contractor.
- The ultimate verification of the jobs was done by a specialized company contracted by the Inspection Department. Three people from this company were involved: one inspector who noticed that the work was not done, one central person who is putting information in the electronic punch list and a second inspector who wrongly said that the job was done.

In order to avoid process safety accidents, a risk analysis has to study the different important two-way interactions. An "interaction" is a kind of action that occurs as two or more subjects have an effect upon one another. These interactions include:

- Communication of any sort (e.g., people talking to each other, communication among groups and organizations);
- The feedback during the operation of a machine or computer.

Figure 5.44 shows the different interactions in the "Operations Protection Layer." A risk study should concentrate on the quality of the different subjects and on the quality of their interactions rather than using a generic frequency of occurrence of an event.

Notes about "OPERATIONS":

- The term OPERATIONS refers to the people that are operating the installation but it includes all activities and interventions needed to ensure safe operations

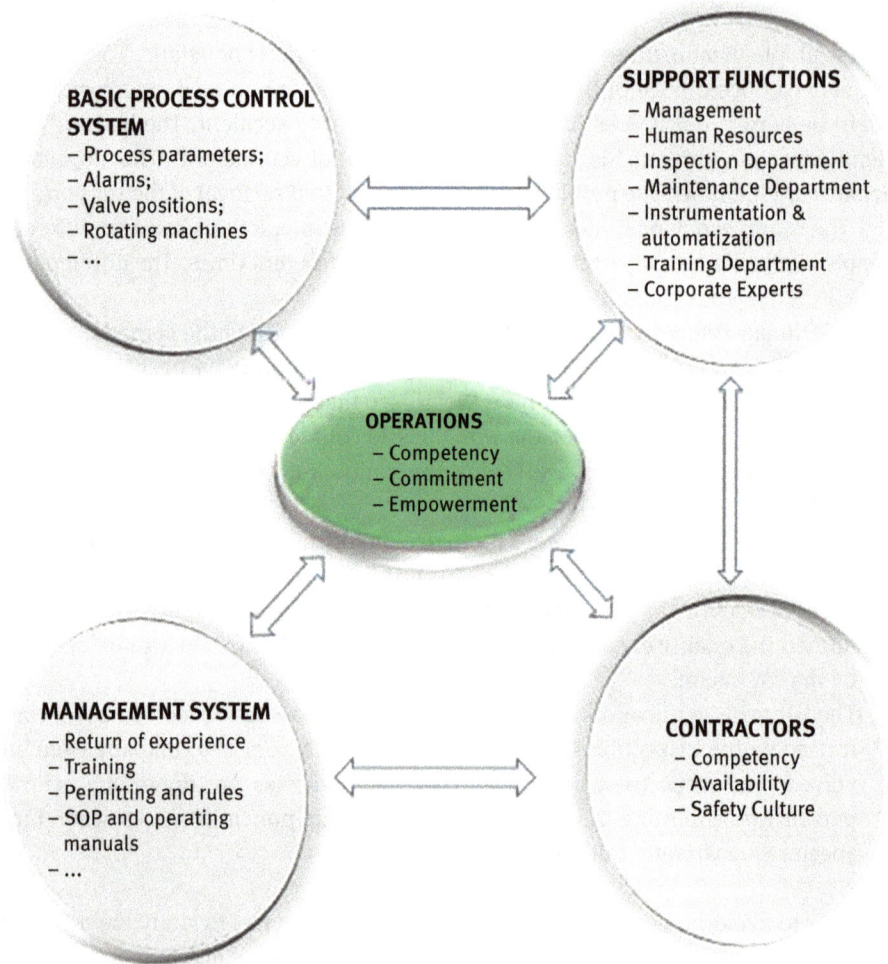

BASIC PROCESS CONTROL SYSTEM
- Process parameters;
- Alarms;
- Valve positions;
- Rotating machines
- ...

SUPPORT FUNCTIONS
- Management
- Human Resources
- Inspection Department
- Maintenance Department
- Instrumentation & automatization
- Training Department
- Corporate Experts

OPERATIONS
- Competency
- Commitment
- Empowerment

MANAGEMENT SYSTEM
- Return of experience
- Training
- Permitting and rules
- SOP and operating manuals
- ...

CONTRACTORS
- Competency
- Availability
- Safety Culture

Figure 5.44: Interaction in the "operations protection layer."

of the installation. This means that it also includes for instance inspection and maintenance activities
- Competency is defined as *the ability to apply validated knowledge, know-how (skills) and behavior in an execution situation* (AFNOR, FDX 50-183- 2002, ISO 10018). This factor is often ignored by risk analysists while it is by far the determining factor in the control of the risks.
- The different arrows show the interrelationship between the different elements of the drawing. The quality of the interrelationship will influence the performance of the operations.

The BPCS will inform the operational about the status of the process parameters (temperature, pressure, flow, etc.). The operational people will have to be trained in order to be able to interpret correctly these parameters and to know how to act in function of these parameters. Designing the BPSC has become increasingly interdisciplinarian in nature. It includes all the hardware (all types of controllers, field instruments, geometric and motion sensors, wiring, valves, servomotors, etc.) as well as operator interface and ergonomics. It is also a very rapid evolving technology. The *Process/Industrial Instruments and Controls Handbook* of McGraw Hill Education was revised 6 times since its first edition in 1957. The sixth edition dates from 2019 (McMillan et al., 2019). Table 5.30 gives an overview of the main undesired events that operational people have to deal with.

Table 5.30: Overview of main undesired events.

Undesired event	Identification	Main risk controls
Mechanical degradation of static equipment	– Visual inspections – Technical inspections	– Competency of operational people – Systematic rounds by operational people and reporting of substandard conditions – Interaction with experts in mechanical degradation mechanisms – Quality of Inspection Program and compliance with the inspection plan – Local expertise in interpretation of inspection results

Table 5.30 (continued)

Undesired event	Identification	Main risk controls
Mechanical degradation of piping and fittings	– Visual inspections – Technical inspections	– Competency of operational people – Systematic rounds by operational people and reporting of substandard conditions – Interaction with experts in mechanical degradation mechanisms of piping and fittings – Quality of Inspection Program and compliance with the inspection plan – Local expertise in interpretation of inspection results
Process deviations and upsets (including utility failures, instrument failures, valve failure, etc.)	– Monitoring of process parameters (pressure, flow, etc.) via BPCS	– Competency of operational people – Hazard identification studies – Process books including trouble shooting – Standard operating procedure (SOP)
Misconduct during operator intervention (draining, sampling, etc.)	– supervision	– Competency of operational people – Operating manuals and SOP – Return of experience – Commitment of operational people – Loss control meetings, discussions – Leadership of supervision and management
Misconduct during intervention of a contractor (lifting, hoisting, etc.)	– Preparation of works – Rules – Supervision	– Competency of operational people – Preparation and planning of works – Selection of contractors – Quality of permitting system – Quality of (first) line supervision
Failure of rotating machines	– Monitoring of process parameters (speed, vibrations, etc.) via BPCS	– Competency of operational people – Monitoring via BPSC – Central monitoring – Expertise in rotating machines – Reporting of substandard conditions

The table shows that the main risk control measure is the competency of the operational people.

Competency is defined as *the ability to apply validated knowledge, know-how (skills) and behavior in an execution situation* (Afnor, FDX 50-183- 2002, ISO 10018). This factor is often ignored by risk analysists while it is by far the determining factor in the control of the risks.

In theory, the operator of an installation must ensure that all parts of the installation are in good condition. That is why inspections and tours are carried out. When a substandard is established, the installation must be repaired or stopped. The protection layer "OPERATIONS" contains includes therefore a large number of safety devices. The difference between the safety devices in this layer and the safety devices in the next layer is the required intervention of the operational people:

- Safety functions (e.g., water curtains and closure of safety valves) that have to be activated by an operator intervention are part of this layer;
- The same safety functions that come into action independent of the decision of the operator (e.g., automatic closure of an emergency shutdown valve) are considered to be part of the next layer of protection.

In large installations (e.g., refineries), operational people will have to deal with a huge number of equipment, for example, thousands of pressure vessels, ten thousands of instruments and so on, a large number of activities and a complexity of different situations. Since it is almost impossible to inspect everything at all times, standards and laws have been enacted over the years that determine the inspection method and inspection frequency.

It is assumed that the operator of an installation acts with due diligence if the inspections are carried out in accordance with the legislation and good industry practices. For example, compliance with standards such as BS, ISO, AFNOR, ASME and API standards are considered as proof of safe operation.

Since some 20 years, industry has gone one step further and lists are drawn up with so-called safety and environmental critical elements (SECE). A ranking of the equipment and instruments as function of their importance (for safety, availability, production, etc.) has always existed in industry, but the terminology "SECE" was introduced in the offshore industry:

- Identification and testing of equipment important for safety on offshore installations was in place since the early 1970s. In June 1974, the API issued Recommended Practice 14C, "Analysis, Design, Installation, and Testing of Basic Surface Safety Systems on Offshore Production Platforms." API-RP-14C uses an associative method for tagging and identifying safety devices whereby a particular safety device is symbolically associated with the process component it protects. A typical device identification (tag number) performs two major functions: (1) it uniquely identifies both the safety device and the type, and (2) it uniquely identifies the process component it protects. The latest revision of API RP 14C dates from May 2018. The recommended practice for subsea applications is API RC 17 V.

– On 6 July 1988, there was a major accident in the North Sea on the Piper Alpha Platform. The explosion and resulting oil and gas fires destroyed Piper Alpha, killing 167 men (61 workers escaped and survived). The Cullen Inquiry was set up in November 1988 to establish the cause of the disaster. It was chaired by the Scottish judge William Cullen. After 180 days of proceedings, it released its report Public Inquiry into the Piper Alpha Disaster (short: Cullen Report) in November 1990 (Cullen et al., 1990). The report made 106 recommendations for changes to North Sea safety procedures from which 37 recommendations covered procedures for operating equipment. The recommendations led to the enactment of the Offshore Safety Act 1992 and the making of the Offshore Installations (Safety Case) Regulations 1992. The Cullen report also recommended to perform safety cases and to apply a risk-based approach.

– The safety case regulations (SCR) were introduced in 1992 and were amended, by the 1996 Offshore Installations and Wells (design and construction) Regulations (DCR), to introduce requirements for the safety critical elements (SCEs) of an offshore installation to be verified as suitable by an independent and competent person (HSE UK Research report 397, 2005). SCEs are any part of the installation, plant or computer programs the failure of which will either cause or contribute to a major accident, or the purpose of which is to prevent or limit the effect of a major accident. The SCR was revised in 2005.

– On 20 April 2010, an explosion and subsequent fire resulted in the sinking of the Deepwater Horizon drilling rig in the Macondo Prospect oil field about 40 miles (64 km) southeast off the Louisiana coast. The accident caused the deaths of 11 workers and a massive offshore oil spill in the Gulf of Mexico, considered the largest accidental marine oil spill in the world, and the largest environmental disaster in US history. Following that accident the European Parliament and the council of the European Union implemented Directive 2013/30 on safety of offshore and gas operations (amending Directive 2004/35/CE). Article 31 of the Directive states "In view of the complexity of offshore oil and gas operations, the implementation of the best practices by the operators and owners requires a scheme of independent verification of safety and environmental critical elements (SECES) throughout the life cycle of the installation, including, in the case of production installations, the design stage." The notion of *safety critical element* became *safety and environmental critical element*. The new Directive requires owners and operators to prevent and mitigate the impact of major accident hazards through the implementation of a systematic and effective approach to risk management. Before exploration or production begins, companies must prepare major hazard reports for offshore installations. These must contain a risk assessment and an emergency response plan. Owners and operators will

needed to comply with the requirements for new builds from 19 July 2016, while existing assets had to brought into line with the new regulations by 19 July 2018.
- The UK Offshore Installations Regulations 2015 modified SCE under SCRs 2005 in SECE. Article 149 states "In practice, it is likely that all safety critical elements (SCEs) under SCR 2005 may be SECEs identified under SCR 2015".
- SECEs are those parts of a system/process which could be engineered (such as fire detectors, alarms, relief systems, emergency shutdown valves and computer programs) or administrative safeguards (such as written safety policies, rules, supervision, schedules and training), or any part thereof:
  - the failure of which could cause or contribute substantially to a major accident hazard;
  - a purpose which is to prevent, or limit the effect of a major accident hazard.

The concept of SECE is applicable to the whole industry but there are some differences. The Directive 2013/30 on safety of offshore and gas operations and the SCRs 2015 requires that:
- The SECEs are identified in the risk assessment for the installations. This must be done prior to the installation being put into operation.
- Independent verification of technical solutions which are critical for the safety of operators' installations must be conducted. This includes independent verification that the SECEs identified in the risk assessment for the installations are suitable.
- A schedule of examination and testing for SECEs must be implemented, including independent verification that this schedule is suitable, up-to-date and operating as intended.

In other words, the SECE's according the Directive 2013/30 on safety of offshore and gas operations and the SCRs 2015 are limited to what has been identified in the risk assessment for the installations. This can be logic from a legislative point of view but not from a safety point of view.

Smith (2001) distinguished between safety-related systems and safety critical systems:
- Safety-related systems are those which, singly or together with other safety-related systems, achieve or maintain a safe state for equipment under their control.
- Safety critical systems are those which, on their own, achieve or maintain a safe state for equipment under their control.

The shortcoming of the definition of Smith is that he does not explicitly consider the potential consequence. While criticality should be related to the potential consequence.

5.11.5 The third layer of protection: automated active protection and passive protection systems

5.11.5.1 Purpose of the third layer

The previous layer was intended to control the situation while this layer as well as the next one are designed to mitigate the potential consequences.

As explained in Figure 5.39, in a number of cases, an undesired event can slip through the control of the operational people. The undesired event is then ready to evolve to a (important) loss of primary containment.

If we compare the situation with driving a car, we are now at the moment that the operator did not brake on time and hence the collision is close. The driver did everything he could and he might be braking but he cannot avoid the collision. The drive will now count on the airbag system and the crash resistance of his car. The air bag system is an active automated protection system while the crash resistance of the car is a passive protection. Both systems are designed with the purpose of mitigating as much as possible the potential consequences.

5.11.5.2 Active versus passive protection systems

This layer comprises automated active protection systems including functional safety systems and passive protection systems.

In engineering, active safety systems are systems activated in response to a safety problem or abnormal event. Such systems may be activated by a human operator, automatically by a computer driven system, or even mechanically. The difference between an active system and a passive system is not always clear-cut.

In the car industry, the difference is when these unique systems come into play. In general, active safety features work to prevent accidents, while passive safety features activate during a collision to protect the driver and passengers. An active safety system works to prevent an accident. These systems always stay active while you drive, and continuously work to keep you from getting into an accident. Most active safety features are electronic and controlled by a computer. They include traction control, electronic stability control and braking systems. These also include advanced driver assist systems that use sensors such as forward collision warning and lane departure warning, along with adaptive cruise control.

Should the unexpected happen, passive safety features come to the rescue by protecting you and your loved ones during an accident. These features are passive because they are only activated once a collision occurs. The best example of a passive safety feature is your airbag system, which triggers upon impact. Similarly, your seatbelts and the overall construction of your car play a role in keeping all passengers safe during a crash, minimizing the overall impact.

In nuclear engineering, active safety contrasts to passive safety in that it relies on operator or computer automated intervention, whereas passive safety systems rely on the laws of nature to make the reactor respond to dangerous events in a favorable manner. This is also the definition that is used for the purpose of this book. A rupture disc or a pressure safety valve is therefore a passive safety system.

5.11.5.3 Automated active protection systems and functional safety systems

Automated active protection systems are the part of the overall safety of a system or piece of equipment that is activated without any human action. It is the part of the overall safety of a system or piece of equipment that depends on automatic protection operating correctly in response to its inputs or failure in a predictable manner (fail-safe).

A special part of automated active protection systems are functional safety systems. The main difference between functional safety and automated active protection system is that functional safety systems are external to the BPCS and they act independently from the operators.

Functional safety was introduced in the 1960s but the technology evolved very fast

1960s	The first of these systems used **hardwired relays and interlock systems.** It was usually only employed where the need for avoiding catastrophe was the greatest. Electromechanical devices were mainly used because electronic systems were still expensive, large and consumed a lot of energy.
1970s	As more **solid-state electronics** became available in the 1970s, they were integrated into the process control systems and protection systems. Solid state began to replace the old relay systems. Engineers began to think more about process hazards, what hazards existed and how they could be controlled. Although reliability seemed better at the outset, solid-state systems were more complex, and had more complex failure modes. These failure modes were not well understood, and little failure history had yet been collected or analyzed.
1980s	**Microprocessor systems** were introduced: the modern PLC was introduced and could be configurable. Software design techniques to make these systems work were developed.
1990s	**Safety PLCs** started to appear. A systematic approach for identifying failure modes and consequences (for both hardware and software). Hardware component failure rates, failure modes and failure distributions were established for use in quantitative analysis. Standards were developed. One of the first functional safety standard available was the VDE 0801 German standard. It was considered law in Germany.

DIN V19250 was published in 1994 (Grundlegende Sicherheitsbetrachtungen für MSR Schutzeinrichtungen: "fundamental safety aspects to consider for control and measurement equipment").

DIN V19250 formed the basis for ANSI/ISA 84.01-1996 that was issued in the USA in 1996. DIN V19250 and ANSI/ISA 84 were the basis for the first draft version of IEC 61508 in 1995 which was approved in 1998.

In 1998, the IEC published a document, IEC 61508, entitled: *Functional Safety of Electrical/Electronic/Programmable Electronic Safety-Related Systems.* This document sets the standards for safety-related system design of hardware and software. IEC 61508 is generic functional safety standard, providing the framework and core requirements for sector specific standard.

2000s By the 2000s, equipment intended for the safety applications were getting certified to the 61508 standard by third party agencies. The safety life cycle, a graphic flowchart included in IEC 61508, was getting used! Design engineers were looking at failure modes and failure rates and mitigation steps to lower the risk of failure in their systems.

International standard IEC 61511 was published in 2003 to provide guidance to end-users on the application of safety instrumented systems (SISs) in the process industries. This standard is based on IEC 61508 for functional safety including aspects on design, construction, and operation of electrical/electronic/ programmable electronic systems. IEC 61511 is a process industry sector specific standard. Other sector specific standards based on IEC 61508 are IEC 61513 (nuclear), IEC 62061 (manufacturing/machineries) and IEC 62525 (railway signaling systems).

In the USA, ANSI/ISA 84.00.01-2004 was issued in September 2004. It primarily mirrors IEC 61511 in content.

The European standards body, CENELEC, has adopted the standard as EN 61511. This means that in each of the member states of the European Union, the standard is published as a national standard (BSI in the UK, AFNOR in France, etc.). The content of these national publications is identical to that of IEC 61511.

2010s In 2010, the second edition of 61508 was published. It filled some gaps that had remained in the first edition. It enhanced some areas of functional safety management, like personnel competency, and strengthened the analysis and calculation of some metrics, like safe failure fraction. HW technology and SW design process continued to improve and advance, but the safety life cycle was still a key ingredient to provide mitigation for risk of failure.

IEC 61511 provides good engineering practices for the application of SISs in the process industry sector. The process industry sector includes many types of manufacturing processes, such as refineries, petrochemical, chemical, pharmaceutical, pulp and paper, and power. IEC 61511 covers the application of electrical, electronic and programmable electronic equipment. IEC 61511 focuses attention on one type of instrumented safety system used within the process sector, the SIS. Before applying the standard IEC 61511, it is important to define whether a SIS is required.

A fallacy that is often made in risk assessments is that all important safety control functions need to be covered by a SIS. The result is that there are too many SIS implemented and that compliance with the IEC 61511 standard is not guaranteed because of the combination of stringent requirements of the standard and the abundance of systems.

What a SIS shall do (the functional requirements) and how well it must perform (the safety integrity requirements) may be determined from HAZOPs, layers of protection analysis (LOPA), risk graphs and so on. All techniques are mentioned in IEC 61511 and IEC 61508. In recent years, LOPA has, one might say, become the method of choice for a number of organizations.

During SIS design, construction, installation and operation, it is necessary to verify that these requirements are met. The functional requirements may be verified by design reviews, such as FMECA and various types of testing, for example factory acceptance testing, site acceptance testing and regular functional testing.

The safety integrity requirements may be verified by reliability analysis. For SIS that operates on demand, it is often the PFD that is calculated. In the design phase, the PFD may be calculated using generic reliability data, for example from OREDA. Later on, the initial PFD estimates may be updated with field experience from the specific plant in question.

A SISs must be independent from all other control systems. It is composed of the same types of control elements (including sensors, logic solvers, actuators and other control equipment) as a BPCS. However, all of the control elements in an SIS are dedicated solely to the proper functioning of the SIS. The specific control functions performed by an SIS are called SIFs. The fundamental concept is that any safety-related system must work correctly or fail in a predictable (safe) way.

The standard defines Safety Integrity Level (SIL) as a relative level of risk-reduction provided by a safety function, or to specify a target level of risk reduction. In simple terms, SIL is a measurement of performance required for a SIF.

For any given design the achieved SIL level is evaluated by three measures:
- Systematic Capability (SC) which is a measure of design quality in order to reasonably justify that the final system attains the required SIL;
- Architecture Constraints which are minimum levels of safety redundancy;
- Probability of Dangerous Failure Analysis.

The probability metric used to determine the SIL level depends on whether the functional component will be exposed to high or low demand:
- **Low demand mode**, as defined in 3.5.16 of IEC 61508-4, is where the frequency of demands for operation made on a safety-related system is no greater than one per year;
- **High demand** or continuous mode, as defined in 3.5.16 of IEC 61508-4, is where the frequency of demands for operation made on a safety-related system is greater than one per year;
- Continuous is regarded as **very high demand**.

Figure 5.45 shows the average PFD (or risk reduction level) and the corresponding SIL level for a low demand mode of operation.

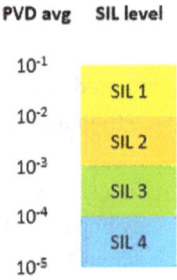

PVD avg   SIL level

10^{-1}	
	SIL 1
10^{-2}	
	SIL 2
10^{-3}	
	SIL 3
10^{-4}	
	SIL 4
10^{-5}	

Figure 5.45: Target failure measures for a safety function operating in a low demand mode of operation.

Within the process industry sector, there has been the tacit assumption that all the scenarios place the SIF being assessed into what the standards describe as low demand mode, and the analysis has been carried out based on this assumption. The other mode of operation for a SIF, high demand mode (or continuous mode) has been seen as something that occurs in other sectors of industry, such as the machinery, manufacturing and the various transport sectors.

However, according to King's (2014) experience, it has identified that there is a type of hazardous event scenario that occurs within the process sector that is not well-recognized by practitioners, and not properly handled by the standard LOPA approach. This occurs when the particular scenario places a high demand rate on the required SIF. King describes in his paper how to recognize a high demand rate scenario and how to handle it.

Figure 5.46 gives the average frequency of a dangerous failure per hour for a high demand of operation.

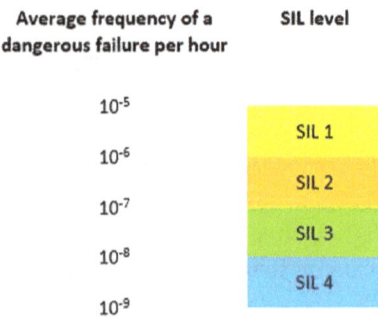

**Average frequency of a   SIL level
dangerous failure per hour**

10^{-5}	
	SIL 1
10^{-6}	
	SIL 2
10^{-7}	
	SIL 3
10^{-8}	
	SIL 4
10^{-9}	

Figure 5.46: Target failure measures for a safety function operating in a high demand mode of operation.

5.11.5.4 Passive protection systems

A passive protection system is the system whose functioning relies on the laws of nature and it does not depend on an algorithm (i.e., a process or set of rules to be followed in calculations or other problem-solving operations) by a human or by a computer. A passive system does not need supply of power to enter in action nor it needs an electronic signal treated.

Table 5.31 gives an overview of some passive protection measures.

5.11.6 The fourth layer of protection: incident management system

Activation of the third layer of protection always result in an incident. The type of incident has a great variety, ranging from a local emergency stop of an equipment to a major accident with multiple casualities. Activation of the third layer will therefore always trigger activation of the fourth layer.

An incident management system is a combination of equipment, personnel, procedures and communications that work together in an emergency to react, understand and respond. Each of the four factors is necessary in order for an incident management system to be effective. Obviously, there will be a degree in the people and resources involved. In case of an emergency stop of one particular equipment the reaction will often be (1) a troubleshooting to understand why the emergency stop happened, (2) repair and correction if needed and (3) a safe restart.

In order to minimize the consequence of unexpected situations companies need to be prepared for emergency situations. The activity to be prepared is called emergency preparedness. An effective incident management system is based upon the organization's identified needs, which establishes evacuation procedures, assigns responsibilities to specific individuals, provides for notification of outside agencies, establishes means of communication, provides for in-house emergency response and prepares the business for other effective actions. Emergency preparedness is a dynamic and continuous activity that considers "on-site" as well as "off-site" unexpected evens (e.g., transport accident).

In the event of an emergency, people and resources will be deployed according to need. In some cases, the size of the emergency situation will increase and intervention effort may increase gradually. Each site will have its internal emergency response plan. This is sometimes called the level 0 of emergency response and recovery. The emergence response is coordinated at the level of the senior management of the site mainly with resources and people from the site.

The authorities will be informed and if needed the authorities will take over the emergency response and recovery. The coordination will be transferred from the senior manager at the site to the authorities. Most countries have a decentralized "emergency response and recovery" for disasters. Most of the time there are three levels:

Table 5.31: Examples of passive protection systems.

Phenomenon or physical subject to be mitigated	Passive protection measure
Limitation of a leak	– Excess flow valves – Orifice in line
Limitation of spread of product	– Collection pit – Bund – Weak weld at top of atmospheric storage tanks – Double wall integrity tanks
Limitation of pressure rise in equipment	– Pressure control valve – Rupture disc – Pressure safety valves to safe location – Explosion vent panels
Evacuation of excess of product	– Drainage sewers and collection pit – Flare system – Blowdown system – Collection pond of firefighting wastewater
Limitation of impact from radiation from fire	– Passive fire protection – Firewalls – Mounded or underground storage
Reduction of overpressure from a vapor cloud explosion	– Low congestion and confinement design
Mitigation of the impact of overpressure from a vapor cloud explosion	– Blast-resistant equipment – Blast-resistant buildings – Blast walls – Low congestion/confinement design – Blast-resistant air-inflated shelters
Runaway	– Maximum overpressure design – Pressure vent with blowdown systems
Mitigation of flame propagation in piping	– Flame arrester – Detonation arrester
Mitigation of explosion and flame propagation in coal mines Decrease of the probability of ignition	– Limestone rock dust on walls – Suspended water bags – ATEX – Earthing – Inerting of vapor space of atmospheric storage tanks

Level Description
1   Level of significant emergency is any disaster with small impact which re-
       quires narrow focus. The crisis will be coordinated at the level of the mayor
       of the town.
2   Level of serious emergency or disaster is any disaster that has wide and pro-
       longed impact. The crisis will be coordinated at the level of the province or
       district.
3   Level of catastrophic emergency or disaster is any disaster that has high
       widespread impact and requires immediate involvement of central govern-
       ment. The crisis will be coordinated at the level of the national government.

As soon as the authorities take over the coordination of the emergency response
and recovery, the civil defense will come into action. Civil defense or civil protection is an effort to protect the citizens of a state from military attacks and natural
disasters.

Programs for prevention, mitigation, preparation, response or emergency evacuation and recovery were implemented in some countries during the 1930s as the
threat of war and aerial bombardment grew. Since the end of the Cold War, the
focus of civil defense has largely shifted from military attack to emergencies and
disasters in general.

In Europe, the European Commission is working on a Union Civil Protection
Mechanism (Decision N° 1313/2013/EU) under which the European Union supports,
coordinates and supplements the action of member states in the field of civil protection to prevent, prepare for and respond to natural and man-made disasters within
and outside the Union.

The Treaty of Lisbon underpins the commitment of the EU to provide assistance,
relief, and protection to victims of natural or man-made disasters around the world
(art. 214), and to support and coordinate the civil protection systems of its Member
States (art. 196). It further mandates the European institutions to define the necessary measures for such actions to be carried out. Information on the Civil Defense in
Europe can be found on the following website: https://ec.europa.eu/echo/who/
about-echo/legal-framework_en

The United Nations International Civil Defense Organization, an intergovern-
mental organization based in Geneva since 1931, serves as a liaison between national civil defense organizations. It counts 59 member states, 16 observer states
and 22 affiliated member states. More information can be found on the website of
ICDO.

6 Process safety risk-based management

6.1 Major disasters are organizational accidents

Process installations and new technologies are complicated and complex. The design, construction and operation of process installations cannot be done by a single person or by a few number of people. Process installations are the result of a combined effort of different companies and within each of these companies via the interrelationship between different people from different departments.

The operation of a process installation will again be done by different people from different organizations.

The organization of a company and the processes to take a decision will have a positive or negative impact on the possibility that a major disaster occurs. This is very well described by, for instance, Hopkins (Hopkins, 2012) for the Macondo accident. Waxin (Waxin, 2015) gives an interesting literature overview of a number of challenges in the logic of decision-making. Some examples are:

- Groupthink, which is *"a mode of thinking that people engage in when they are deeply involved in a cohesive in group, when the members' striving for unanimity overrides their motivation to realistic appraise alternative courses of action."*
- Group polarization, which occurs *"when the position that is held on an issue by the majority of the group members is intensified as a result of discussion."*
- Over optimism, which refers to *"decisions based on delusional optimism rather than on a rational weighting of gains, losses and probabilities."*

Lavallo and Kahneman (2003) writes that *"the optimistic biases of individual employees become mutually reinforcing and unrealistic views of the future are validated by the group."* An example is the Bhopal accident. According to the initial studies, the market in India required about 3,000 tons of pesticide. But Union Carbide built in the late 1970s, a unit for the production of 5,000 tons of pesticide because the economics of such a unit were more attractive. The production started in 1980 and the sales in 1981 (2,704 tons) confirmed a market of 3,000 tons. From 1982, the factory made losses and despite communication campaigns, farmers bought less Sevin than expected. The situation became worse in 1983 due to drought while the sales decreased to 2,308 tons. Cost saving plans were put in place and a disaster was in the making (Lapierre and Moro, 2001).

Organizational structures and clear definition of the responsibilities and the interrelationship between the different parts of an organization are vital in the prevention of major process accidents.

https://doi.org/10.1515/9783110632132-006

6.2 The pillars of a risk-based management strategy

The basis of our risk management strategy is competency (including knowledge and experience), which is shared via communication and networks. Competency is defined as "the ability to apply validated knowledge, know-how (skills) and behavior in an execution situation" (Afnor, FDX 50-183- 2002, ISO 10018).

An acceptable risk is achieved by a combination of technical measures, organization and procedures and appropriate human behavior. This is shown in Figure 6.1.

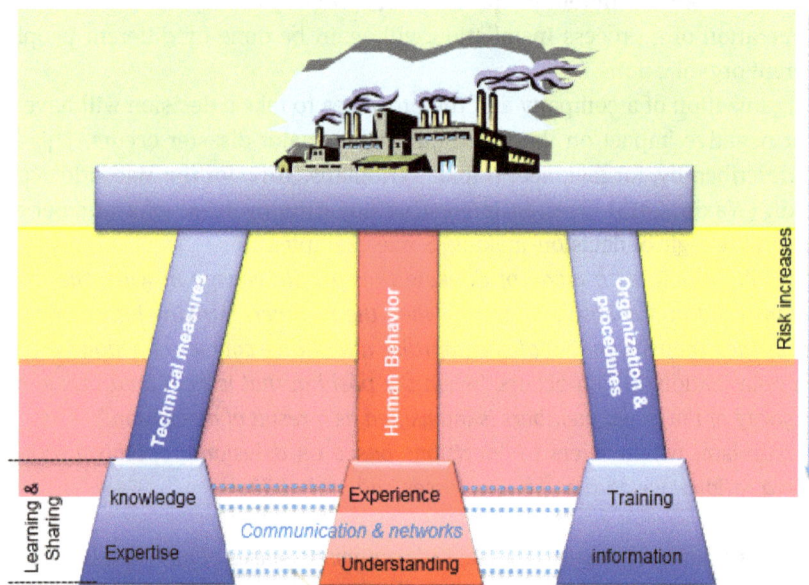

Figure 6.1: The pillars of a risk-based management strategy.

In many textbooks one will read that safety is achieved via technical measures, organizational measures and human behavior. It is my belief that two important points are often overlooked and not enough highlighted:

– The importance of competency. Competency is defined as the "*Ability to apply validated knowledge, know-how (skills) and behavior in an execution situation*" (Afnor, FDX 50-183- 2002, ISO 10018).
– Human behavior is often the result of the organization.

CCPS published the following pillars and focal points for risk-based process safety (Table 6.1) management system. I have put in red and underlined the pillars and focal points, which are impossible without in depth competency. It should be obvious that without in-depth operational process safety competency that it is impossible to deploy an efficient process safety management system.

Table 6.1: Pillars and focal points for risk-based process safety.

Commit to process safety	
– Ensure management cares and provide adequate resources and proper environment	– Process safety culture
	– Compliance with standards
– Ensure employees care	– Process safety competency
– Demonstrate commitment to stakeholders	– Workforce involvement
	– Stakeholder outreach

Understand hazards and risk	
– Know what you operate	– Understand residual risk
– Identify means to reduce or eliminate hazards	– Process Knowledge Management
	– Hazard identification and risk analysis
– Identify means to reduce risk	

Manage risk	
– Know how to operate processes	– Contractor management
– Know how to maintain processes	– Training and performance assurance
– Control changes to processes	– Management of change
– Prepare for, respond to, and manage incidents	– Operational readiness
	– Conduct of operations
– Operating procedures	– Emergency management
– Safe work practices	
– Asset integrity and reliability	

Learn from experience	
– Monitor and act on internal sources of information	– Measurement and metrics
	– Auditing
– Monitor external sources of information	– Management review and continuous improvement
– Incident investigation	

Human behavior programs are back in fashion these days. In fact, they've returned from never being gone. The reasons for this are mainly that experts are constantly looking for new ideas, reinventing old ideas over time. In my opinion, most human behavior programs underestimate the influence of the organization. The culture is also determined to a large extent by the organization. The term "HRO" (High Reliability Organization) refers in the first place to the organization. The importance of the organization becomes very obvious when we observe people from contractor companies. The behavior of an individual or a team from a contractor company A will behave differently when they work for client X or client Z. The behavior will even be different when they work for different sites of the same client.

6.3 Can we use the recipes of management of occupational health and safety?

The management of occupational health and safety has a history of more than 200 years. Many concepts and ideas have been developed in this period (Swiss cheese model, Bird pyramid, safety management systems, human behavior programs, leadership principles, Bradley curve, the cultural ladder, etc.) and it is legitimate to question whether we cannot simply use the same recipes for process safety. In essence, both cases involve management of loss prevention and there is a general consensus that proper management techniques can be applied to production, business, safety, etc.

The main principles of management will of course be the same. The devil is however in the details. Two aspects are very dominant in the management of process safety: "competency and uncertainty."

In the United States of America, the Occupational Safety and Health Administration (OSHA) is a large regulatory agency of the United States Department of Labor that was established by Congress under the Occupational Safety and Health Act (OSH Act), which President Richard M. Nixon signed into law on December 29, 1970.

OSHA's mission is to "assure safe and healthy working conditions for working men and women by setting and enforcing standards and by providing training, outreach, education and assistance."

Because of the particularities of major accident prevention, OSHA published in the Federal Register (55 FR 29150) on July 17, 1990, a proposed standard, "*Process Safety Management of Highly Hazardous Chemicals*" – containing requirements for the management of hazards associated with processes using highly hazardous chemicals to help assure safe and healthful workplaces. Table 6.2 gives an overview of some of the important requirements of the Process Safety Management standard. Reading the table leaves no doubt that it is all about knowledge, competency and organization.

Many of the recipes that have worked for the management of occupational health and safety will only have some added value when the basis (Figure 6.1) is solid. This is less the case for occupational health and safety because there is far less uncertainty involved and the situation in most cases is not complicated. For instance, the measures to be taken to avoid "to fall from height" are not complicated. Hence soft skills (leadership, commitment, etc.) will play an important role. Operating a refinery in a safe manner will, in the first place, require technical understanding and competency, and soft skills will be of a second order.

Table 6.2: Some OSHA requirements for a Process Safety Management System (OSHA 1910.119).

1	**Process Safety Information**	Information on the hazards of the highly hazardous chemicals in the process shall consist of at least the following: toxicity, permissible exposure limits, physical data, reactivity data, corrosivity data, and thermal and chemical stability data, and hazardous effects of inadvertent mixing of different materials.
		Information on the technology of the process must include at least the following: a block flow diagram or simplified process flow diagram, process chemistry, maximum intended inventory, safe upper and lower limits for such items as temperatures, pressures, flows or compositions and an evaluation of the consequences of deviations, including those affecting the safety and health of employees.
		Information on the equipment in the process must include the following: materials of construction, piping and instrument diagrams (P&IDs), electrical classification, relief system design and design basis, ventilation system design, design codes and standards employed, material and energy balances and safety systems (e.g., interlocks, detection or suppression systems).
2	**Process Hazard Analysis**	The process hazard analysis is a thorough, orderly, systematic approach for identifying, evaluating and controlling the hazards of processes involving highly hazardous chemicals. The employer must perform an initial process hazard analysis (hazard evaluation) on all processes covered by this standard. The process hazard analysis methodology selected must be appropriate to the complexity of the process and must identify, evaluate and control the hazards involved in the process.
3	**Operating Procedures**	The employer must develop and implement written operating procedures, consistent with the process safety information, that provide clear instructions for safely conducting activities involved in each covered process. OSHA believes that tasks and procedures related to the covered process must be appropriate, clear, consistent and most importantly, well communicated to employees. The procedures must address at least the following elements: Steps for each operating phase: initial startup; normal operations; temporary operations; emergency shutdown, including the conditions under which emergency shutdown is required, and the assignment of shut down responsibility to qualified operators to ensure that emergency shutdown is executed in a safe and timely manner;
		safety and health considerations: properties of, and hazards presented by, the chemicals used in the process; precautions necessary to prevent exposure, including engineering controls, administrative controls and personal protective equipment; control measures to be taken if physical contact or airborne exposure occurs; quality control for raw materials and control of hazardous chemical inventory levels; and any special or unique hazards.
		Safety systems (e.g., interlocks, detection or suppression systems) and their functions.

Table 6.2 (continued)

4	**Employee Participation**	Employers must develop a written plan of action to implement the employee participation required by PSM. Under PSM, employers must consult with employees and their representatives on the conduct and development of process hazard analyses and on the development of the other elements of process management, and they must provide to employees and their representatives access to process hazard analyses and to all other information required to be developed by the standard.
5	**Training**	OSHA believes that the implementation of an effective training program is one of the most important steps that an employer can take to enhance employee safety. Accordingly, PSM requires that each employee presently involved in operating a process or a newly assigned process must be trained in an overview of the process and in its operating procedures. The training must include emphasis on the specific safety and health hazards of the process, emergency operations including shutdown, and other safe work practices that apply to the employee's job tasks. Those employees already involved in operating a process on the PSM effective date do not necessarily be given initial training. Instead, the employer may certify in writing that the employees have the required knowledge, skills and abilities to safely carry out the duties and responsibilities specified in the operating procedures.
6	**Contractors**	Many categories of contract labor may be present at a jobsite; such workers may actually operate the facility or do only a particular aspect of a job because they have specialized knowledge or skill.
		PSM, therefore, applies to contractors performing maintenance or repair, turnaround, major renovation or specialty work on or adjacent to a covered process.
7	**Pre-Startup Safety Review**	It is important that a safety review takes place before any highly hazardous chemical is introduced into a process. PSM, therefore, requires the employer to perform a pre-startup safety review for new facilities and for modified facilities when the modification is significant enough to require a change in the process safety information. Prior to the introduction of a highly hazardous chemical to a process, the pre-startup safety review must confirm that the following: Construction and equipment are in accordance with design specifications; Safety, operating, maintenance, and emergency procedures are in place and are adequate; A process hazard analysis has been performed for new facilities and recommendations have been resolved or implemented before startup, and modified facilities meet the management of change requirements; and training of each employee involved in operating a process has been completed.

Table 6.2 (continued)

8	**Mechanical Integrity**	OSHA believes, it is important to maintain the mechanical integrity of critical process equipment to ensure it is designed and instaled correctly and operates properly. PSM mechanical integrity requirements apply to the following equipment: pressure vessels and storage tanks; piping systems (including piping components such as valves); relief and vent systems and devices; emergency shutdown systems; controls (including monitoring devices and sensors, alarms and interlocks) and pumps.
9	**Management of Change**	OSHA believes that contemplated changes to a process must be thoroughly evaluated to fully assess their impact on employee safety and health and to determine needed changes to operating procedures. To this end, the standard contains a section on procedures for managing changes to processes. Written procedures to manage changes (except for "replacements in kind") to process chemicals, technology, equipment and procedures, and change to facilities that affect a covered process, must be established and implemented. These written procedures must ensure that the following considerations are addressed prior to any change: the technical basis for the proposed change, impact of the change on employee safety and health, modifications to operating procedures, necessary time period for the change, and authorization requirements for the proposed change.
10	**Incident Investigation**	The investigation must be by a team consisting of at least one person knowledgeable in the process involved, including a contract employee if the incident involved the work of a contractor, and other persons with appropriate knowledge and experience to investigate and analyze the incident thoroughly.

6.4 The senior process safety engineer

6.4.1 The role and positioning of the senior process safety engineer

Around the end of the nineteenth century, it became clear for most large companies that there was a need for a safety engineer for a full-time position in the company. The safety engineer was a knowledgeable person in the aspects of occupational health and safety and should be the advisor of the management. Ensuring safety is a duty of the management, but it was not possible for the line management to oversee all aspects of the rapid evolution of the profession of occupational health and safety.

The management of process operations is facing a similar situation by the end of the twentieth century in relation to the process safety of their operations.

Technological risk management in general and process safety management in particular, have evolved into a professional field in the last 50 years. Risk analysis is an engineering discipline with a semi-scientific basis which is defined as "having or using certain aspects of science." Understanding the engineering part of a process installation is crucial but not enough. The senior process safety engineer needs to have knowledge of all aspects in relation to the process installation and he has to be knowledgeable in the different risk analysis and assessment techniques. Figure 6.2 shows the positioning of the senior process safety engineer in relation to the different disciplines.

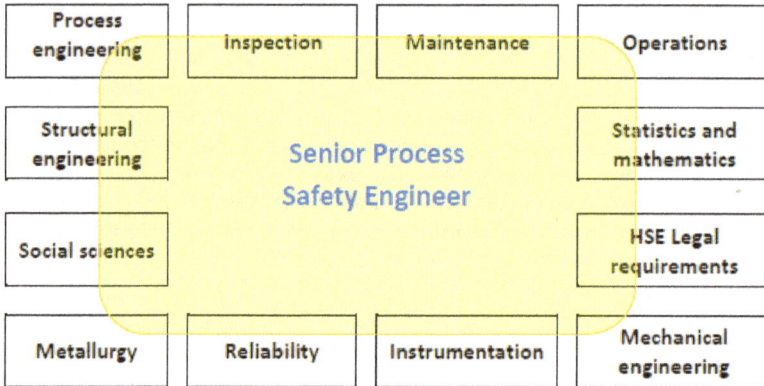

Figure 6.2: Positioning of the senior process safety engineer.

6.4.2 Relation with the process engineer

The process engineer is very knowledgeable about a limited number of processes (e.g., steam cracking processes, styrene production processes, alkylation processes). He will be consulted whenever a process needs to be improved, modified etc.

The senior process safety engineer has a good understanding about thermodynamics, chemistry and physics and process technologies. He is very familiar with PFD's and P&IDs. He will concentrate on all possible hazards related to a particular process and products. He will be assisted by a process engineer for the in depth understanding. The senior process safety engineer will in addition have:
- experience with a broad spectrum of hazard identification techniques (HAZOP, What If, Swift, FMEA, FTA, etc.) which will enable him to identify hazards;
- good knowledge about the databases that can be consulted for the study of previous accidents.

6.4.3 Relation with technical inspection

Inspection services are knowledgeable about degradation mechanisms, inspection methods and inspection plans.

The senior process safety engineer will have a good understanding about corrosion mechanisms, characteristics of materials, inspection techniques, risk-based inspection and degradation mechanisms.

The senior process safety engineer will systematically identify hazards related to mechanical integrity using for this, the in depth knowledge of the inspectors. His knowledge will enable him to ask the right questions which will help him to identify the hazards. Examples of questions: do we have a program for small bore management, do we have the right material, is there a risk of corrosion, etc.

6.4.4 Relation with maintenance

The senior process safety engineer will have a good understanding of maintenance, reliability-centered maintenance, hazards related to maintenance activities, etc.

He will closely work together with maintenance people in order to identify the hazards related to maintenance activities and the interference between maintenance and operations.

The senior process safety engineer is very knowledgeable about techniques such as FMEA and FMECA. He is also familiar with reliability techniques (see further).

6.4.5 Relation with operations

The senior process safety engineer has a good understanding of the organization of operations and the hazards related the operations. He is very knowledgeable in using techniques (e.g., HAZID, HAZOP, audits) to analyze an organization and its potential failures. He is knowledgeable about best practices for among others: writing SOPs, loss control team meetings, personal communication techniques, shift handover, management of organizational change, etc.

6.4.6 STEM

STEM stands for Science, Technology, Engineering and Mathematics.

Risk has been defined in industry as a combination of likelihood of occurrence and impact.

The senior process safety engineer will therefore need an in depth understanding about mathematics and statistics. All models used for consequence estimations make extensive use of mathematics.

Probabilistic modeling and consequence modeling makes extensive use of computer science and computer models.

Mechanical integrity of static and rotating equipment is a very important contributor in the major accident prevention. The senior process safety engineer will have a good understanding about the mechanical engineering of static as well as rotating equipment in order to identify the hazards related to loss of mechanical integrity. He will lead FMEA and FMECA meetings in order to identify the critical points.

6.4.7 Legal requirements

In many regions of the world the legislation imposes a number of legal obligations for operators of high-risk installations (e.g., safety case for offshore industry, safety studies and QRA's for onshore facilities). The company can decide to make use of consultants or to perform all the work in-house. In both cases, the senior process safety engineer will need to fully understand the legal requirements.

6.4.8 Automatization and instrumentation

The senior process safety engineer will use instrumentation in the prevention and mitigation barriers. He will have a good understanding about the design principles of a process (distributed) control system, the architecture of a control loop, the basic requirements of control determinism, fault tolerance, performance, etc., different techniques for measurement and principles for selection of instruments.

The senior process safety engineer will be knowledgeable about the failure mechanisms of instrumentation and the concept of Safety Instrumented Systems. He will be very knowledgeable in LOPA calculations. He will look for assistance from instrumentation experts.

6.4.9 Reliability

The senior process safety engineer will have significant expertise in reliability of mechanical equipment and instrumentation. Reliability calculations or failure probability calculations are part of the basic competences of the senior process safety engineer.

6.4.10 Metallurgy

The senior process safety engineer will have an understanding of metallurgy, iron-carbon phase diagrams, degradation mechanisms, physical characteristics of materials and so on.

He will be able to ask the right questions to metallurgist in order to identify the hazards related to specific metallurgy topics (e.g., heat treatment of metals, hardening processes of metals, influence of alloying elements).

6.4.11 Social science

There are two main reasons why the senior process safety engineer needs to have understanding about social sciences:
- Risk communication with own people and with the communities;
- Human behavior is an important contributor in major accident prevention.

The senior process safety engineer will need to have basic knowledge in human factor fundamentals and Human Factor analysis techniques.

6.4.12 Structural engineering

Major accidents can be devastating for buildings and process structures due to heat load and/or blast load. The senior process safety engineer will have to evaluate the resistance of buildings and installations in case of a major accident in order to estimate properly the risk. He will make use of structural engineering experts but he needs to understand the basics (and limits) of, for instance, single degree freedom calculations, multi degree freedom calculations, dynamic impact calculations, FEM techniques, etc. The senior process safety engineer will have to decide when to ask for which type of expertise in structural resistance calculations.

6.4.13 Specific knowledge

The specific professional core competences (knowledge and expertise) of the senior process safety engineer are:
- Extensive experience and knowledge about accidents that happened in the past and the lessons learned;
- Use of hazard identification techniques (What If, Swift, HAZOP, FMEA, etc.);
- Use of consequence modeling techniques simple models as well as CFD modeling;
- Reliability studies and calculations of probabilities;
- Use of Fault Tree Analysis, Event Tree Analysis;
- In depth understanding of the physics and thermodynamics involved in accidental release of hazardous products;
- Impact calculations (from fire, explosion, toxic emissions and collisions);
- Risk calculations (e.g., matrix, PLL, PLM, Risk Contours);

- Accident investigation techniques;
- Knowledge in process safety management systems.

The function is very technical. An excellent base for the senior process safety engineer is mechanical engineering, chemical engineering or STEM.

The causes and sources of accidents are multiple (technical, organizational, human behavior, cultural, etc.), and hence every experience in one of these fields is useful. However, it should be understood that process safety is a discipline in itself and hence the best is to start as soon as possible in the discipline.

The senior process safety engineer makes extensive use of the expertise of people in the different disciplines (engineering, inspections, operations, etc.) and it is therefore not necessary at all that he/she works for a long period of time in one of these disciplines.

6.4.14 Main tasks of the senior process safety engineer

Management of major risk is about questions:
- What can go wrong?
- How can it impact my activities?
- What can I do to avoid it?
- Is it worthwhile to take additional actions?

The senior process safety engineer is the central person who is permanently monitoring these questions. The main tasks of the senior process safety engineer therefore are:
- Identification of the major hazards;
- Evaluation of the risks of these hazards;
- Report to the management the status of the main hazards and the risk management of these risks;
- Elaboration of bow-tie for each of the major hazards;
- Training the staff about the barriers;
- Reporting to the site management on the status of the barriers.

7 Conclusion

7.1 Occupational Health and Safety

Occupational Health and Safety refers to the relationship between the employee and the employer and is regulated by labor law, which evolved over a period of about 200 years and which continue to evolve. The main driving forces behind the rules that regulated Occupational Health and Safety at work are a quest for justice to ensure the well-being of the employee and the creation of a fair and equitable basis for the national and international economy. The burdens and benefits of the economy must remain fairly distributed within a society.

The Occupational Health and Safety regulations are result oriented, which means that injuries (even very small injuries) are by definition not acceptable. The worker uses his or her physical and mental capacity to produce goods and/or to deliver services and help the employer and the society to create an added value for the economy. The worker gets a monetary compensation but the agreement is that his/her well-being should not be affected. Employee well-being is a holistic concept. It includes all aspects that can affect the employee's well-being. For some 25 years, great importance has been attached to the psychosocial factors that can influence the well-being of the employee.

The work should be prepared and checks have to be done so that the work can be executed without any uncertainty about the occurrence of an accident. The worker and the employer have to work together in order to perform the work such a way that his/her well-being of the worker is not negatively affected. It is a result-oriented objective, which means that the only acceptable result is zero accidents and zero harm. If an accident happens, lessons are learned and measures are taken to avoid recurrence. The worker receives compensation for the suffering he had. The objectives of occupational health practice that were originally defined in 1950 by the Joint International Labor Organization and the World Health Organization Committee on Occupational Health stated that:

> Occupational health should aim at the promotion and maintenance of the highest degree of physical, mental and social well-being of workers in all occupations; the prevention amongst workers of departures from health caused by their working conditions; the protection of workers in their employment from risks resulting from factors adverse to health; the placing and maintenance of the worker in an occupational environment adapted to his physiological and psychological capabilities; and, to summarize: the adaptation of work to man and of each man to his job.

Since about 1900, it became obvious that managing occupational health and safety is an integral part of managing a business.

It became more and more obvious that for a very large number of activities and jobs, the objective of zero harm is realistic. Performing a work at height or a work in a confined space can be done without any risk for the health of the worker. There is

https://doi.org/10.1515/9783110632132-007

no rocket science involved to identify the hazards and to define the conditions and to implement the measures in order to perform the work safely.

The situation can be complex because many contributing factors are involved (technique, organization and humans) but the problem is not complicated.

In order to achieve the objective of zero harm:

1. Companies should train their managers in the basics of Occupational Health and Safety legislation. Many managers are convinced that managing Occupational Health and Safety is a soft skill and that charismatic leadership will be largely sufficient. They sometimes ignore the scientific basis for managing Occupational Health and Safety.

2. The position of the chief executive of the company is vital. Where he/she is personally involved, the whole organization is aware of it.

3. Creating a safe working environment is about "doing and not only about talking." Managers at all levels have to be in contact with their staff daily to check the efficiency and effectiveness of the safety activities that are in place, and safety achievements must be celebrated. It is therefore important to embed all activities in an Occupational Health and Safety management system.

4. Companies must introduce "rituals" and "taboos." Rituals are since a number of years being investigated experimentally both in the laboratory and in the field, through the use of techniques common to social psychology, cognitive psychology, neuroscience, behavioral economics and experimental anthropology, they are predefined sequences characterized by rigidity, formality and repetition that are embedded in a larger system of symbolism and meaning, which partially lack direct instrumental purpose. A typical example is to start all meetings with a "safety moment."

5. Companies have to define the appropriate methods, concepts and models for the management of Occupational Health and Safety and to stick to these methods, concepts and models for a very long time (several decades). Most of the methods and concepts have been developed in the twentieth century, and these methods, concepts and models are effective. A current problem is that companies regularly reorganize and that new managers start with new ideas that are often "old ideas" but presented differently. The result within the workforce is a combination of (1) resistance to change and (2) confusion. Too often and in particular when accidents happen companies tend to try "new theories," while success is often the result of continuing what has been started.

6. Develop in-house expertise in the field of Occupational Health and Safety. This expertise should be focused on the main situations that can lead to fatal accidents or high potential accidents. For most industries, these situations are: (1) work on energized systems, (2) work at height, (3) lifting and hoisting activities, (4) circulation on site, (5) work above water, (6) excavation, (7) confined space and (8) moving parts. An underlying cause of accidents is what is called "line of

fire." It means that people are putting themselves in a position that make them a target in case the energy is released.

7. The main challenges for the twenty-first century in the field of Occupational Health and Safety in the Western world are mental health and physical fitness of the workers. Managers have to become manager coaches, and the managers need to focus on mental health of the workers, and companies have to put in place health programs that support the workers to improve their physical fitness. A good physical fitness supports the mental health.

7.2 Process safety

The introduction of new technologies and new processes led to the emergence of what is called technological risk management and process safety. A better term would be "process safety engineering" because these activities are very close to process engineering.

Uncertainties are inherent to the new technologies and complicated processes, and in many cases, it becomes impossible to guarantee that these new technologies and processes can be operated with zero uncertainties.

In order to deal with these uncertainties, the notion of "uncertainty management" or "risk management" was introduced. The notion of risk was studied by several academic disciplines that came up with different definitions. Psychologists and sociologists have written a whole library of books about their approach toward these uncertainties.

People who design, construct and operate these technologies and processes have principally an engineering or scientific background. They developed since the 1940s a whole arsenal of methods and theories to analyze and to assess the risk in an engineering framework. Risk analysis and assessment is an engineering discipline on a scientific basis.

Engineers apply science and mathematics to develop technical and organizational solutions to solve all types of problems. Engineers usually invent, design, build and explore all types of things (machines, process installations, bridges, etc.). The starting point is the required functionality or process output, and the engineer will then perform the required calculations in order to design and construct the machines, the process installations or any other good. The engineer uses standards and codes that have proven to be reliable.

Historical evidence, however, taught us that new technologies can sometimes fail in an unexpected way. The unexpectedness finds its origin in lack of knowledge and the variability of parameters. Engineers developed methods to approach a design from a new point of view. During the design, they started to consider the possibility that a component would fail or that the intended purpose would not be achieved. They then looked in the potential causes.

Uncertainties were grouped under the term of "risks," and a mathematical expression was developed in order to make risks quantifiable. The approach was more and more structured and process safety engineering emerged over the last 70 years as a mature engineering discipline.

It is important that management of companies understand that:

1. Process safety is an engineering discipline and it is different from Occupational Health and Safety.
2. The structure of the organization and the decision-making process are key aspects in the prevention of major process accidents. Major disasters are mainly organizational accidents which are different from occupational accidents that are mainly personal accidents.
3. Technical and engineering competency is crucial to ensure process safety and that is why process safety engineering is a discipline of engineering. The only argument to house process safety engineering in an health, safety and environment (HSE) department is to guarantee independency from the engineering and the business.
4. The many concepts, ideas and methods that have been developed to manage Occupational Health and Safety are not necessarily applicable for process safety. Many ideas and concepts (Heinrich's domino concept, Bird's pyramid, etc.) were developed by studying occupational accidents. The typology of technological disasters is, however, very different. Major disasters (e.g., Macondo and Bhopal) are often the result of the coincidental convergence of very different minor failures, which on the first sight have nothing to do with each other and which on themselves are not very dangerous. It is very difficult to identify these so-called weak signals very early because as long as they do not all come together they can be considered as small marginal deviations. The Swiss Cheese model of Reason represents this very well. The linearity between substandard conditions/acts and the accident is not as obvious as what one can find for occupational accidents.
5. Process safety accidents and major disasters find their origin much deeper in the organization. A simple and logic decision at a certain moment in time (e.g., suppression of a function) or a latent failure (e.g., failure of a level measurement) can be a detail for a long time until it becomes a major contributor of a major accident. It is therefore important that the organization is built around a very strong technical competency and that the installations with high consequence potential are managed by senior management who has a strong technical background.
6. The methods that are used to assess the residual risks (consequence calculations, probability estimates, risk matrices, etc.) need to be standardized and applied by people who have been trained for these methods. These methods seem to be very simple but in reality they require a lot of experience to be effective.
7. Senior management must be trained about the potential major accidents and the measures that are in place to avoid occurrence of these major accidents. Senior management has to be regularly informed about the health status of these measures, and every degradation of the important measures needs to be reported to and validated by the senior management.

Appendix I: SWIFT categories and questions

Category	Questions
Material problems	– Flammability?
	– Thermal instability?
	– Flash points?
	– Explosive powder?
	– Static electrical charge buildup and discharge can occur (grounding/bonding)?
	– Storage temperature?
	– Storage pressure?
	– Reactivity (with process chemicals, process service side fluids, firefighting agents)?
	– Toxicity?
	– Exposure limits (personal protective equipment requirements)?
	– Physical properties (freezing point, boiling point, viscosity, etc.)?
	– Corrosivity?
	– Contamination from outside sources?
	– Contamination though process connections?
	– Mixing hazards?
	– Settling hazards?
	– Pyrophoric potential?
	– The process/storage temperature is sufficiently high that unexpected or unwanted water can cause a hazard?
	– Wrong material is unloaded and introduced into system?
	– What did we miss from a material view point which could impact EHS, quality, etc.?
External effects or influences	– (Natural) High winds?
	– (Natural) Flooding?
	– (Natural) Extreme cold?
	– (Natural) Extreme heat?
	– (Natural) High humidity?
	– (Natural) Low humidity?
	– (Natural) Earthquake?
	– (Natural) Hurricane?
	– (Natural) Tornado?
	– (Man-made) A crane drops its load on safety critical piping, equipment or instrumentation?
	– (Man-made) A crane upsets and impacts?
	– (Man-made) Maintenance activity disrupts (vessel lining, instrumentation lines, etc.)?
	– (Man-made) A forklift, truck, rail car or aircraft impacts?
	– (Man-made) Arson?
	– (Man-made) Over pressure or shrapnel impact from a nearby explosion?
	– (Man-made) sabotage?

https://doi.org/10.1515/9783110632132-008

(continued)

Category	Questions
	– (Man-made) Civil disturbance?
	– (Man-made) Poor access to equipment or instrumentation results in adverse effects (inadequate maintenance, inability to operate, emergency response situations)?
	– (Man-made) An external fire impinges on equipment?
Operating errors and other human factors	– (Action) Too little?
	– (Action) Wrong location?
	– (Action) Incorrect chemical used?
	– (Action) Substitution because of unavailability or convenience?
	– (Action) Makes an unauthorized change in procedure?
	– (Action) Procedure not being followed?
	– (Action) Makes an unauthorized change in conditions?
	– (Action) Makes an unauthorized change in equipment (integral parts, bearings, etc.)?
	– (Action) A control or safety system is bypassed or otherwise disabled?
	– (Action) Controls are in manual?
	– (Action) Poor communication (especially between shifts)?
	– (Ergonomics) Controls are confusing?
	– (Ergonomics) Procedure is confusing?
	– (Ergonomics) Feedback is inadequate?
	– (Ergonomics) Piping (manifold, etc.) is confusing?
	– (Ergonomics) Necessary equipment is hard to access for maintenance or difficult to operate?
	– (Ergonomics) Necessary equipment is unavailable?
	– (Ergonomics) Drawings incomplete, misleading or confusing?
	– (Training) Insufficient knowledge?
	– (Training) Insufficient training (initial and refresher)?
	– (Training) Inadequate skill level (performance not adequately monitored)?
	– (Training) Inadequate instruction?
	– (Surroundings) Real hazard adversely affects judgment?
	– (Surroundings) Perceived hazard adversely affects judgment?
	– (Surroundings) Operator discomfort adversely affects performance (noise, humidity, temperature, HVAC, protective gear cumbersome)?
	– (Surroundings) Weather prevents operator from taking necessary action (lightning, ice storm, etc.)?
	– (Surroundings) Personnel hazards adversely affect operator performance (exposure to hot pipes, missing hand rails and close quarters)?
	– (Work organization) Task responsibility is not assigned?
	– (Work organization) Insufficient motivation?
	– (Work organization) Afraid of making a mistake?
	– (Work organization) Number of personnel to perform the operations?
	– (Task characteristics) Task too complex?
	– (Task characteristics) Task is too boring?

(continued)

Category	Questions
	– (Information) Too much information?
	– (Information) Too little information?
	– (Information) Incorrect information?
	– (Information) Incomplete information?
	– (Time/sequence) Not enough time to respond?
	– (Time/sequence) Too soon?
	– (Time/sequence) Too late?
	– (Time/sequence) Wrong sequence?
	– (Time/sequence) Too often?
	– (Time/sequence) Too infrequently?
	– (Action) Too much?
Analytical or sampling errors	– Sample is not taken?
	– Difficulty in obtaining a representative sample?
	– Sample is not analyzed?
	– Significance of test results not understood by operator?
	– Test results are delayed?
	– Test results are incorrect?
	– Sample is thermally unstable?
	– Sample is pressure sensitive?
	– Sampling procedure is unsafe?
	– In-line analytical device is out of calibration?
	– Lab-based analytical device is out of calibration?
	– Sample point left open or leaking?
	– Person performing sample analysis is not trained?
	– Calibration equipment/weights not available?
	– Calibration/reference documents lost or missing?
	– Correct PPE not available or used?
	– Results of testing do not meet process requirements?
	– Sample frequency inadequate or not followed?
Equipment or instrumentation malfunctions	– (Process color coding) Does your plant have a specified color coding requirement?
	– (Process color coding) Are your piping systems properly color coded and labeled?
	– (Process color coding) Are your employees trained on the color codes?
	– (Pumps) Pump fails to pump?
	– (Pumps) Pump seal leaks?
	– (Pumps) Pump cavitation?
	– (Pumps) Pump is deadheaded?
	– (Pumps) Pump runs backward following maintenance?
	– (Pumps) Pump leaks through?
	– (Blowers) Blower fails to pump?
	– (Blowers) Blower seals leak?
	– (Blowers) Blower is in surge?

(continued)

Category	Questions
	– (Blowers) Blower runs backward following maintenance?
	– (Blowers) Blower has excessive lubrication oil present in outlet?
	– (Compressors) Compressor fails to compress?
	– (Compressors) Compressor seals leak?
	– (Compressors) Compressor is in surge?
	– (Compressors) Compressor is choked?
	– (Compressors) Liquid reaches compressor?
	– (Compressors) Other problems (runs hot, high vibration, etc.)?
	– (Valves) Valve fails to close?
	– (Valves) Valve fails open?
	– (Valves) Valve leaks by?
	– (Valves) Valve packing leaks?
	– (Valves) Valve cannot be opened on demand?
	– (Valves) Valve cannot be closed on demand?
	– (Valves) Valve does not control properly (flow/pressure)?
	– (Other components) Line plugs?
	– (Other components) Strainer plugs?
	– (Other components) Control orifice plugs?
	– (Other components) Corrosion (internal and external)?
	– (Other components) Erosion?
	– (Agitators) Agitator stops?
	– (Agitators) Agitation is insufficient?
	– (Agitators) Agitation starts at an inappropriate time?
	– (Control instrumentation) Flow control loop fails?
	– (Control instrumentation) Temperature control loop fails?
	– (Control instrumentation) Pressure control loop fails?
	– (Control instrumentation) Level control loop fails?
	– (Control instrumentation) Process computer/distributed control system fails?
	– (Control instrumentation) Failure condition is inappropriate?
	– (Electrical) Area classification is impaired?
	– (Electrical) Component integrity is not adequately maintained (equipment, cables, conduits, cabinets and connections)?
	– (Electrical) Emergency power fails (not tested or maintained)?
	– (Vessels) Pressure design rating is exceeded?
	– (Vessels) Vacuum design rating is exceeded?
	– (Vessels) Temperature design rating is exceeded (high or low)?
	– (Structures) Loadings have been increased significantly since the original design?
	– (Structures) Critical piping or equipment shifts because of inadequate support as the result of aging, or settling, etc.?
	– (Structures) Fire proofing inadequate (critical equipment and piping supports, critical control and safety instrumentation, and wiring)?

(continued)

Category	Questions
	– (Structures) Piping supports are not adequately designed for emergency vent thrust, thermal expansion?
	– (N_2 standards) N_2 (nitrogen) inerting – is N_2 inerting required to prevent safety or quality issues?
	– (N_2 standards) What happens if inerting fails?
	– (N_2 standards) How are personnel protected from N_2 exposure?
	– (Safety instrumentation) Flow shutdown fails?
	– (Safety instrumentation) Temperature shutdown fails?
	– (Safety instrumentation) Pressure shutdown fails?
	– (Safety instrumentation) Level shutdown fails?
	– (Safety instrumentation) Vibration shutdown fails?
	– (Safety instrumentation) Hydrocarbon detection/shutdown fails?
	– (Safety instrumentation) Toxic vapor/gas detection shutdown fails?
	– (Safety instrumentation) Shutdown interlock fails to work when needed (inadequate testing)?
	– (Safety instrumentation) Shutdown interlock activates when it is not needed?
	– (Safety instrumentation) Failure condition is inappropriate?
	– (Safety instrumentation) Are common cause failures possible (process side – plugging or fouling, etc.; external side – critical instrument lines destroyed by fire, etc.)?
Process upsets of unspecified origin	– (Flow) Low or no flow occurs?
	– (Flow) High flow occurs?
	– (Flow) Reverse flow occurs?
	– (Temperature) High temperature occurs?
	– (Temperature) Low temperature occurs?
	– (Pressure) High pressure occurs?
	– (Pressure) Low pressure occurs?
	– (Pressure) A vacuum is drawn?
	– (Pressure) Required vacuum is not established?
	– (Phase) Foaming occurs?
	– (Phase) Separation is poor?
	– (Phase) A phase inversion occurs?
	– (Phase) Flashing occurs?
	– (Chemistry) pH is too high?
	– (Chemistry) pH is too low?
	– (Chemistry) Reaction rate is sluggish?
	– (Chemistry) Reaction rate is rapid?
	– (Chemistry) An unstable material is produced?
	– (Chemistry) An incompatible material is introduced?
	– (Explosive) An explosive dust concentration is produced within the equipment?

(continued)

Category	Questions
	– (Explosive) An explosive dust concentration is produced external to the process?
	– (Explosive) An explosive vapor concentration is produced within the equipment?
	– (Explosive) An explosive vapor concentration is produced external to the process?
	– (Explosive) Static charge generated by falling or moving nonconductive liquid?
	– (Explosive) Flammable substance in contact with electrical components or hot surface?
Utility failures	– Cooling water fails?
	– Process water fails?
	– Refrigeration fails?
	– Power fails?
	– Different voltages of electrical failures?
	– Steam fails?
	– Air fails?
	– Inert fails?
	– HVAC system fails?
	– Communications system fails?
	– Fire water system fails?
	– Effluent removal/treatment fails?
	– Vacuum fails?
	– Is the power distribution in the building putting an excessive load on the UPS or any main load breaker?
Integrity failure or loss of containment	– (Process caused) Runaway decomposition reaction?
	– (Process caused) Runaway polymerization reaction?
	– (Process caused) Component melts?
	– (Process caused) Cryogenic failure as the result of exceeding the lower temperature specification of material of construction (including unintended flash cooling possible from relief devices)?
	– (Process caused) A heat exchanger tube fails?
	– (Process caused) Pressure leakage from adjacent process sections?
	– (Process caused) Static head?
	– (Process caused) Repeated extreme cycling of conditions (pressure, temperature, concentration, etc.)?
	– (Inherent to the materials) Either internal or external corrosion failure occurs (include stress corrosion cracking)?
	– (Inherent to the materials) Hydrogen embrittlement occurs?
	– (Inherent to the materials) Erosion failure occurs?
	– (Inherent to the materials) Fatigue occurs?
	– (Inherent to the materials) Incorrect material of construction for the service (including piping, vessels, pumps, seals, flanges, gaskets, valve internals, etc.)?

(continued)

Category	Questions
	– (Inherent to the materials) Normal wear and tear?
	– (Inherent to the materials) Inadequate design rating for the service (temperature and pressure)?
	– (Inherent to the materials) Inadequate design rating for an upset in service (temperature and pressure)?
Emergency operations	– An internal fire occurs?
	– An internal explosion occurs?
	– A physical over/underpressure occurs?
	– A fire occurs in a nearby unit?
	– An external fire occurs within the unit?
	– An explosion occurs in a nearby unit?
	– An internal explosion occurs within the unit?
	– A toxic release occurs in a nearby unit?
	– A toxic release occurs within the unit?
	– Combination failures (i.e., a fire or explosion ruptures the fire main which cannot be isolated using section valves, etc.)?
	– Emergency system inaccessible because of incident?
	– Emergency system inoperative (not tested, operator not trained, etc.)?
	– Clear responsibilities not assigned (either on-site or off-site: response, declaring an emergency, public notification and public relations)?
Environmental release	– This event or scenario is not covered by the emergency plan?
	– An event of this magnitude is not addressed by the emergency plan?
	– The emergency plan is out of date (personnel roles, authority, training, communications and specific response requirements)?
	– The resources for coping with this particular emergency are inadequate (PPE for responders and operators, emergency medical treatment and first aid, decontamination, fire fighting)?
	– The response team has not been trained to deal with this type of scenario?
	– Any unacceptable fugitive emissions during normal operations or during intermittent operations (truck loading)?
	– Leak/release detection capability is not adequate?
	– The sources of all potential releases cannot be isolated in a timely manner?
	– The potential effect from a leak/release have not been adequately considered?
	– The likely impact zone or effects of a leak/release are not been adequately considered?
	– Mitigation of a leak/release is not possible or is impractical?
	– Site security and control is inadequate?
	– The alarm system fails, is inadequate, or misleading?
	– Is there a potential for causing an impact to soil, groundwater, wastewater, surface water and/or the air pathway and are existing safeguards in place to prevent such occurrences?

(continued)

Category	Questions
	– Have there been releases that caused environmental impacts to one of the above pathways? – Which existing safeguard (if any) failed and why? What has been done to upgrade the current safeguard? What is done to maintain the integrity of existing safeguards? – Are any additional safeguards required? – Could the environmental release lead to a potential shutdown of the unit? If yes, how and what are the business interruption implications? – Could releases lead to permit exceedance? Have there been past exceedances? Should anything else be done to reduce the chances of permit exceedances? – Is there a mechanism to measure and monitor environmental releases and a means to measure and monitor the impact? – Are there potential conflicts between environmental safeguards and other impacts (safety or economic)? – What key parameters should be measured and monitored?
Safety devices	– Emergency relief opens lower than the set point? – Emergency relief opens higher than the intended set point (pressure trapped between leaking rupture disk and emergency relief valve, incorrect set point, excessive back pressure downstream, etc.)? – Emergency relief fails to open? – Emergency relief capacity is inadequate (design basis, throughput increased from original design, additional users on the header)? – Emergency relief discharge location is inappropriate? – Fire protection system fails (automation failure, manually activated equipment inaccessible due to proximity of fire, etc.)? – Fire protection capacity is inadequate (undersized, incorrect agents, etc.)? – Fire (or release) detection system inadequate in remote or unattended areas? – Area monitor detection? – Identify possible conflicts in the design of safeguards for addressing different issues. For example, systems designed to prevent overpressurization (a relief valve) may lead to an environmental air impact? – Identify key parameters that should be reviewed as potential precursors for the performance standard triangles?
Operability concerns	– What dexterity/agility level is required to operate the system? – What vision requirements (color and depth perception) are required for safe operations? – Is physical strength a requirement to perform the operations? – Are there physical size limitations or requirements to perform the required operations? – Are there access area requirements for safe operations?

(continued)

Category	Questions
	– What staffing levels are required to maintain a safe and productive level of operations?
	– Is the physical environment such that safe work practices can be conducted?
Quality factors	– Is there a contamination source into the process?
	– Is material produced to proper specifications?
	– Is material produced to proper color standards?
	– Is material produced to proper concentrations?
	– What off gases are produced in production operations?
Process control	– Is the process under proper control?
	– Are there adequate numbers of controls/operators monitoring the process?
	– Are process critical interlocks in place and operational?
	– Are operating ranges properly set and being adhered to?
	– Are there adequately experienced technical operators or are there technical deficiencies?
	– Is training up to date?
	– Are all software/hardware requirements in place?
	– Have all applicable measurements been considered for the process?
	– Does the alarm function work, when required?
	– Should technicians be able to access controls/valves from DCS as well as on the floor?
	– Are there times when excessive alarming occurs?
	– Should instrument fail conditions cause shutdowns?
	– Have the fail-safe positions for the valves been documented on P&IDs?
	– Should the interlocks be DCS or hardwired?
	– Can safety or quality issues result from improper or unreliable metering (flow, temperature, composition, pressure and limit switch)?
	– Is there an operational UPS (uninterruptible power supply)?
	– Is UPS properly designed and functionally tested?
Reliability factors	– What is the historical status of the unit or process being reviewed?
	– What measurement systems have been utilized?
	– What is the maintenance history?
	– What spares are available/critical?
	– Has an inspection process been established?
	– What training has been provided? Is this training up to date?
	– What testing parameters are instituted?
	– Are materials of construction used considered to be in line with good engineering practices?
	– Have unplanned failure contingencies been addressed?
Facility siting	– Are toxic and/or flammable chemicals processed in this unit?
	– Is there a potential explosion hazard due to either: could releases from any of the piping/equipment containing toxic liquids drain toward occupied buildings if released?

(continued)

Category	Questions
	– Could flammable liquids within any of the piping and equipment in this unit drain toward occupied buildings if released?
	– Are drainage systems sufficient to handle large spills and water used for firefighting without the runoff endangering buildings in the vicinity?
	– Does a release into the drainage system in one area create potential for mixing incompatible materials and creating fumes in any occupied building?
	– Are toxic and/or flammable chemicals processed in this building?
	– Combustible dusts handled or stored in piping/equipment located inside occupied buildings?
	– Any part of the process within the building subject to a runaway reaction?
	– Are rooms within buildings containing the explosion hazard designed with explosion relief panels to minimize structural damage and risk to personnel?
	– Is there explosion venting for the equipment if an explosion should occur? If provided, does it meet NFPA 68 requirements? Is the vent directed to a safe location?
	– Are building occupants not involved directly with the process separated from the hazardous operations by solid walls?
	– Do the walls and wall openings separating personnel from hazardous operations involving flammable materials have at least a 1-h fire rating to allow safe evacuation?
	– Has a building blast risk study been performed (e.g., API 753)?
	– Building name/number?
	– Do buildings have solid walls with no windows facing process equipment that presents an explosion potential?
	– Are air intakes for buildings that could be subject to gas or vapor clouds of either toxic or flammable material equipped with appropriate detection and alarm systems to warn occupants of hazardous releases?
	– Are the air intakes equipped with a means to isolate ducts if there is a release of hazardous vapor or gas in the area?
	– Is there a safe breathing air supply available for critical personnel who must remain in building during emergencies?
	– Are communication systems provided to alert affected building personnel of the emergency conditions?
	– Are exits placed to minimize exposure of personnel evacuating the building during the emergency?
	– Does the electrical equipment in the unit and buildings meet the requirements of the national electrical code?
	– Have the temporary and contractor buildings been included in this analysis?
	– Does the emergency plan cover this building?

(continued)

Category	Questions
	– Are routing emergency drills conducted for building occupants who may be affected by the emergency?
	– Is the emergency plan adequate and well understood?
	– Is the spacing/structural design of this building adequate to protect building personnel from the known process hazards?
	– Was the spacing based on accepted standards such as NFPA Code 30, Industrial Risk Insurers spacing guidelines and electrical codes?
	– Does the building have adequate protective features to protect occupants from hazardous chemical intrusion, such as positive pressure?
	– Does the team judge this building to be properly sited?
	– Are the plant buildings near this process properly sited from the hazards identified or should the team recommend a further siting study be conducted?

References

ACSNI, Human Factors Study Group: Third Report – Organizing for Safety, HSE Books, 1993.

AIHA, Emergency Response Planning Guidelines, AIHA, Akron, 1989.

Aldrich M, Safety First: Technology, Labor and Business in the Building of American Work Safety, 1870–1939, The John Hopkins University Press, Baltimore and London, 1997.

Akhmedjanov F, Reliability databases: State-of-the-art and perspectives, Forskningscenter Risoe, Risoe report No. 1235(EN), Denmark, 2001.

Aleksandrova AY, Timofeeva S, Analyzing and Assessing the State of Safety Culture at the Mining Industry Facilities in the Irkutsk Region, IOP Conf. Series: Earth and Environmental Science 459, 2020.

Angus M, The World Economy: Historical Statistics, Statistical Appendix, 2007.

Anon, De versierde mens: mode van 1700 tot 1930, Openbaar Kunstbezit in Vlaanderen eenentwintigste jaargang april/mei/juni 1983.

ANSI/ISA-S84.01-1996, Application of Safety Instrumented Systems for the Process Industries, ISA, Instrument Society of America, 1996.

Anyakora SN, Engel GFM, Lees FP, Some Data on the Reliability of Instruments in the Chemical Plant Environment, Chemical Engineering, London, 1971.

Arabian-Hoseynabadi H, Oraee H, Tavner P, Failure modes and effects analysis (FMEA) for wind turbines. International Journal of Electrical Power & Energy Systems, 32, 817–824, 2010.

Arbous AG, Kerrich JE, Accident statistics and the concept of accident proneness. Biometrics, 4(4), 340–432, 1951.

Arulanantham DC, Lees FP, Some data on the reliability of pressure equipment in the chemical plant environment. International Journal of Pressure Vessels and Piping, 9, 327–338, 1981.

Ashelford J: The Art of Dress: Clothing and Society 1500–1914, Abrams, 1996.

Assailly J-P, La psychologie du risqué, Sciences du Risque et Dangers, Collection dirigé par Franck Guarnieri, Lavoisier, 2010.

Atherton J, Gil F, Incidents that Define Process Safety, CCPS and John Wiley & Sons, Inc., 2008.

Terje A, Three influential risk foundation papers from the 80s and 90s: Are they still state-of-the art? Reliability Engineering and System Safety, 193, 2020.

Back KC, Thomas AA, MacEwen JD, Reclassification of materials listed as transportation health hazards. Wright-Patterson Air Force Base, OH: 6570th Aerospace Medical Research Laboratory, Report no. TSA-20-72-3, pp. A-182 to A-183, 1972.

Barber R, Burns M, A Systems Approach to Risk Management, Presented at ANZ Systems, Mooloolaba, 2002.

Barnett A, Wang A, Passenger mortality risk estimates provide perspectives about flight safety. Flight Safety Digest, 19(4), 1–12, 2000. https://flightsafety.org/fsd/fsd_apr00.pdf.

Baschel S, Koubli E, Roy J, Gottschalg R, Impact of component reliability on large scale photovoltaic systems' performance. Energies, 11, 2018.

Beerens HI, Post JG, Uijt de Haag PAM, The use of generic failure frequencies in QRA: the quality and use of failure frequencies and how to bring them up-to-date. Journal of Hazardous Materials, 130(3), 265–270, 2006.

Bellamy LJ, Ale BJM, Whiston JY, Mud ML, Baksteen H, Hale AR, Papazoglou IA, Bloemhoff A, Damen M, Oh JIH, The software tool Storybuilder and the analysis of the horrible stories of occupational accidents. Safety Science, 46(2), 186–197, 2008. 2008.

Beyer DS, Industrial Accident Prevention, Houghton Mifflin Company, 1916.

Bigelow P, Robson L, Occupational Health and Safety Audit Instruments: A Literature Review, Toronto, Institute for Work and Health, 2005.

https://doi.org/10.1515/9783110632132-009

Bird EF Jr, George GL, Practical Loss Control Leadership, the Conservation of the Environment, Property, Process and Profits, International Loss Control Institute, 1985.

Blake RP, Industrial Safety, Prentice-Hall Inc., Englewood Cliffs, New York, 1943.

Bliss CI, The methods of probits, Science 79, pp. 38–39, 1934.

Blustein D, Masdonati J, Rossier J, Psychology and the International Labor Organization, The Role of Psychology in the Decent Work Agenda, June 2017.

Bolt J, Timmer M, van Zanden JL, GDP per capita since 1820, van Zanden JL, eds., How Was Life? Global Well-bing since 1820, OECD Publishing, 2014.

Bowerman H, Owens GW, Rumley JH, Tolloczko JJA, Interim Guidance Notes for the Design and Protection of Topside Structures Against Explosion and Fire, The Steel Construction Institute Document SCI-P-112/487, January 1992.

Bradbury JA, The Use of Social Science Knowledge in Implementing the Nuclear Waste Policy Act, PhD, Graduate School of Public and International Affairs, University of Pittsburgh, 1989.

Brasie WC, Simpson DW, Guidelines for estimating damage explosion. Loss Prevention, 2(91), 1968.

Breakwell GM, Psychology of Risk, Cambridge University Press, 2014.

Bronstein JL, Caught in the Machinery, Stanford University Press, California, 2008.

Buffinton A, The Second Hundred Years' War, 1689–1815, 1929.

Burgan B, Resistance of Structures to Fires and Explosions, Lecture at South Chine University of Technology, 2020.

Burham J, The syndrome of accident proneness (Unfallneigung): why psychiatrists did not adopt and medicalize it. History of Psychiatry, 19(3), 251–274, 2008.

Bush SH, Pressure vessel reliability. Journal of Pressure Vessel Technology, 54–69, 1975.

Busch C, Heinrich's Local Rationality: Shouldn't "New View" Thinkers Ask Why Things Made Sense to Him? Thesis/Project Work Submitted in Partial Fulfilment of the Requirements for the MSc in Human Factors and System Safety, Lund University, 2018.

Cadier-Rey G, Les Français de 1900, Editions Garnier, ISBN 97823541525, 2015.

Cadwallader LC. Selected component failure rate values from fusion safety assessment tasks, Report INEEL/EXT-98-00892, Idaho National Engineering and Environmental Laboratory Nuclear Engineering Technologies Department Lockheed Martin Idaho Technologies Company Idaho Falls, ID 83415–3860, 1998

Carey LM, Anderson HR, Atkinson RW, Beevers SD, Cook DG, Strachan DP, Dajnak D, Gulliver J, Kelly FJ, Are noise and air pollution related to the incidence of dementia? A cohort study in London, England, BMJ Open, 2018.

Carr KE European food history – Renaissance to today. Quatr.us Study Guides, August 4, 2017 (Web. June 4, 2019).

CCPS, Guidelines for Process Equipment Reliability Data with Data Tables, American Institute for Chemical Engineers, New York, 1989.

CCPS, Guidelines for Engineering Design for Process Safety, American Institute of Chemical Engineer, 1993.

CCPS, Inherently Safer Chemical Processes – A Life Cycle Approach, American Institute of Chemical Engineer, A John Wiley & Sons Inc. publication, 1996, 2nd edition 2009.

CCPS, Guidelines for Design Solutions for Process Equipment Failures, A John Wiley & Sons Inc. publication, 1998.

CCPS, Guidelines for Chemical Process Quantitative Risk Analysis, American Institute of Chemical Engineers, 2000.

CCPS, Guidelines for Hazard Evaluation Procedures, 3rd edition, A John Wiley & Sons Inc. publication, 2008.

CCPS, A Practical Approach to Hazard Identification, for Operations and Maintenance Workers, Published by John Wiley & Sons, Inc., Hoboken, New Jersey, 2010.

CEOC, Anniversary Book, Booklet, edited by, Grunert A, Kotová D, Marc Van Overmeire for the 50th anniversary of CEOC International, 2011.

Chaplin Z, Howard K, Update of pipeline failure rates for land use planning assessments, HSE UK, RR1035 Research Report, 2015.

Chernin L, Vilnay M, Shufrin I, Cotsovos D, Pressure-Impulse Diagram Method – A Fundamental Review, Engineering and Computational Mechanics, 2019.

Clark C, The Railroad Safety Movement in the United States: Origin and Development, 1869 to 1893, PhD Dissertation, University of Illinois, 1966.

Clarke L. Acceptable Risk? Making Choices in a Toxic environment, University of California Press, Berkeley, 1989.

Colli A, Failure mode and effect analysis for photovoltaic systems. Renewable and Sustainable Energy Review, 50, 804–809, 2015.

Conklin T, The 5 Principles of Human Performance: A contemporary update of the building blocks of Human Performance for the new view of safety, 2019.

Covello VT, The perception of technological risks: a literature review. Technological Forecasting and Social Change, 23, 285–297, 1983.

Cozzani V, Salzano E, Threshold values for domino effects caused by blast wave interaction with process equipment. Journal of Loss Prevention in the Process Industries, 17, 437, 2004.

COVO Study, Risk Analysis of Six Potentially Hazardous Industrial Objects in the Rijnmond Area, a Pilot Study, A Report to the Rijnmond Public Authority, 1981.

CPR, Committee for the Prevention of Disasters, Guidelines for Quantitative Risk Assessment – "Purple Book", CPR 18E, The Hague, 1999.

CPR, Committee for the Prevention of Disasters, Methods for the Determination of Possible Damage to People and Objects Resulting from Release of Hazardous Materials, CPR 16E, The Hague, 2005.

Cramer JS, The Origins of Logistic Regression, Tinbergen Institute Discussion Paper TI-2002-119/4, Faculty of Economics and Econometrics, University of Amsterdam and Tinbergen Institute, 2002.

Crawley F, Tyler B, Hazard Identification Methods, EPSC and IChemE, 2003.

Crawley F, Tyler B, HAZOP: Guide to Best Practice, 3rd Edition, IChemE, 2015.

Cresswell WL, Frogatt P, Accident proneness or variable accident tendency?, Journal of the Statistical and Social Inquiry Society of Ireland, 1962.

Crickmer JR, Cave L, The data required for fast reactor safety assessment by probability methods, Proceedings of a Symposium on Fast Reactor Physics and related Safety Problems held by the International Atomic Energy Agency in Karlsruhe, 30 October – 3 November 1967, 1968.

Cross R, Youngblood R, Probabilistic Risk Assessment Procedures Guide for Offshore Applications, Report JSC-BSEE-NA-24402-02, 2018.

Cullen, The Hon. Lord William Douglas. The Public Inquiry into the Piper Alpha Disaster, Presented to Parliament by the Secretary of State for Energy by Command of Her Majesty. London: H.M. Stationery Office. 488 pages, 2 volumes, November 1990.

Czujko J, Design of Offshore Facility to Resist Gas Explosion Hazard, CorrOcean ASA, Oslo, 2001.

Danesh N. Response of Building Structure and its Components to Blast Loads, PhD, Carleton University Ottawa, Ontario, 2017

De Fruyt F, Jesus S, Applied personality psychology: Lessons learned from the IWO field. European Journal of Personality, 17, S123–S131, Wiley InterScience, 2003.

DeBlois LA, Industrial Safety Organization for Executive and Engineer, McGraw-Hill, 1926.

Descazeaux M, Jean-Claude R, Camille B, Damien S-M, Serious Injury and Fatality Prevention, Focusing on the Essential, ICSI, 2019.

Deferme J, Alles strijdt wat naar vrijheid haakt. Theorievorming over de staking in de Belgische politiek, 1884–1914. Bijdragen En Mededelingen Betreffende De Geschiedenis Der Nederlanden, 117(2), 145–167, 2002.

Dekker S, The Field Guide to Human Error Investigations, Ashgate Publishing, 2002, Reissued by Routledge.

Dekker S, Ten Questions about Human Error, A New View of Human Factors and Safety Systems, Lawrence Erlbaum Associates, Publishers, London, 2005.

Dekker S, Safety Differently: Human Factors for a New Era, 2nd edition, CRC Press, 2014.

Dekker S, The Foundations of Safety Science, A Century of Understanding Accidents and Disasters, CRC Press, 2019.

Department of Defence, Reliability Prediction of Electronic Equipment, MIL-HDBK-217E, Washington, 1982.

Departement Omgeving, Richtlijnen Voor Kwantitatieve Risicoanalyse, Indirecte Risico's En Milieurisicoanalyse – Versie 2.0, Administratieve Diensten Van De Vlaamse Overheid, Beleidsdomein, Omgeving, 2019.

de Weger D, Pietersen CM, Reuzel PGJ, Consequences of exposure to toxic gases following industrial disasters. Journal of Loss Prevention in the Process Industries, 4, 1991.

DNV, OREDA – Offshore Reliability Data Handbook, 3rd Edition, Norway, 1997.

DOE, Department of Energy, Chemical Process Hazards Analysis, DOE-HDBK-1100-2004, 2004.

Rayner M, Wildavsky AB, Risk and Culture: An Essay on the Selection of Technical and Environmental Dangers, Berkeley, University of California Press, 1983.

Drago JP, Pike DH, Borkowski RJ, Goldberg FF, The In-Plant Reliability Data Base for Nuclear Power Plant Components: Data Collection and Methodology Report, NUREG/CR-2641, 1982.

Drury HB, Scientific Management; A History and Criticism, PhD diss., Columbia University, 1915.

Dussault Heather B, The Evolution and Practical Applications of Failure Mode and Effects Analysis, Rome Air Development Center, Air Force Systems Command, RADC-TR-83-72, 1983.

Eastman C, Work-Accident and the Law, New York, Volume 2 of the Pittsburgh Survey, Russell Sage Foundation Publications, 1910.

Edmonds J, Human Factors in the Chemical and Process Industries, 1st edition, Elsevier, 2016.

EASA ATPL Training, Human Performance, Jeppesen Sanderson Inc., 2014.

Eide SA, Calley MB, "Generic Component Failure Rate Database," Proceedings of the International Topical Meeting on Probabilistic Safety Assessment, PSA '93, American Nuclear Society, Clearwater Beach, FL, January 26–29, pages 1175–1182, 1993.

Eisenberg NA, Lynch CJ, Breeding RJ, Vulnerability Model: A simulation of Damage Resulting from Marine Spills, US Coast Guard, Report CG-D-136-75, 1975.

Engels F, Die Lage Der Arbeitenden Klasse in England, Leipzig, 1845.

EPA, The History of Drinking Water Treatment, EPA-816-F-00-006, February 2000.

Epstein R, The historical origins and economic structure of Workers' Compensation Law. Georgia Law Review, 16(4), 1982.

European Agency for Safety and Health at Work, Economic incentives to improve occupational safety and health: a review from the European perspective, 2010.

Evans RJ, The Pursuit of Power: Europe 1815–1914, Viking, 2016.

FABIG, Explosion Resistant Design of Offshore Stuctures, Technical Note 4, The Steel Construction Institute, 1998.

FABIG a, Simplified Methods for Analysis of Response to Dynamic Loading, Technical Note 7, The Steel Construction Institute, 2002.

FABIG b, An Advanced SDOF Model for Steel Members Subject to Explosion Loading: Material Rate Sensitivity, Technical Note 10, The Steel Construction Institute, 2002.

FABIG, Design of Low to Medium Rise Buildings against External Explosions, Technical Note 14, The Steel Construction Institute, 2018.

Factory Mutual, Handbook of Industrial Loss Prevention, 2nd edition, McGraw-Hill Book Company, 1967.

Fairlie and Sumner, An independent scientific evaluation of health and environmental effects 20 years after the nuclear disaster providing critical analysis of a recent report by the International Atomic Energy Agency (IAEA) and the World Health Organisation (WHO), 2006.

Farris Engineering Services, Understanding Causes for RV Chatter, Valve User Magazine, Issue 46, page 59, Autumn 2018.

Farmer E, The study of personal differences in accident liability. Journal of the National Institute of Industrial Psychology, 3, 1927.

FEMA, The Federal Emergency Management Agency, Continuity Risk Toolkit, US Department of Homeland Security, 2017.

Fine W, Mathematical Evaluation for Controlling Hazards, Naval Ordonnance Laboratory, White Oak, Maryland, March 1971.

Fine W, Mathematical evaluation for controlling hazards. Journal of Safety Research, 3(4), 1971.

Finney DJ, Probit Analysis, Cambridge University Press, Cambridge, UK, 1947.

Fiske ST, Taylor, Social Cognition, Reading, 1978.

Forty S, 100 Innovations of the Industrial Revolution from 1700 to 1860, Haynes Publishing, 2019.

Gadd S Dr., Keeley D Dr, Balmforth H Dr, Health & Safety Laboratory, Good Practice and pitfalls in risk assessment, HSE, Research report 151, 2003.

Gallivan S, Taxis K, Franklin BD, Barber N, Is the principle of a stable Heinrich ratio a myth? Drug Safety, 31(8), 637–642, 2008.

Gilbert C, Journé B, Laroche H, Bieder C, Safety Cultures, Safety Models, Taking Stock and Moving Forward, Springer Open, 2018.

Glasstone S, Hirschfelder JO, Parker D, Kramish A, Smith RC, The Effects of Atomic Weapons, US Atomic Energy Commission, Los Alamos Scientific Laboratory, 1950.

Glasstone S, Dolan PJ, The Effects of Atomic Weapons, United States Department of Defense and the United States Department of Energy, 3rd revision, 1977.

Glynis B. The Psychology of Risk, Cambridge University Press, 1st edition, 2007.

Gould J, Glossop M, Ioannides A, Review of Hazard Identification Techniques, Health and Safety Laboratory, HSL/2005/58, 2000.

Green AE, Quantitative Assessments of System Reliability, IChemE Symposium Series No 33, 1972.

Greenwoods M, Woods HM, The Incidence of Industrial Accidents with Special Reference to Multiple Accidents, Industrial Fatigue Research Board, report No. 4, London, 1919.

Graham K, Kinney G, A Practical safety analysis system for hazards control. Journal of Safety Research, 12(1), 1980.

Hämäläinen P, Takala J, Saarela K, Global estimates of occupational accidents. Safety Science, 44, 137–156, 2006.

Heinrich HW, Granniss ER, Industrial Accident Prevention, A Scientific Approach, 4th edition, McGraw-Hill Book Company, 1959.

Heikkilä A-M, Inherent Safety in Process Plant Design, An Index-Based Approach, VTT, Technical Research Center of Finland, ESPOO, 1999.

Hellemans M, The Safety Relief Valve Handbook, Design and Use of Process Safety Valves to ASME and International Codes and Standards, IChemE, Elsevier Ltd, 2009.

Herzberg F, Mausner B, Peterson RO, Capwell DF, Job Attitudes: Review of Research and Opinion, Psychological Service of Pittsburgh, 1957.

Hewitt M, Relative Culture Strength, A Key to Sustainable World-Class Safety Performance, DuPont Safety Resources, 2011.

Hibbert AJ, Turnbull CJ, Measuring and Managing the Economic Risks and Costs of With-profit Business, Institute of Actuaries and Faculty of Actuaries, 2003.

Hill D, Technical Support for APS related to McMicken Thermal Runaway and Explosion, McMicken Battery Energy Storage System Event Technical Analysis and Recommendations, DNV-GL report, Document 10209302-HOU-R-01, 18 July 2020.

Hill KM, Knott H, The Design of Plants for Handling Hydrofluoric Acid, Symposium of Chemical Process Hazards, 1960.

Hill RR, Stinebaugh JA, Briand D, Wind Turbine Reliability: A Database and Analysis Approach, Sandia National Laboratories, Sandia Report SAND 2008-0983, 2008.

Hills RL, Power from Steam: A History of the Stationary Steam Engine, Cambridge University Press, 1993.

Homans GC, Social Behavior: Its elementary forms, 1961.

Hoorelbeke P, Het gebruik van dispersiemodellen in milieu- en veiligheidsstudies, Veiligheidsnieuws No. 96, 1992.

Hoorelbeke P, Handboek Kanscijfers ten behoeve van het opstellen van een veiligheidsrapport. AMINAL, Dienst Gevaarlijke Stoffen en Risicobeheer, 1994.

Hoorelbeke P, Vapor Cloud Explosion Hazard in a Petrochemical Installation, VUB, 2004.

Hoorelbeke R, Brewerton, Explosion Analysis of an Onshore Plant with Worked Examples, Quantitative Risk Analysis, Probabilistic Explosion and MDOF Response Analysis, FABIG Technical Meeting on Protection of Onshore Oil and Gas Plants Against Fire and Explosion, 6–7 April 2005.

Hoorelbeke P, Roosendans D, Societal Risk Criteria for Industrial Activities, Analysis of 1657 Accidents over a 70 Year Post WW II Period, FABIG technical newsletter, 2015.

Hoorelbeke P, Van Overmeire M, Roosendans D, Boogaerts G, Master of Safety Engineering – Qualitative Risk Analysis Techniques, KUL, 2019.

Howden-Chapman P, Mackenbach J, Poverty and painting: Representations in the nineteenth century Europe. BMJ, 325, 2003.

Hobsbawm E, Industry and Empire, the Birth of the Industrial Revolution, The New Press, New York, 1999.

HSC Advisory Committee on Dangerous Substances, Major Hazard Aspects of the Transport of Dangerous Substances, HMSO, London, 1992.

HSE, Health And Safety Executive UK, Canvey: An Investigation of Potential Hazards from Operations in the Canvey Island/Thurrock Area, London, HM Stationery, 1978.

HSE, Health And Safety Executive UK, Canvey: A Second Report. A Review of the Potential Hazards from Operations in the Canvey Island/Thurrock Area Three Years after Publication of the Canvey Report, London, HM Stationery, 1981.

HSE, Health And Safety Executive UK, Workplace fatal injuries in Great Britain, 2019, Annual statistics, July 2019.

HSE, Health And Safety Executive UK, An evaluation of current legislative requirements for verification of elements critical to the safety of offshore installations, Bomel Limited, Crown Copyright, Research Report 397, 2005.

Hopkins A, Disastrous Decisions: The Human and Organisational Causes of the Gulf of Mexico Blowout, 2012.

Hudson P, Safety Management and Safety Culture, The Long Hard and Winding Road, Pearse W, Callagher C, Bluff L, Eds., Occupational Health and Safety Management Systems Crown, 2001.

Human Engineering, A review of safety culture and safety climate literature for the development of the safety culture inspection toolkit, HSE UK research report 367, 2005.

Hunther Regina L, Layton DW, Anspaugh LR. Opportunities and impediments for risk based standards: Some views from a workshop. Risk Analysis, 14, 863–68, 1994.

IAEA, Summary Report on the Post-accident Review Meeting on the Chernobyl Accident, Report by the International Nuclear Safety Advisory Group, Safety Series N° 75 – INSAG-1, Vienna, 1986.

IAEA, International Atomic Energy Agency, Basic Safety Principles for Nuclear Power Plants, A report by the International Nuclear Safety Advisory Group, INSAG-3, Vienna, 1988.

IAEA, International Atomic Energy Agency, Defense in depth, A report by the International Nuclear Safety Advisory Group, INSAG-10, Vienna, 1996.

Ibbetson D, A Historical Introduction to the Lax of Obligations, ISBN-13:9780198764113, 2001.

IEC. International Electrotechnical Commission Standard 60812, Analysis techniques for system reliability – Procedure for failure mode and effects analysis (FMEA), Geneva, 2006.

IEEE Std. 500-1984, IEEE Guide to the Collection and Presentation of Electrical, Electronic, Sensing Component and Mechanical Equipment Reliability Data for Nuclear-Power Generation Stations, IEEE, New York, 1984.

ILO, Model Code of Safety Regulations for Industrial Establishments for the Guidance of Governments and Industry, First Printed in 1949, Geneva, Switzerland, 1954.

Institute of Makers of Explosives (IME), The American Tables of Distance, 2011.

ISO 17776, Petroleum and natural gas industries – offshore production installations – Guidelines on tools and techniques for hazard identification and risk management, 2000.

IPO, Guidelines for the Preparation of Off-site Safety Industrial Sites, Report IPO Project A-73, The Hague, 1994.

Jaeger CC, Webler T, Rosa EA, Renn O, Risk, Uncertainty and Rational Action, Taylor & Francis, 2013.

John C. The Prevention of Factory Accidents, Longmans, Green and Co, New York and Bombay, 1899.

Jonkman SN, van Gelder PHAJM, Vrijling JK, An overview of quantitative risk measures for loss of life and economic damage. Journal of Hazardous Materials, 99, 1–30, 2003.

Julie B, Holroyd J. Review of human reliability assessments. HSE Research Report RR, 679, Crown, 2009.

Kaems D, Construction – The Real Safety of Punch Press Work. NSC Proceedings, 8, 316–327, 1919.

Kaplan S, Garrick BJ, On the quantitative definition of risk. Journal of Risk Analysis, 1, 11–27, 1981.

Karlos V, Solomos G, Calculation of Blast Loads for Application to Structural Components, JRC Technical Reports, JRC 32253-2011, 2013.

Kemeny J, The President's Commission on The Accident at Three Mile Island, 1979.

King AG, SIL Determination and High Demand Mode, IChemE 24th Symposium on Hazards, Series N° 159, Edinburgh, 2014.

King E, Safety engineering in the textile plant. NSC Proceedings, 10, 819–823, 1921.

Kinney GF, Wiruth AD, Practical Risk Analysis for Safety Management, Naval Weapons Center, NWC TP 5865, 1976.

Kirigia JM, Muthuri RNDK, The fiscal value of human lives lost from coronavirus disease (COVID-19) in China, Kirigia and Muthuri, BMC Research Notes, 2020.

Klein JA, Two centuries of process safety at DuPont. Process Safety Progress, 28(2), 2009.

Kletz TA, "Hazard Analysis – A Quantitative Approach to Safety", I Chem E Symposium Series No 34, p 75, 1971.

Kletz TA, What you don't have can't leak, Chemistry and Industry, pp. 287–292, 1978.

Kletz T, An engineering's view of human error, 1nd edition, IChemE, 1985 (2nd edition was published in 1991 and 3rd edition in 2001)

Kletz TA, HAZOP – Past and Future, Reliability Engineering and System Safety, Vol. 55, Elsevier, 263–266, 1997.

Kletz T, Process Plants: A Handbook for Inherently Safer Design, CRC Press, 1998.

Knowlton AE, Shipley DK, Introduction to Hazard and Operability Studies, HO/SD/760003, 1976.

Kogan N, Wallach MA, Risk Taking as a Function of the Situation, the Person and the Group, New Directions in Psychology III, New York, 1967.

Koslin D, "Value-Added Stuffs and Shifts in Meaning: An Overview and Case-Study of Medieval Textile Paradigms", in Koslin and Snyder, Encountering Medieval Textiles and Dress.

Kouabenan DR, Cadet B, Hermand D, Sastre MTM, Psychologie du risqué: identifier, évaluer, prévenir, De Boeck, 2006.

Lagadec P, Le Risque Technologique Majeur, Pergamon, 1981.

Landefeld JS, Seskin EP, The economic value of life: linking theory to practice. American Journal of Public Health, 72, 555–566, 1982.

Lapierre D, Moro J, Il était minuit cinq à Bhopal, Pocket, 2001.

Lavallo D, Kahneman D, Delusion of success: How Optimism Undermines Executives' Decisions, Harvard Business Review, 2003.

Law F, Newell W, The Prevention of Industrial Accidents, General Pamphlet, The Fidelity and Casualty Company of New York, 1909.

Lawler EE, Pay and Organizational Effectiveness: A Psychological View, New York, McGraw Hill, 1971.

Lawley HG, Operability studies and hazard analysis. Chemical Engineering Progress, 70(4), 45, 1974.

Lees FP, A Review of Instrument Failure Data, Process Industry Hazards, Vol. 6, Institution of Chemical Engineers, 1976.

Lees F, Loss Prevention in the Process Industries, Butterworths, 1980.

Le Bon G, Psychologie De Foules, Alcan, 1895.

Léoni L, Histoire de la prévention des risques professionnels, EN3S – Ecole nationale supérieure de Sécurité sociale, N°51, ISSN 0988-6982, 2017.

Libre J-M, Collin-Hansen C, Kjeilen-Eilertsen G, Rogstad TW, Stephansen C, Brude OW, Bjorgesaeter A, Brönner U, ERA Acute-Implementation of a New Method for Environmental Risk Assessment of Acute Offshore Oil Spills, Society of Petroleum Engineers, SPE-190540-MS, 2018.

Lifar AS, Brom AE, FMECA use for the equipment reliability analysis in hydro-power engineering, Proceedings of the IOP Conference Series: Earth and Environmental Science, Ota, Nigeria, 18–20, 2019.

Licht DM, Polzella DJ, Human Factors, Ergonomics, and Human Factors Engineering: An Analysis of Definitions, Technical Report of the Crew System Ergonomics Information Analysis Center (CSERIAC), 1989.

LG Chem, Report to the Arizona Corporation Commission In the Matter of the Commission's Inquiry of Arizona Public Service Battery Incident at McMicken Energy Storage Facility, 30 July 2020.

Linssen W, De Jonge K, Belgische Ingenieurs in de negentiende eeuw, 2013.

Lowrance W, Of Acceptable Risk: Science & the Determination of Safety, Published by William Kaufmann, 1976.

Luhmann N. Technology, environment and social risk: A systems perspective. Industrial Crisis Quarterly, 4, 223–31, 1990.

Lyman OR, The History of the Quantity Distance Tables for Explosive Safety, Memorandum Report ARBL-MR-02925, Ballistic Research Laboratory, 1979.

Malchaire J, Stratégie générale de gestion des risques professionnels, Cahiers de notes documentaires – Hygiëne et sécurité de travail – N° 186, 2002.

Mannan S, Lees'Loss Prevention in the Process Industries, Hazard Identification, Assessment and Control, Third Edition, Elsevier, 2005.

Mansfield D, Poulter L, Kletz T, Improving Inherent Safety, Health and Safety Executive, OTH 96 521, 1996.

Marbe K, Ueber Unfallversicherung und Psychotechnik, Prakt. Psychol. IV, 1923.

Marshall VC, Major Chemical Hazards, Ellis Horwood Ltd. and John Wiley & Sons, 1987.

Mary D. Risk Acceptability according to the Social Sciences, Russell Sage Foundation, New York, 1985.

Mather J, The Coal Mines: The Dangers and Means of Safety, London, Longman, 1853.

McMillan KG, Hunter Vegas P, Process/Industrial Instruments and Controls Handbook, 6th edition, McGraw Hill, 2019.

Miller RL, Doyle WH, Loss Prevention, Volume 1, A CEP technical manual published by the American Institute of Chemical Engineers, New York, 1967.

Mitchell C, Mitchell P, Landmark Cases in the Law of Tort, Bloomsbury, 2010.

Morris T, Southern Slavery and the Law, 1619 – 1860, University of North Carolina, 1996.

Nassani DE, Simple A, Model for calculating the fundamental period of vibration in steel structures. Asia-Pacific Chemical, Biological & Environmental Engineering Society, APCBEE Procedia, 9, 339–346, 2014.

NEA, OECD, Chernobyl: Assessment of Radiological and Health Impacts, Nuclear Energy Agency, Organisation for Economic Co-operation and Development, 2002.

Newbold EM, A Contribution to the Study of the Human Factor in the Causation of Accidents, Industrial Fatigue Research Board, report No. 34, London, 1926.

Nielsen DS, The Cause/Consequence Diagram Method as a Basis for Quantitative Accident Analysis, Danish Atomic Energy Commission Research Establishment Risö, Report Risö-M-137A, 1971.

NORSOK Standard N-001, Structural design, Rev 3, August 2000.

NRC (National Research Council). Risk Assessment in the Federal Government: Managing the Process, Committee on the Institutional Means for Assessment of Risks to Public Health, Washington, 1983.

OGP, international Association of Oil and Gas Producers, Storage Incident Frequencies, Report No. 434–3, 2010.

Ones DS, Viswesvaran C, Empirical and theoretical considerations in using conscientiousness measures in personnel selection. Paper presented at the Fifth European, Congress of Psychology, Dublin, 1997.

Oppert F, On Melonosis of the Lungs and Other Lung Diseases from Inhalation of Dust, John Churchill and Sons, London, 1866.

Orymowska J, Sobkowicz P, Hazard identification methods. Scientific Journal of Silesian University of Technology. Series Transport, 95, 145–158, 2017.

Otway HJ, Maurer D, Thomas K, Nuclear power: the question of public acceptance. Journal Futures, 10(2), 109–118, 1978.

Parliamentary Papers, Reports from Committees, Select Committee on Accidents in Coal Mines, 1854.

Parliamentary Papers, Reports from Committees, Report from the Selected Committee on Employers' Liability for Injuries to their servants, 1876.

Parry CF, Relief Systems Handbook, Institution of Chemical Engineers, 1992.

Pasman HJ, History of Dutch process equipment failure frequencies and the Purple Book. Journal of Loss Prevention in the Process Industries, 24, 208–213, 2011.

Pershing JA, Handbook of Performance Technology, 3rd Edition, John Wiley & Sons, Inc, 2006.

Pidgeon N, O'Leary M, Man-Made Disasters: why technology and organizations (sometimes) fail. Safety Science, 34, 15–30, 2000.

Piketty, Capital in the Twenty-First Century, 2013.

Philips CAG, Warwick RG, A Survey of Defects in Pressure Vessels Built to High Standards of Construction and its Relevance to Nuclear Primary Circuit Envelopes, Authority Health and Safety Branch, Risley, Warrington, Lancashire, UK, AHSB R162, UKAEA Report, 1969.

Price CW, How to organize for safety. NSC Proceedings, 6, 92–98, 1917.

Prugh RW, Application of Fault Tree Analysis, Loss Prevention, Vol. 14, American Institute of Chemical Engineers, 1981.

RADC Reliability Notebook, Volume I, Technical Report No. RADC-TR-67-108, November 1968.

Raj PK, Hydrogen Fluoride and Fluorine Dispersion Models Integration Into the Air Force Dispersion Assessment Model (ADAM), Final Report GL-Ta-90-0321, 1990.

Rasmussen S, Rules and Knowledge; Signals, Signs, and Symbols, and Other Distinctions in Human Performance Models IEEE Transactions, Man and Cybernetics, N° 3, May 1983.

Rayner S, Risk in Cultural Perspective, Acting under Uncertainty: Multidisciplinary Conceptions. Theory and Decision Library (Series A: Philosophy and Methodology of the Social Sciences), Vol. 13, Springer, Dordrecht, 1990.

Rawles C, Benchmark Minerals' Lithium ion Battery Megafactory Assessment, 2019 https://www.benchmarkminerals.com/who-is-winning-the-global-lithium-ion-battery-arms-race/

Rawson AJ, Accident proneness. Psychosomatic Medicine, 6(1), 88–94, 1944.

Reason J, Managing the Risks of Organizational Accidents, Ashgate Publishing Company, London, 1997.

Reitynbarg D, Makarow, Psychol. Abstracts, VII, N° 6002, 1933.

Rémond R, Le XIXe siècle 1815–1914, Introduction à l'histoire de notre temps, Editions du Seuil, 1974.

Renn O, Risk perception and risk management: A review. Risk Abstracts, 7, 1990.

Renn O, Concepts of Risk: A classification, Social Theories of Risk, ed, Krimsky S, Golding D, Praeger Publishers, 1992.

Renn, O. Concepts of Risk: A classification, Krimsky, S, Golding, D (Ed.), Social Theories of Risk, Praeger Publishers, 1992.

RIVM, Rijksinstituut voor Volksgezondheid en Milieu, Handleiding Risicoberekeningen Besluit externe veiligheid inrichtingen, Ministerie van Volksgezondheid, Welzijn en Sport, 2020.

Roberts D, Wind Farm Reliability Estimates, Seattle, WA Global Energy Concepts, 2007.

Robson, Ship/platform collision incident database (2001), HSE UK Research report 053, Crown, 2003.

Rodgers MD, Blanchard RE, Accident Proneness: A Research Review, Federal Civil Aeromediical Institute Report DOT/FAA/AM-93-9, 2003.

Rogovin M, Three Mile Island, A Report to the commissioners and to the public, Nuclear Regulatory Commission Special Inquiry Group, 1980.

Roosendans D, Safety Critical Measures EPSC Report Number 33, 2012.

Roosendans D, Mitigation of Vapor Cloud Explosions by Chemical Inhibition using Alkali Metal Compounds, Ph.D. Thesis, Univ. of Brussels, ISBN 978-9-49231-294-5, 2018.

Royal Commission on the Employment of Children in Factories, Reports from the Commissioners XX, 1833.

Rynes SL, Gerhart B, Minette KA, The importance of pay in employee motivation: discrepancies between what the environment say and what they do. Human Resource Management, Wiley Periodicals, 43(4), 381–394, 2004.

Salzano E, Hoorelbeke P, Khan F, Amyotte P, Overpressure Effects, Domino Effects in the Porcess Industries, editors, Reniers H, Cozzani V, Elsevier, 2013.

Samuel G. The Effect of Nuclear Weapons, prepared by the United States of Defense, published by the United States Atomic Energy Commission, June 1957.

Sayed A, El-Shimy M, El-Metwally M, Elshahad M, Reliability, availability and maintainability analysis for grid-connected solar photovoltaic systems. Energies, 12, 2019.

Savage I, The Economics of Railroad Safety, Kluwer Academic Publishers, 1998.

Salvendy G, Handbook of Human Factors and Ergonomics, John Wiley and Sons, 2012.

Schmitt E, Unfallaffinität und Psychotechnik im Eisenbahndienst. Industrielle Psychotechnik, 3, 144–153, 1926.

Scott HM, Review: The Second "Hundred Years War" 1689–1815. The Historical Journal, 35, 443–469, 1992.

Shi Y, Hao H, Li Z-X, Numerical derivation of pressure-impulse diagrams for prediction of RC column damage to blast loads. International Journal of Impact Engineering, 35, 1213–1227, 2008.

Simonds RH, Grimaldi JV, Safety Management, Edited by, Erwin R, Inc. Homewood III, 1956.

Shah J, Qureshi N, Chlorine Gas vs. Sodium Hypochlorite: What's the Best Option? Opflow July 2008.

Shafiee M, Dinmohammadi F, An FMEA-based risk assessment approach for wind turbine systems: A comparative study of onshore and offshore. Energies, 7, 619–642, 2014.

Sherman K, "Electoral lists of France's July Monarchy, 1830–1848". French Historical Studies, 1971.

Shimmin S, Wallis D, Fifty Years of Occupational Psychology in Britain, the Division and Section of Occupational Psychology, Leicester, 1994.

Schmidt M, Investigating Risk Perception: A Short Introduction, Chapter 3 In: Schmidt M. 2004. Loss of Agro-biodiversity in Vavilov Centers, with A Special Focus on the Risks of Genetically Modified Organisms (Gmos), PhD Thesis, Vienna, 2004.

Single J, Schmidt J, Denecke J, State of research on the automation of HAZOP studies. Journal of Loss Prevention in the Process Industries, 62, 2019.

Shell International Exploration & Production B.V., Overview Hazards and Effects Management Process, EP 95-0300, 1995.

Shewhart's book Statistical Method from the Viewpoint of Quality Control, The Graduate School, The Department for Agriculture, Washington, 1939.

Slovic P, Fischhoff B, Lichtenstein S, Perceived risk: Psychological factors and social implications. Proceedings of the Royal Society of London, 374, 17–34, 1981.

Slovic P, Perception of risk. Science, 236, 1987.

Slovic P, The Psychology of Risk, Saude e Sociedade, Vol 19, no. 4, São Paulo, 2010 paper is based on SLOVIC, P. "Trust, emotion, sex, politics, and science: Surveying the risk-assessment battlefield", Bazerman MH, et al., eds., Environment, Ethics, and Behavior, San Francisco, New Lexington, 277–313, 1997, 2010.

Smith JD, Reliability Maintainability and Risk, 6th edition, Butterworth Heinemann, 2001.

Smith TA, Warwick RG, The second Survey of Defects in Pressure Vessels Built to High Standards of Construction and its Relevance to Nuclear Primary Circuit Envelopes, United Kingdom Atomic Energy Authority, Warrington, Lancashire, UK, AHSB R162, UKAEA Report SRD-R-30, 1974.

Saloniemi A, Oksanen HE, Accidents and fatal accidents – some paradoxes. Safety Science, 29(1), 59–66, 1998.

SRI, Critical Industry Repair Analysis – Petroleum Industry, prepared for the Office of Civil Defense under Contract OCD-PS-64-201, OCD Subtask 3311 A, Report No. CIRA-4, Advance Research, Inc., DDC No. AD 482 909L, 1965.

Starr C, Social Benefit versus Technological Risks – What is our society willing to pay for safety. Science, 19, 165, 1232–1238, 1969.

Stephensen MM, Minimizing Damage to Refineries, US Department of Interior, Office Oil and Gas, 1970.

Stewart M, Lewis OT, History and Organization of Codes, Pressure Vessel Field Manual, 2013.

Stuster J, The Human Factors and Ergonomics Society: Stories from the first 50 years, HFES, 2006.

Swain AD, Guttmann HE, Handbook of Human Reliability Analysis with Emphasis on Nuclear Power Plant Applications, Final Report, Sandia National Laboratories, U.S. Nuclear Regulatory Commission, NUREG/CR-1278, August 1983.

Swuste P, van Gulijk C, Zwaard W, Ongevalscausaliteit in de negentiende en in de eerste helft van de twintigste eeuw, de opkomst van de brokkenmakertheorie in de Verenigde Staten, Groot-Brittanië en Nederland, Tijdschrift voor toegepaste Arbowetenschap, nr. 2, 2009.

Takeda SS, Spilker KD, Fashioning Fashion: European Dress in Detail, 1700–1915, LACMA/Prestel USA, 2010.

Tausseef SM, Abbasi T, Suganya R, Abbasi SA, A critical assessment of available software for forecasting the impact of accidents in chemical process industry. International Journal of Engineering, Science and Mathematics, 6(7), 2017.

Taylor FW, The Principles of Scientific Management, New York, Harper & Brothers, 1911.

Ten Berge WF, Vis van Heemst M, Validity and Accuracy of a Commonly Used Toxicity Assessment Model in Risk Analysis, 4th Int. Symp. Loss Prevention, Harrogate, 1983.

Ten Berge WF, Zwart A, Appelman LM, Concentration – time mortality response relationship of irritant and systemically acting vapours and gases. Journal of Hazardous Materials, 13, 301–309, 1986.

The Guardian, 2005 https://www.theguardian.com/business/2005/feb/07/politics. economicpolicy

The International Association of Engineering Insurers. IMIA Working Group Paper, 112(19), 2019.

Tomei PA, Russo GM, Workplace Safety Culture Model (WSCM): presentation and validation. International Journal of Advanced Engineering Research and Science, 6(4), 2019.

TOTAL, The H.O.F. Approach, Human and Organizational Factors for Safety, Internal brochure of ONE HSE, October 2019.

Perrin T, Working today: understanding what drives employee engagement. The Tower Perrin Report, 2003.

Trbojevic VM, Optimising hazard management by workforce engagement and supervision, Health and Safety Executive Research Report RR 637, 2008.

UFIP, Guide Méthodologique UFIP pour la Réalisation des Etudes de Dangers en Raffineries, Stockages et Dépôts de produits liquides et Liquéfiés, 2002.

United States Atomic Energy Commission, Theoretical Possibilities and Consequences of Major Accidents in Large Nuclear Power Plants, A Study of the Possible Consequences if Certain Assumed Accidents, Theoretically Possible but Highly Improbable, Were to Occur in Large Nuclear Power Plants, WASH 740, 1957.

United States Atomic Energy Commission, Reactor Safety Study, An Assessment of Accident Risks in US Commercial Nuclear Power Plants, WASH 1400, 1975.

U.S. Department of Labor, Handbook for Analyzing Jobs, Washington D.C., U.S. Government Printing Office, 1991.

U.S. Department of Health and Human Services, Public Health Service Agency for Substances and Disease Registry, Toxicological Profile for chlorine, 2010.

USNRC, United States Nuclear Regulatory Commission, Historical Review and Observations of Defense-in-Depth, NUREG/KM-0009, April 2016.

Van den Bosse E, Arbeidsomstandigheden en gezondheid, 23 juli 2012.

van Heemst V, Estimating Chlorine Toxicity Under Emergency Conditions, Chlorine Safety Seminar, Brussels, November 14–15, 1990.

Van Horn WH, Foget CR, Staackman M, Damage to the Basic Chemical Industry from Nuclear Attack and Resultant Requirements for Repair and Reclamation, 'u7LS 687-4, prepared for Stanford Research Institute and Office of Civil Defense, Contract Number 12475 (6300A-300), Work Unit 3311 B, June 1968.

Van Onacker E, Vakbondswerking in het België van de 19de eeuw en in het hedendaagse China: een vergelijking, Masterproef, Faculteit Economie en bedrijfskunde, 2009.

Van Schaack D, Editor, Safeguarding for the Prevention of Industrial Accidents, Aetna Life Insurance Company, 1910.

Vernon HM, Accidents and Their Prevention, Cambridge, University Press, 1936.

Vlek C, Stallen PJM, Rational and personal aspects of risk. Acta Psychologica, 45, 273–300, 1980.

Vlek C, Stallen PJM, Persoonlijke Beoordeling Van Risico's: Over Risico's, Voordeligheid En Aanvaardbaarheid Van Individuele, Maatschappelijke En Industriële Activiteiten, University of Groningen, Institute for Experimental Psychology, 1979.

Vlek CAJ, Stallen PJM. Judging risks and benefits in the small and in the large. Organizational Behaviour and Human Performance, 28, 235–71, 1981.

VRB, Energy Storage Safety Monitor, June 2020.

VROM, Methods for the Calculation of Physical Effects Due to Releases of Hazardous Materials (Liquids and Gases), Yellow Book, CPR 14E, 3rd edition, The Hague, 2005.

Waddell G, Burton AK, Is Work Good for Your Health and Well-being?, London, The Stationary Office, 2006.

Wang B, Wu C, Shi B, Huang L, Evidence-based safety (EBS) management: A new approach to teaching the practice of safety management. Journal of Safety Research, 2017.

Walker FE, "Estimating Production and Repair Efforts in Blast damaged petroleum refineries", Stanford Research Institute, SRI Project MU 6300-620, July 1969.

Waxin Raphaël, Facteurs Humains et Organisationnels du Management de la Sécurité Industrielle, ESCP Europe, Executive Mastères Spécialisés, Thèse professionnelle, Promotion 2013–2015, 2015.

Weigman K, The consequence of errors. European Molecular Biology Organization Report, 6(4), 306–309, 2005.

Wertenbach HG, Spread of Flames on Cylindrical Tanks for Hydrocarbon Fluids, Gas and Erdgas 112, 1971.

Westley F, Zimmerman B, Patton Q, Getting to Maybe: How the World is Changed, 2006.

Westrum R, Adamski AJ, Organizational Factors Associated with Safety and Mission Success in Aviation Environments, Garland DJ, Wise JA, Hopkin VD, Eds., Handbook of Aviation Human Factors, Lawrence Erlbaum, Mahwah, NJ, 67–104, 1999.

Weyman AK, Kelly CJ, "Risk Perception and Risk Communication: a review of literature", HSE Research report 248/1999, 1999.

Whittingham RB, The Blame Machine, Why Human Error Causes Accidents, Elsevier, 2004.

WHO, Mental Health and Well-being at the Workplace, – Protection and Inclusion in Challenging Times, Edited by, Baumann A, Muijen M, World Health Organization, 2010.

Wikipedia: several searches were performed in Wikipedia.

Winsemius W, De Psychologie Van Het Ongevalsgebeuren, Stenfert Kroese's Uitgeverij, Leiden, 1953.

Withers RMJ, Lees FP, The assessment of major hazards. The lethal toxicity of chlorine, Part 2. Model of toxicity to man. Journal of Hazardous Materials, 12, 283–302, 1985.

WHAZAN, Theory Manual, 1988.

Yorio PL, Moore SM, Examining factors that influence the existence of Heinrich's safety triangle using site-Specific H&S data from more than 25,000 establishments. Risk Analysis, 38(4), 839–852, 2018.

Young R, Accident prevention in steel plants. Iron Age, 89, 30–40, 1912.

Zerga JE, Job Analysis: A Resume and Bibliography. Journal of Applied Psychology, 249–267, 1943.

Zinn JO, Social Theories of Risk and Uncertainty: An Introduction, Blackwell Publishing Ltd, 2008.

Zwart A, Woutersen RA, Acute inhalation toxicity of chlorine in rats and mice: Time – concentration mortality relationships and effects on respiration. Journal of Hazardous Materials, 19, 195–208, Elsevier Science Publishers B.V., Amsterdam, 1988.

Index

https://doi.org/10.1515/9783110632132-010

www.ingramcontent.com/pod-product-compliance
Lightning Source LLC
Chambersburg PA
CBHW080903220326
41598CB00034B/5459